JAPANESE NATIONAL LARGE TELESCOPE
AND RELATED ENGINEERING DEVELOPMENTS

JAPANESE NATIONAL LARGE TELESCOPE AND RELATED ENGINEERING DEVELOPMENTS

Proceedings of
the International Symposium on Large Telescopes,
held in Tokyo, Japan, 29 November – 2 December, 1988

Edited by

T. KOGURE
Department of Astronomy, Faculty of Science, Kyoto University, Japan

and

A. T. TOKUNAGA
Institute for Astronomy, University of Hawaii, U.S.A.

Reprinted from Astrophysics and Space Science
Volume 160, 1989

Springer-Science+Business Media, B.V.

Library of Congress Cataloging-in-Publication Data

International Symposium on Large Telescopes (1988 : Tokyo, Japan)
    Japanese national large telescope and related engineering
  developments : proceedings of the International Symposium on Large
  Telescopes, held in Tokyo, Japan, 29 November–2 December, 1988 /
  edited by T. Kogure and A.T. Tokunaga.
        p.   cm.
    "Reprinted from Astrophysics and space science, volume 160/1–2."

    1. Telescope--Congresses.   2. Telescope--Japan--Congresses.
  3. Very large array telescopes--Congresses.   I. Kogure, T.
  (Tomokazu)   II. Tokunaga, Alan Takashi, 1949-   .   III. Title.
  QB88.I597   1988
  681'.412--dc20   89-24623

Printed on acid-free paper

All Rights Reserved

ISBN 978-94-017-2005-2   ISBN 978-94-017-2003-8 (eBook)
DOI 10.1007/978-94-017-2003-8
© 1989 Springer Science+Business Media Dordrecht
Originally published by Kluwer Academic Publishers in 1989.
Softcover reprint of the hardcover 1st edition 1989

JAPANESE NATIONAL LARGE TELESCOPE AND RELATED ENGINEERING DEVELOPMENTS

*Proceedings of the International Symposium on Large Telescopes,
held in Tokyo, Japan, 29 November–2 December, 1988*

Edited by

T. KOGURE

Department of Astronomy, Faculty of Science, Kyoto University, Japan

and

A. T. TOKUNAGA

Institute for Astronomy, University of Hawaii, U.S.A.

TABLE OF CONTENTS

THIRD SESSION
THE JNLT PROJECT

FOURTH SESSION
TELESCOPE ENGINEERING AND FABRICATION OF
LARGE MIRRORS

FIFTH SESSION
NEW PLANS AND ACHIEVEMENTS WITH EXISTING TELESCOPES

SIXTH SESSION
INSTRUMENTATION

SEVENTH SESSION
INTERFEROMETRY

SUMMARY OF THE SYMPOSIUM

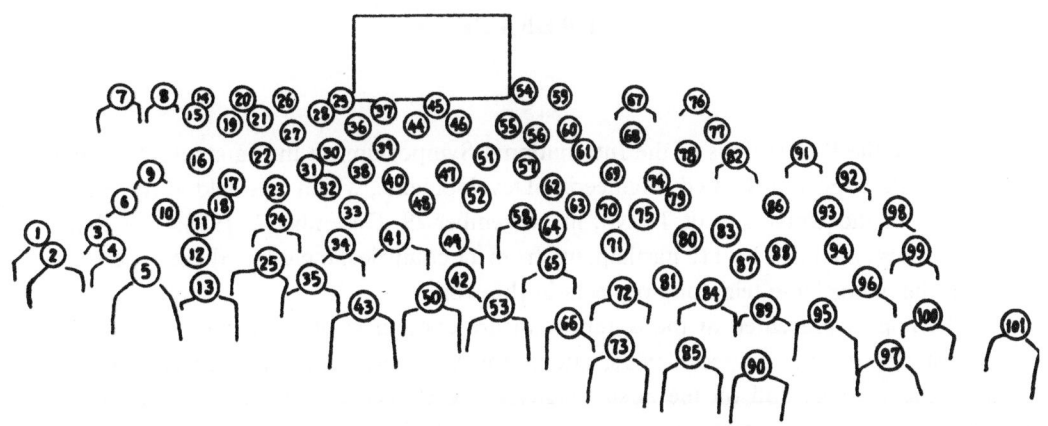

IDENTIFICATION TABLE

1. R. K. Smith
2. S. J. Medwadowski
3. R. Tsukada*
4. W. A. Siegmund
5. H. Raff
6. Y. Yamashita
7. K. Koide*
8. A. Ohmura*
9. N. Itoh
10. T. Kawaguchi
11. H. A. Miska
12. B. Lhenry
13. J.-P. Rozelot
14. T. Hasegawa*
15. M. Itoh*
16. I. Mikami
17. M. Kitamura
18. S. Nishimura
19. K. F. Beckstette
20. M. Doi*
21. J. Noumaru*
22. S. Hayakawa
23. S. Tamura
24. J. Spangenberg-Jolley
25. H. J. Smith
26. H. Sugai*
27. C. Barbieri
28. K. Sorimachi
29. H. Ando
30. N. Okada
31. S. Sato
32. T. Nagata
33. C. R. Blackwell
34. K. Kodaira
35. S. G. Wang
36. S. Shectman
37. S. Isobe
38. T. Yamashita
39. S. Yamashita
40. Ch. Kinoshita
41. D. Enard
42. K. Ishida
43. B. Hidayat
44. R. Angel
45. T. Yamagata
46. Y. Nakai
47. R. Muller
48. M. Watanabe
49. P. J. Lena
50. T. Kogure
51. A. Chelli
52. J. M. Beckers
53. S. Okamura
54. A. Miyashita
55. K. Iwasaki
56. M. Sekiguchi*
57. M. Shimizu
58. T. Hirayama
59. J. Morimoto
60. N. Baba
61. T. Aoki*
62. R. A. McLaren
63. R. Davies
64. W. van Citters
65. L. D. Barr
66. M. Iye
67. T. Noguchi
68. E. H. Zhang
69. A. T. Tokunaga
70. S. Hayashi
71. K. Torii
72. B. Mack
73. M. Nakagiri
74. T. Maihara
75. D. N. B. Hall
76. K. Tomita
77. K. Sato
78. T. Kohno
79. W. Y. Wong
80. W. S. Smith
81. R. D. Cannon
82. Y. Funakoshi
83. S. Kikuchi
84. R. Ellis
85. K. Okita
86. Y. Matsui
87. J.-P. Swings
88. R. Hirata
89. T. J. Lee
90. H. Kimura
91. S. Araya
92. T. Nishimura
93. H. Ohtani
94. M. Sasaki*
95. H. Hugenell
96. M. C. Morris
97. A. Labeyrie
98. N. Kaneko
99. P. R. Gillingham
100. R. Laing
101. K. Nariai

* Assistant member of the Symposium.

PREFACE

These are the Proceedings of the International Symposium on the Japanese National Large Telescope and Related Engineering Developments, which was held at the Sanjo Kaikan of the University of Tokyo, in November 29–December 2, 1988. The Symposium was attended by 116 participants from 11 countries, including the key persons from the major large telescope projects in the world.

The papers presented at the Symposium are composed of 18 invited reviews, 23 contributed papers, 5 special talks, and 5 poster papers. They were presented in 7 scientific sessions and on the poster panels. The Proceedings contains most of the papers with some rearrangement of the sessions and papers for the convenience of readers.

In the course of the editorial work, we had helpful cooperation from Dr T. Maihara and Mr R. Hirata of Kyoto University to whom we express our hearty thanks. Thanks are also due to Miss N. Sakon for her kind secretarial assistance, and to Mr J. Noumaru for assistance in the editorial work.

The Editors
T. KOGURE
A. T. TOKUNAGA

Fig. 1. Conceptual model of JNLT. (Kodaira, p. 138).

Fig. 5a. First 3.5 m trial casting, now being polished for the ARC telescope. (Angel, p. 66).

Fig. 5b.   Second 3.5 m trial casting to be polished for the WIN telescope. (Angel, p. 66).

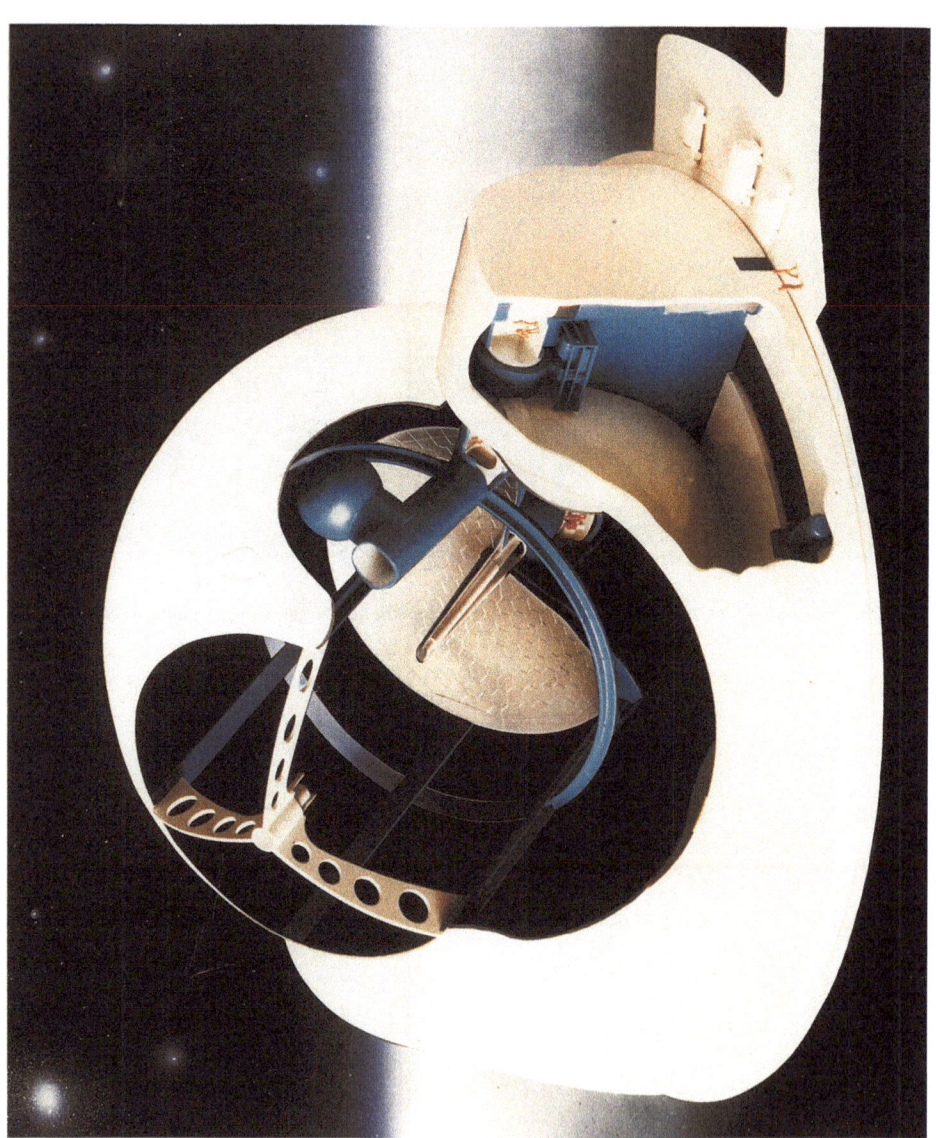

Fig. 15. The Central-Axis Reflector (ZAS). A super telescope of the new generation which, ground-based or stationed in orbit, allows observation of a universe never seen before. (Hügenell).

Fig. 3. A view of the VLT. This mock-up shows at the forefront the interferometric tunnel where beam combination occurs. (Lena, p. 367).

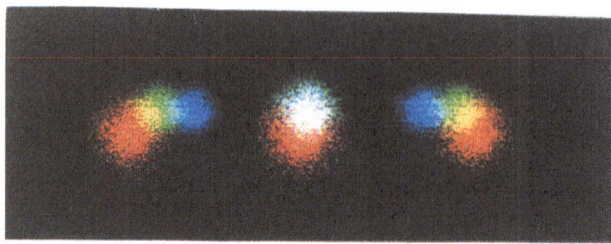

Fig. 2. Wideband speckle image and dispersed speckle image of two point sources (He–Ne laser and Hg lamp). (Baba, p. 374).

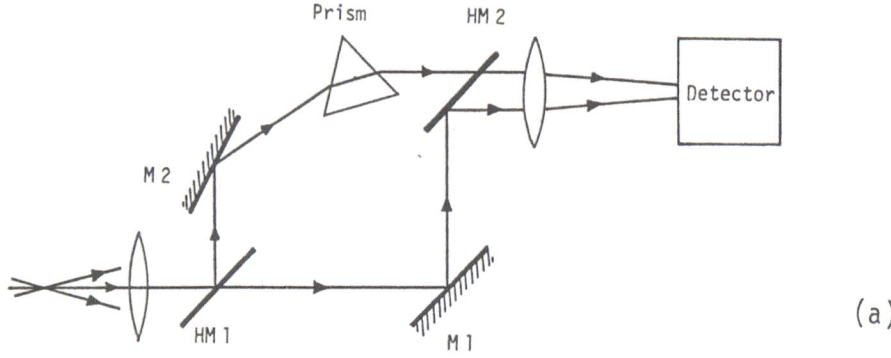

(a)

(b)

Fig. 6. New version of speckle spectrometer and a sample image. (Baba, p. 376).

ORGANIZING COMMITTEE

T. KOGURE, Department of Astronomy, Kyoto University, Kyoto (Chairman)

K. KODAIRA, National Astronomical Observatory, Mitaka, Tokyo

S. ISOBE, National Astronomical Observatory, Mitaka, Tokyo

M. IYE, National Astronomical Observatory, Mitaka, Tokyo

S. OKAMURA, Kiso Observatory, Institute of Astronomy, Faculty of Science, University of Tokyo, Kiso, Nagano

W. TANAKA, Department of Astronomy, University of Tokyo, Tokyo

R. HIRATA, Department of Astronomy, Kyoto University, Kyoto

K. TAKAKUBO, Department of Astronomy, Tohoku University, Sendai

S. NISHIMURA, National Astronomical Observatory, Mitaka, Tokyo

S. SATO, National Astronomical Observatory, Mitaka, Tokyo

T. TSUJI, Institute of Astronomy, Faculty of Science, University of Tokyo, Mitaka, Tokyo

PARTICIPANTS

H. ANDO, National Astronomical Observatory, Mitaka, Tokyo, Japan

R. ANGEL, Steward Observatory, University of Arizona, Tucson, U.S.A.

S. ARAYA, Goto Optical MFG Co., Tokyo, Japan

N. BABA, Department of Applied Physics, Faculty of Engineering, Hokkaido University, Sapporo, Japan

C. BARBIERI, Department of Astronomy, University of Padova, Italy

L. D. BARR, National Optical Astronomy Observatories, Tucson, U.S.A.

J. M. BECKERS, European Southern Observatory, Garching by München, F.R.G.

K. F. BECKSTETTE, Carl Zeiss O-Labor, Oberkochen, F.R.G.

C. R. BLACKWELL, Aoer Freeman Fox Ltd., London, England, U.K.

A. ROSENBERG, Royal Greenwich Observatory, Herstmonceux Castle, East Sussex, England, U.K.

R. D. CANNON, Anglo-Australian Observatory, Epping, Australia

A. CHELLI, Instituto de Astronomia, Observatorio Astronomico Nacional, Mexico, Mexico

R. DAVIES, Department of Astrophysics, University of Oxford, Oxford, England, U.K.

M. G. EDMUNDS, Department of Physics, University of Wales, Wales, U.K.

R. S. ELLIS, Physics Department, Durham University, Durham, England, U.K.

D. ENARD, European Southern Observatory, Garching bei München, F.R.G.

Y. FUNAKOSHI, Hida Observatory, Kyoto University, Gifu, Japan

P. R. GILLINGHAM, Anglo-Australian Observatory, Coonabarabran, Australia

D. N. B. HALL, Institute for Astronomy, University of Hawaii, Honolulu, Hawaii, U.S.A.

S. HAYAKAWA, University of Nagoya, Nagoya, Japan

S. HAYASHI, Joint Astronomy Centre, Hilo, Hawaii, U.S.A.

B. HIDAYAT, Bosscha Observatory, Lembang, Indonesia

R. HIRATA, Department of Astronomy, Faculty of Science, Kyoto University, Kyoto, Japan

T. HIRAYAMA, National Astronomical Observatory, Mitaka, Japan

M. HONMA, Ministry of Education, Science and Culture, Tokyo, Japan

M. HÜGENELL, Carl F. Angstenberger, Consultants, Ludwigshafen-am-Rhein, F.R.G.

K. ISHIDA, Kiso Observatory, Institute of Astronomy, Faculty of Science, University of Tokyo, Nagano, Japan

M. ISHIGURO, Nobeyama Radio Observatory, National Astronomical Observatory, Nagano, Japan

S. ISOBE, National Astronomical Observatory, Mitaka, Tokyo, Japan

N. ITOH, Mitsubishi Electric Co., Amagasaki, Japan

K. IWASAKI, Kwasan Observatory, Kyoto University, Kyoto, Japan

M. IYE, National Astronomical Observatory, Mitaka, Japan

N. KAIFU, Nobeyama Radio Observatory, National Astronomical Observatory, Nagano, Japan

N. KANEKO, Faculty of Science, Hokkaido University, Sapporo, Japan

A. KATSUNUMA, NIKON Co., Tokyo, Japan

T. KAWAGUCHI, Corning Japan, Tokyo, Japan

S. KIKUCHI, National Astronomical Observatory, Mitaka, Japan

H. KIMURA, Purple Mountain Observatory, Nanjing, China

C. KINOSHITA, Mitsubishi Electric Co., Amagasaki, Japan

M. KITAMURA, Akatsuki-machi, Hachioji, Japan

Y. KOBAYASHI, Institute of Astronomy, Faculty of Science, University of Tokyo, Mitaka, Japan

K. KODAIRA, National Astronomical Observatory, Mitaka, Japan

T. KOGURE, Department of Astronomy, Faculty of Science, Kyoto University, Kyoto, Japan

T. KOHNO, Tokyo Metropolitan Institute of Technology, Tokyo, Japan

S. KOKAJI, Mechanical Engineering Laboratory, Tsukuba, Japan

Y. KOZAI, National Astronomical Observatory, Mitaka, Japan

H. KUROKAWA, Mechanical Engineering Laboratory, Tsukuba, Japan

A. LABEYRIE, CERGA, Observatoire de Calern, St-Vallier de Thiey, France

R. LAING, Royal Greenwich Observatory, Herstmonceux Castle, East Sussex, England, U.K.

T. J. LEE, Royal Observatory, Blackford Hill, Edinburgh, Scotland, U.K.

P. J. LENA, Observatoire de Paris, Meudon, France

B. LHENRY, G.I.E. Telas, Cannes La Bocca Cedex, France

B. MACK, Royal Greenwich Observatory, Herstmonceux Castle, East Sussex, England, U.K.

T. MAIHARA, Department of Physics, Faculty of Science, Kyoto University, Kyoto, Japan

Y. MATSUI, CANON Co., Tokyo, Japan

S. MATSUMOTO, Carl Zeiss Co., Tokyo, Japan

R. A. McLAREN, Canada–France–Hawaii Telescope Corporation, Kamuela, Hawaii, U.S.A.

S. J. MEDWADOWSKI, Consulting Structure Engineer, San Francisco, U.S.A.

I. MIKAMI, Mitsubishi Electric Co., Amagasaki, Japan

H. S. MISKA, Corning Glass Works, Advanced Products Department, New York, U.S.A.

A. MIYASHITA, National Astronomical Observatory, Mitaka, Japan

J. MORIMOTO, NIKON Co., Tokyo, Japan

M. C. MORRIS, Royal Greenwich Observatory, Herstmonceux Castle, East Sussex, England, U.K.

R. MUELLER, Schott Glaswerke, Mainz, F.R.G.

T. NAGATA, Department of Physics, Faculty of Science, Kyoto University, Kyoto, Japan

M. NAKAGIRI, National Astronomical Observatory, Mitaka, Japan

Y. NAKAI, Kwasan Observatory, Faculty of Science, Kyoto University, Kyoto, Japan

T. NAKAGAWA, Institute of Space and Astronautical Science, Sagamihara, Japan

T. NAKAMURA, KEK, Tsukuba, Ibaraki, Japan

K. NAKAO, Space Engineering Development Co. Ltd., Tokyo, Japan

M. NAKAYAMA, Hamamatsu Photonics Co., Hamamatsu, Japan

K. NARIAI, National Astronomical Observatory, Mitaka, Japan

S. NISHIMURA, National Astronomical Observatory, Mitaka, Japan

T. NISHIMURA, Steward Observatory, University of Arizona, Tucson, U.S.A.

T. NOGUCHI, National Astronomical Observatory, Mitaka, Japan

M. ODA, Institute of Physical and Chemical Research, Wako, Saitama, Japan

K. OGURA, Kokugakuin University, Tokyo, Japan

H. OHTANI, Department of Astronomy, Faculty of Science, Kyoto University, Kyoto, Japan

S. OKAMURA, Kiso Observatory, Institute of Astronomy, Faculty of Science, University of Tokyo, Nagano, Japan

K. OKITA, Okayama Astrophysical Observatory, National Astronomical Observatory, Okayama, Japan

H. OKUDA, Institute of Space and Astronautical Science, Sagamihara, Japan

H. RAFF, Carl Zeiss, Oberkochen. F.R.G.

J. P. ROZELOT, G.I.E. Telas, Cannes La Bocca Cedex, France

K. SATO, Mizusawa Astrogeodynamics Observatory, National Astronomical Observatory, Mizusawa, Japan

S. SATO, National Astronomical Observatory, Mitaka, Japan

S. A. SHECTMAN, Mount Wilson and Las Campanas Observatories, Pasadena, California, U.S.A.

M. SHIMIZU, Wakaba-cho, Tachikawa, Japan

W. A. SIEGMUND, University of Washington, Telescope Project, Seattle, Washington, U.S.A.

H. J. SMITH, McDonald Observatory at Mount Locke, University of Texas, Austine, U.S.A.

R. K. SMITH, Corning Glass Works, Advanced Products Department, New York, U.S.A.

W. S. SMITH, Contraves Goerz Co., Pennsylvania, U.S.A.

K. SORIMACHI, CANON Co., Tokyo, Japan

J. SPANGENBERG-JOLLEY, Corning Glass Works, Advanced Products Department, New York, U.S.A.

A. SUZUKI, Mechanical Engineering Laboratory, Tsukuba, Japan

J.-P. SWINGS, Institute d'Astrophysique, Université de Liège, Cointe-Ougrée, Belgium

M. TAKEUCHI, Department of Astronomy, Faculty of Science, Tohoku University, Sendai, Japan

S. TAMURA, Department of Astronomy, Faculty of Science, Tohoku University, Sendai, Japan

M. TANAKA, Institute of Astronomy, Faculty of Science, University of Tokyo, Mitaka, Japan

W. TANAKA, Department of Astronomy, Faculty of Science, University of Tokyo, Tokyo, Japan

M. TARENGHI, European Southern Observatory, Garching bei München, F.R.G.

A. T. TOKUNAGA, Institute for Astronomy, University of Hawaii, Honolulu, U.S.A.

K. TOMOTA, A.E.S., Tsukuba, Ibaraki, Japan

T. TSUJI, Institute of Astronomy, Faculty of Science, University of Tokyo, Mitaka, Japan

W. VAN CITTERS, Division of Astronomical Science, National Science Foundation, Washington D.C., U.S.A.

S. G. WANG, Beijing Astronomical Observatory, Chinese Academy of Sciences, Beijing, China

J. WATANABE, National Astronomical Observatory, Mitaka, Japan

M. WATANABE, Ad-In Research Inc. Co., Tokyo, Japan

W. Y. WONG, National Optical Astronomy Observatory, Tucson, Arizona, U.S.A.

N. YAJIMA, Institute of Space and Astronautical Science, Sagamihara, Japan

T. YAMAGATA, National Astronomical Observatory, Mitaka, Japan

S. YAMASHITA, NIKON Co., Tokyo, Japan

T. YAMASHITA, National Astronomical Observatory, Mitaka, Japan

Y. YAMASHITA, National Astronomical Observatory, Mitaka, Japan

Y. YONEDA, ULVAC Japan Ltd., Kanagawa, Japan

E. H. ZHANG, Beijing Astronomical Observatory, Beijing, China

WELCOME ADDRESS OF THE DIRECTOR OF THE NATIONAL ASTRONOMICAL OBSERVATORY*

Y. KOZAI

National Astronomical Observatory, Tokyo, Japan

Distinguished Guests, Ladies and Gentlemen,

It is a great honour and pleasure for me to welcome you, all the participants, on behalf of local organizers and the National Astronomical Observatory.

The National Astronomical Observatory is a new organization which was founded on July 1, this year. And the title of the meeting is, as all of you know, the JNLT and Related Engineering Development and JNLT stands for Japanese National Large Telescope. One of the most important initiatives to found the National Astronomical Observatory in place of the Tokyo Astronomical Observatory, which was one of research institutes of the University of Tokyo, was to strengthen the administrative power and to make inter-university cooperation easier to build JNLT at Mauna Kea as well as astronomical instruments.

The largest optical telescope we have now in Japan is the 188 cm telescope at Okayama, dedicated in 1960. During the 30 years after its dedication nearly 30 large optical-infrared telescopes of 2 to 4 m sizes have been dedicated in the world and contributed to the rapid and revolutionary progress of astronomy.

During this period Japanese astronomers have not been idle, however, as you will hear this morning from the first three speakers. In fact, several X-ray astronomy satellites have made excellent contributions to high-energy astrophysics, millimeter wavelength radio astronomy has been developed with the Nobeyama radio astronomy facilities, infrared astronomy has been developed, theoretical astrophysicists have been very active, and the detectors at Kamioka caught neutrino bursts from SN 1987a.

Despite such progress we are afraid that our optical-infrared astronomy facilities are far behind the frontiers in the world. To overcome this situation the idea of JNLT, and, therefore, the National Astronomical Observatory, came out and already some design studies of the telescope and site tests at Mauna Kea have been made, although we have not yet succeeded in obtaining funds for the telescope.

There are in the world several projects to build large telescopes of 7 to 10 m sizes and even more have been started and planned. I am very happy to see delegates of almost all such projects as well as directors, eminent astronomers, and engineers of existing observatories. Therefore, for me this meeting is really an international meeting for exchanging ideas about the new technology of large telescopes.

I expect that JNLT as well as other projects and all of the participants will benefit by attending the meeting. And in every sense I wish the success of the meeting.

* Presented at the Symposium on the JNLT and Related Engineering Developments, Tokyo, November 29–December 2, 1988.

OPENING ADDRESS OF THE CHAIRMAN OF THE ORGANIZING COMMITTEE*

TOMOKAZU KOGURE

Department of Astronomy, Faculty of Science, Kyoto University, Kyoto, Japan

Distinguished Participants,
Ladies and Gentlemen,

On behalf of the Organizing Committee of the Symposium and also of GOPIRA, the Group of Optical and Infrared Astronomers in Japan, I extend our hearty welcome to all of the participants coming from many countries and from many parts of our country.

We, the Japanese optical and infrared astronomers, have been organized into GOPIRA in 1980, and since then, we have eagerly worked out on focusing the idea of a future telescope and adopted the 7.5 m telescope plan in 1984, which is now called JNLT, Japanese National Large Telescope. We then started the studies for the engineering feasibility of the telescope and instrumentation, and for the scientific objectives to be carried out with this large telescope.

The present JNLT project has two outstanding significance for our Japanese astronomers. One is that the 7.5 m monolithic-mirror telescope is really an engineering challenge we have never experienced. The largest telescope in Japan has long been the 1.88 m telescope at the Okayama Astrophysical Observatory for more than 25 years. The construction of such a large 7.5 m telescope requires engineering development in every part of its construction.

The second is that JNLT will be the first Japanese facility to be constructed in an international site which is the summit of the Mauna Kea. This is also the first experience for us and requires a new step in the form of international cooperation.

This meeting is primarily aimed to discuss and to exchange information on the telescope engineering of the JNLT and of other major large-telescope projects in the world. On our side, we offer the present state of our studies of the JNLT and its instrumentation for the critical discussions in the meeting. Through the 4-day discussions we hope we shall get many new ideas and new information which are directly related to the promotion of our JNLT project. We feel thankful to the presence of outstanding participants from many countries and in many fields of astronomy and engineering. We particularly express our gratitude to the invited speakers and the speakers of the special talks and of contributed papers.

The present Symposium is organized by GOPIRA, under the cooperation of the National Astronomical Observatory, which has been newly formed in July 1 of this year as the host institute of the JNLT project, and under the support of the Astronomical

* Paper presented at the Symposium on the JNLT and Related Engineering Developments, Tokyo, November 29–December 2, 1988.

Society of Japan. The Symposium is also supported financially by the following scientific foundations:

Asahi Glass Foundation for Industrial Technology,
The Inamori Foundation,
Inoue Foundation for Science,
Kajima Science Foundation,
Mitsubishi Foundation,
Toray Science Foundation,
Yoshida Foundation for Science and Technology.

We have also accepted the financial support from many companies, and it is clear that without such wide support, the organization of the Symposium should not be possible. In this occasion I extend our hearty gratitude to all of these foundations and companies.

Finally I repeat my hope for the success of this meeting, and my thanks to all of the participants.

FIRST SESSION

ASTRONOMY AND LARGE TELESCOPES

COSMOLOGY AND LARGE TELESCOPES*

SATIO HAYAKAWA

Nagoya University, Nagoya, Japan

(Received 23 December, 1988)

Abstract. Assuming a large collecting area, a good angular resolution and a large field of view expected for the Japanese National Large Telescope (JNLT), we demonstrate that JNLT will provide a useful means of studying cosmological objects of interest. Among them I discuss how cosmological parameters and evolutionary effects can be obtained from redshift-magnitude relations, galaxy counts, distant supernovae, quasar properties, and large-scale structures. An advantage of near infrared observations is emphasized.

1. Introduction

The design of large telescopes has been often guided by the advancement of cosmological observation. The Hale telescope was designed for extending the redshift-magnitude relation to large redshifts. Space Telescope aims at deep space surveys, achieving a large limiting magnitude which is not attainable with ground-based telescopes. The Japanese National Large Telescope (JNLT) is not exceptional and is, therefore, designed for the observation of faint objects such as distant galaxies in the optical and near infrared ranges. We will discuss how the JNLT will contribute to cosmological studies.

To be quantitative, we assume that the JNLT has a collecting area with a diameter of 7.5 m, a field of view with a diameter of 30′, and an angular resolution of 0.3″. The fluctuation of thermal radiation from the optical system is a limiting factor at long wavelengths but is assumed to be unimportant in the K-band, with which we are mainly concerned in the present paper. A good angular resolution enables us to distinguish a galaxy from a point source, since the angular diameter of an ordinary galaxy is larger than the minimum angular size resolved. A large field of view allows us to observe a distant cluster of galaxies as a whole. A large collecting area makes it possible to observe almost all galaxies without being limited by the photon number collected. The capability in the infrared range favours the observation of highly redshifted objects, since photons received therefrom are mostly at near infrared wavelengths.

In what follows we will present several salient examples in which the JNLT will play a powerful role.

2. Hubble Diagram and Galaxy Count

Cosmological models can be tested by a number of observations (Sandage, 1988). Among several methods we here discuss the redshift-magnitude relation and the galaxy count which have been extensively studied. The redshift-magnitude relation, the Hubble

---

* Paper presented at the Symposium on the JNLT and Related Engineering Developments, Tokyo, November 29–December 2, 1988.

diagram, for galaxies, has been recently calculated by Yoshii and Takahara (1988) with and without evolution. The $V - z$ and $K - z$ relations for three values of the deceleration parameter $q_0 = 0.02$, 0.5, and 1.0 and for the redshifts of galaxy formation $z_F = 3$ and 5 with and without evolution are reproduced in Figures 1 and 2, respectively.

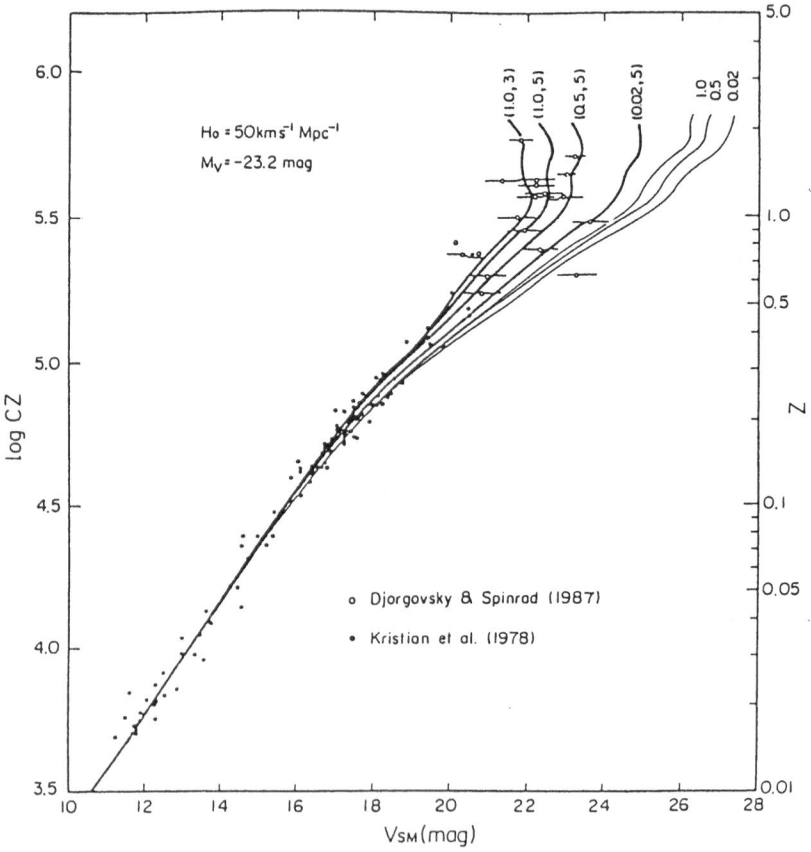

Fig. 1. The redshift-magnitude relations in the V-band. Thick and thin lines represent those with and without evolution, respectively, calculated by Yoshii and Takahara (1988). The numbers attached to the lines represent the values of q_0 and z_F. Observed results are shown by solid and open circles.

Comparison of the calculated $V - z$ relations with observed data indicates that the evolution models are favoured. The values of q_0 and z_F may be obtainable if the V-magnitude can be measured to $V \simeq 25$. The distinction between the evolution and non-evolution models can be more easily made with the $K - z$ relation if the magnitudes at $z > 2$ are measured to $K \simeq 20$.

The magnitudes of $V \simeq 25$ and $K \simeq 20$ give photon counting rates of about 10 s^{-1} and 10^2 s^{-1}, respectively, with conventional detectors attached to JNLT. It is, therefore, not difficult to measure the magnitudes of individual galaxies faint enough to distinguish between models. However, comparison of a model with observed data is subject to the luminosity distribution which results in a wide distribution of magnitude

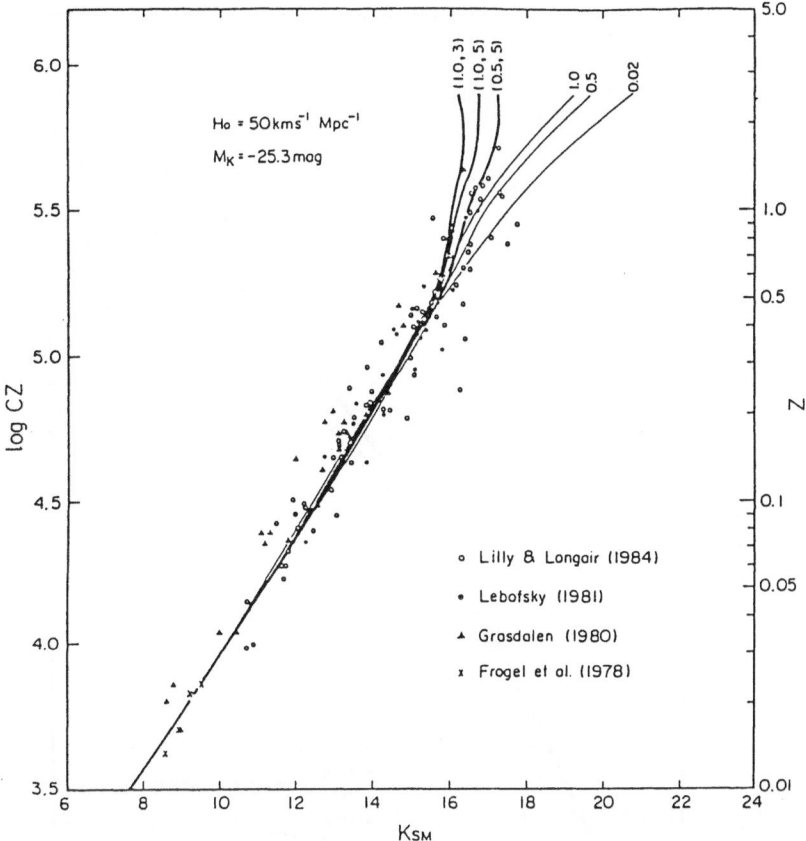

Fig. 2. The redshift-magnitude relations in the K-band represented in the same way as in Figure 1.

for a given redshift. Hence, the comparison requires the data of many galaxies to enable us to obtain the distribution of magnitude and the average magnitude for given z. The number of galaxies for which these data are available to date is not enough for accurate comparison but will be increased by utilizing large telescopes such as JNLT.

Models with and without evolution and of different q_0 can be distinguished from each other by reference to the differential number count of galaxies, the number of galaxies per solid angle per magnitude interval. The relations for the B- and K-bands calculated by Yoshii and Takahara (1988) are reproduced in Figures 3 and 4, respectively.

Comparison of the calculated relation with observed data in the B-band favours the evolution model, but the observation to $B \simeq 28$ is necessary to obtain q_0 and z_F. In the K-band both the evolution and non-evolution models give similar results for faint magnitudes. Hence, the K-band observation is useful for obtaining q_0 and will be performed with the JNLT.

Expected results of the galaxy count have been illustrated by Chokshi and Wright (1988) in their simulation calculations. In their Figures 3–5 the simulated images of galaxies with $z_F = 10$ in the K-band within a $100'' \times 100''$ field of view are given for selected values of density parameter Ω_0 and cosmological constant Λ_0. One can readily

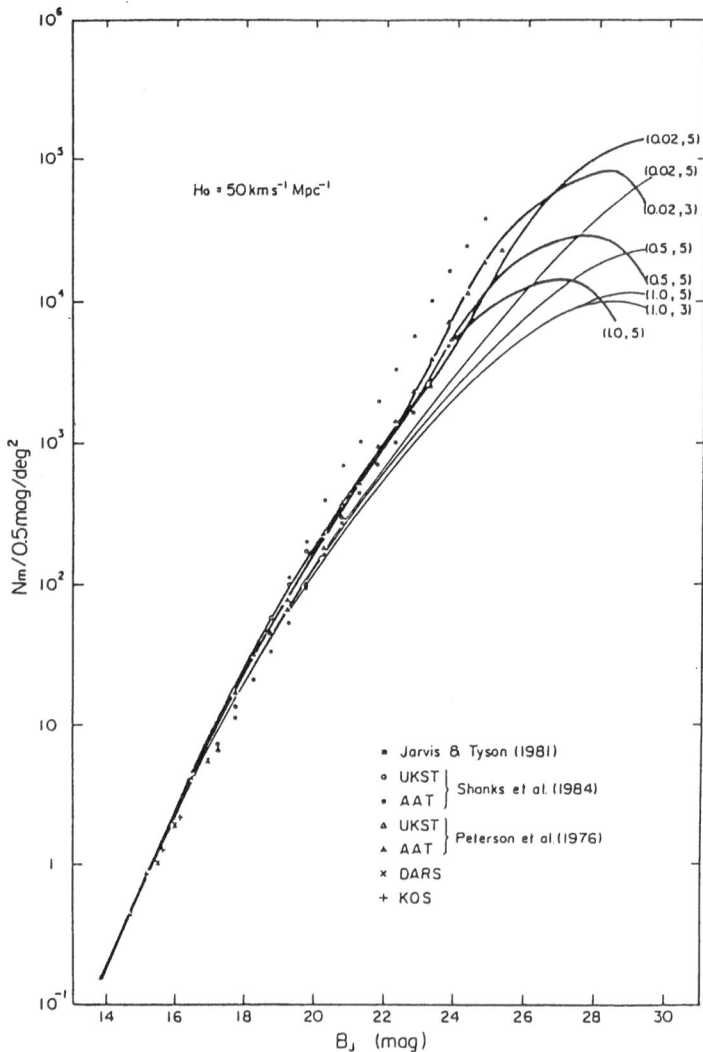

Fig. 3.   The numbers of galaxies in the 0.5 mag range per deg² versus *B* mag. Thick and thin lines are those calculated by Yoshii and Takahara (1988) with the same models as in Figures 1 and 2. Observed results are compared.

see that the images of $\Omega_0 = 0$ and 1 for $\Lambda_0 = 0$ are clearly distinguishable. It is also easy to distinguish between the images of $\Lambda_0 = 0$ and 1 for $\Omega_0 = 1$. An advantage of near-infrared observations can be observed by the comparison between optical and 4 μm images for $z_F = 5$, as shown in their Figures 6–8. More numerous galaxies are observable at 4 μm than at optical wavelengths.

JNLT will be capable to obtain such images, since it will achieve the band photometry to $m \simeq 28$ and 25 in the optical and near infrared ranges, respectively. The spectroscopic observations to $m \simeq 23$ and 20 in the respective ranges will be attainable to obtain redshifts.

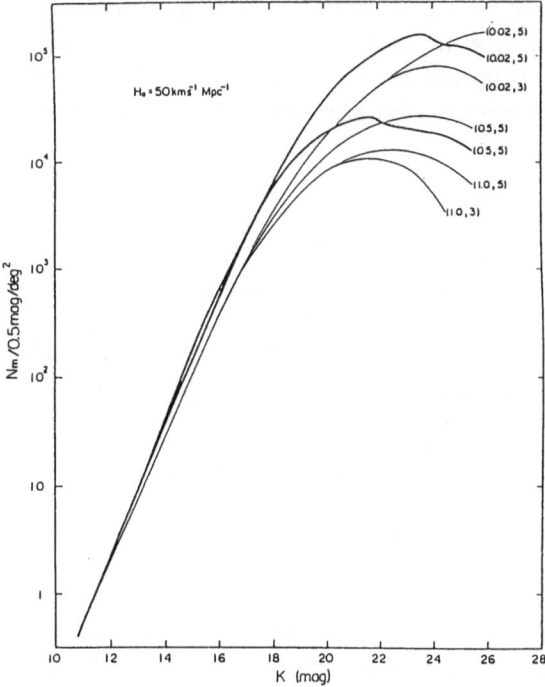

Fig. 4.   The numbers of galaxies in the 0.5 mag range per deg^2 versus K magnitude represented in the same way as in Figure 3.

3. Distant Supernovae

The Hubble constant H_0 is one of the most important quantities in cosmology and can be, in principle, obtained with the methods described in the preceding section as well as with other methods. However, its value has been given in a wide range 40–100 km s^{-1} Mpc^{-1}. One way of determining H_0 is the observation of Type 1 supernovae since they are regarded as standard candles with a peak luminosity of about 10^{43} erg s^{-1}, approximately equal to the luminosity at which the luminosity function of galaxies peaks. With this method the value of H_0 has been determined to be 50–70 km s^{-1} Mpc^{-1} (Arnett *et al.*, 1985; Wooseley *et al.*, 1986).

Distant supernovae can be efficiently observed by the JNLT owing to its good resolution and wide field of view. Since the angular diameter of a galaxy is at least several times the minimum resolvable angle, a supernova which appears off the galactic center can be identified. Then the light curve gives us the redshift parameter z and the apparent magnitude, through which we can obtain H_0 and possibly the decleration parameter q_0. Since the photometric observation covers the near infrared range, we can obtain the bolometric magnitude which is free from the complexity associated with radiative transfer. The probability of finding a supernova is a few tenth in a cluster of galaxies, on account of that a cluster contains a few hundred galaxies, the supernova rate is about one in 100 years, and its high luminosity lasts for 0.2 years. Since a cluster of galaxies lies in a field of view if $z > 0.2$, the detection of distant supernovae is feasible. If $z > 1$, one can always find supernovae in a thousand or more galaxies in a field of view.

The observation of supernovae also provides information on the evolutionary effects of galaxies. The supernova rate is a signature of the star formation rate. The probability of finding distant supernovae may be higher than that estimated by reference to the supernova rate obtained from nearby galaxies, since the star formation rate in early galaxies is likely to be higher. The luminosity and light curve of a Type II supernova depend on the metallicity, as has been discussed in connection with SN 1987A in LMC. Type II supernovae may occur also in elliptical galaxies in their early stage when they contained gas. If the types of galaxies are distinguished from each other and distant supernovae in each type of galaxies are studied, a large step forward will be attainable on the evolution of galaxies.

4. Quasar Evolution

The quasar is the brightest object which provides information in the early stage of galaxy formation. A decrease in the population of quasars beyond $z \simeq 3$ has been considered to indicate that quasars became bright at about $z \simeq 3$. A different interpretation is, however, to attribute the decrease in the quasar population to the obscuration by dust.

Assuming that dust in foreground galaxies is responsible for the extinction of distant quasars, Heisler and Ostriker (1988) have calculated the density of quasars as functions of z and magnitude. Figure 5 represents the comoving number densities $n(z)$ of quasars

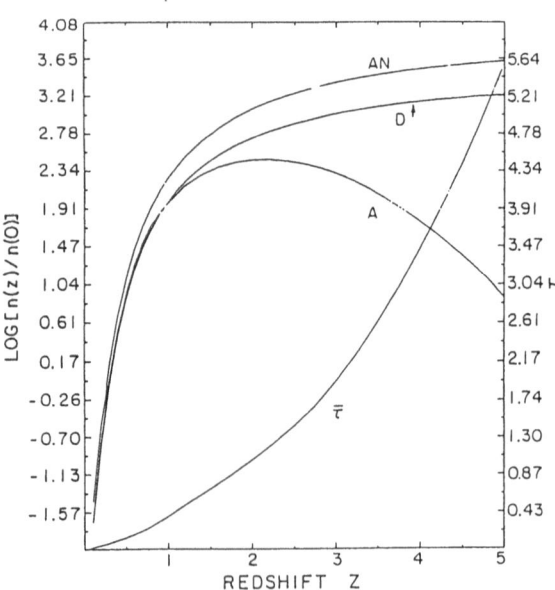

Fig. 5.   The relative numbers of quasars to be observed in the B-band versus redshift z calculated by Heisler and Ostriker (1988) with (A) and without (AN and D) extinction by foreground galaxies. The curve $\bar{\tau}$ represents the optical depth for B extinction.

brighter than the present absolute B magnitude of -25 for different models. The model A takes into account the luminosity evolution, the luminosity dependent density and the average optical depth $\bar{\tau}$ to a redshift z for extinction in the B-band. The model AN is the same as the model A but $\bar{\tau} = 0$. The model D is also dust-free, but the model parameters for the luminosity evolution and the luminosity dependent density are slightly different from those in the model A. From this figure they argue that the observed density decreases towards high z beyond the maximum at $z \simeq 2$ as shown by curve A, while the real density continues to increase towards high z.

The dust obscuration model can be easily tested by infrared observations. The optical depth of $\tau_B \simeq 5$ gives an insignificant extinction in the K-band. Hence, the density $n(z)$ in the K-band or at longer wavelengths would behave like that in the model AN or D. Since $B - K \simeq 4.5$ for the quasar frequency spectrum of $v^{-0.5}$ without extinction, a deep survey with the JNLT will easily obtain $n(z)$ in the K-band. In a field of view of the JNLT the number of quasars with $B < 25$ is estimated to be about 10 for the model A, whereas $n(z)$ with $K < 20$ would be about 300 for the model AN.

The dust obscuration may alternatively be due to dust surrounding the active nucleus. It has been argued that the Type II Seyfert is an active galactic nucleus partially obscured by an accretion disk (see, for example, Krolik, 1988). In early elliptical galaxies a greater amount of gas than in the present ellipticals may be accreted by massive nuclei. Then the colour would be somewhat different from that of nuclei obscured by foreground dust, because dust close to the nuclei should melt, and the metal abundance of infalling gas would be different.

If more distant quasars are observed in the near infrared range, the probability of gravitational lensing would be higher than that for quasars observed at present. The angular resolution of JNLT is good enough to obtain lensed images The resolution of a fraction of an arc-sec is capable of giving the distribution of mass responsible for lensing. If quasar images are observed mainly in the near infrared range, a foreground galaxy responsible for lensing could be observed in the optical range owing to weak optical intensities of the quasar images. In other words, the foreground galaxy is conspicuous in the optical range, while the quasar images can be distinguished therefrom in the near-infrared range.

The necessity of higher resolution and larger collecting area is demonstrated by the observation of quasars with the 3.6 m CFHT at Mauna Kea (Le Fèvre and Hammer, 1988). Under good seeing conditions ($0''.75$ and $0''.87$ in R), 3C 194 and 3C 225A were resolved into multiple components. Some of them may be images lensed by foreground galaxies, and some others may not be physically associated. Unfortunately, however, the spectra of individual components have not yet been obtained to give their redshifts. JNLT will be able to perform photometric observations in various bands including the near-infrared range and to determine the redshifts of individual components thus resolved. The result will shed light on the evolution of quasars.

5. Large-Scale Structure

A non-uniform distribution of galaxies to form superclusters and voids has been observed on the scale of 100 Mpc (Oort, 1983; Bahcall, 1988). A motion of galaxies with a velocity of several hundreds of km s^{-1} has been found in a region within about 100 Mpc. If these inhomogeneities are traced back to past epochs, we may be able to know the growth of density and velocity fluctuations in the course of evolution of the Universe.

The angular scale of such inhomogeneities is on the order of one degree at $z \sim 1$. Hence, large-scale structures at $z > 1$ can be studied by the observation of galaxies in several fields of view of JNLT in conjunction with the Hubble diagram and the galaxy count discussed in Section 2. In contrast to the non-uniform distributions of normal galaxies and clusters of galaxies, intergalactic gas clouds responsible for the Lα absorption lines of quasars show a uniform distribution. Since the mass of such a cloud is estimated to be on the order of $10^8 M_0$, a similar uniform distribution may be expected for dwarf galaxies of comparable mass. Dwarf galaxies of $m < 25$ can be observed by JNLT, and their distribution within a distance of 100 Mpc will be studied.

The large-scale structure has been studied mainly for $z < 0.03$ or for the linear scale smaller than 200 Mpc (de Lapparent *et al.*, 1986). The two-point correlation functions for galaxies and clusters of galaxies summarized in Figure 6 show a unique dependence on the distance r, $r^{-1.8}$, but their strengths or the correlation lengths r_c as defined by

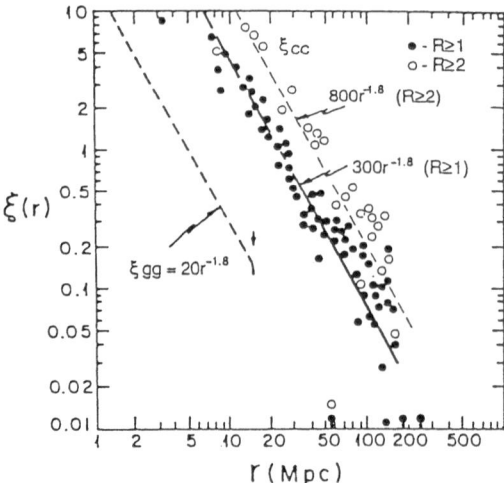

Fig. 6.   The two-point correlation functions for galaxies and clusters of galaxies with richness classes $R \geq 1$ and $R \geq 2$, summarized by Bahcall (1988).

$(r_c/r)^{1.8}$ are different for galaxies and clusters of galaxies, as shown in Figure 6 (Bahcall, 1988). For clusters of galaxies the strength increases with the cluster richness or the number of galaxies in a cluster; the relation between the correlation strength and the mean richness is shown in Figure 7 (Bahcall, 1988). This would suggest the universality of the correlation function if normalized by the mean distance between objects or the

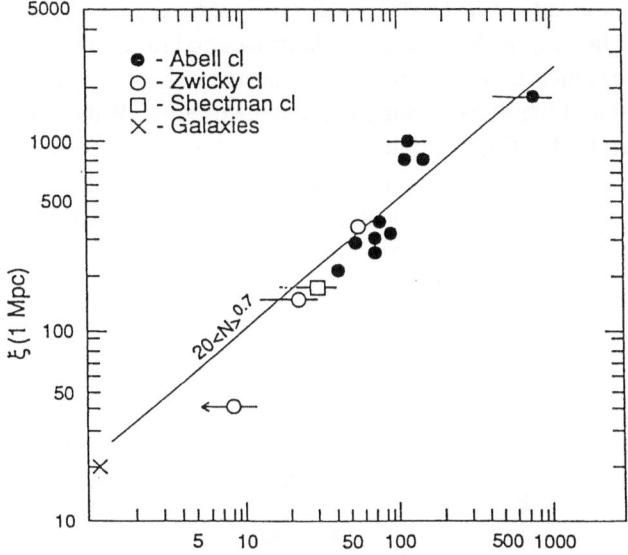

Fig. 7.   The correlation strength versus the average richness given by Bahcall (1988).

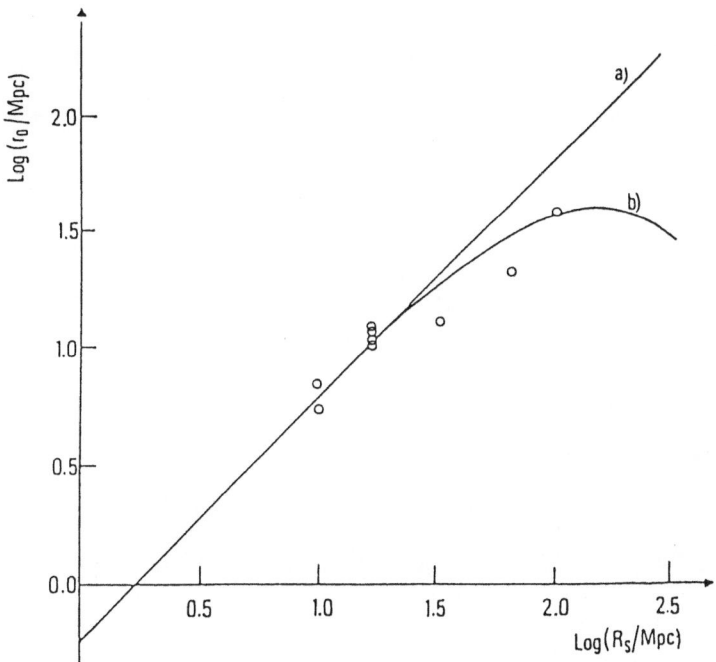

Fig. 8.   The correlation length versus the sample scale given by Calzetti *et al.* (1988). The curve (a) shows
a linear relation, while curve (b) results from a mixture of correlated and random components. Observed
data for clusters of galaxies are shown by circles.

sample radius. Calzetti *et al.* (1988) have demonstrated that the correlation length increases nearly linearly as the sample size increases and then tends to saturate, as seen in Figure 8. The turn-over of this relation is expected if there are two components, one being correlated and the others being randomly distributed. If the correlation function is scale invariant, the fractal dimension is $3 - 1.8 = 1.2$, suggesting that a one-dimensional structure such as cosmic strings may play a dominant role in clustering. In order to see the evolution of correlations, it is important to study correlations at high redshifts, at which the distance is small and a large comoving scale size is attainable.

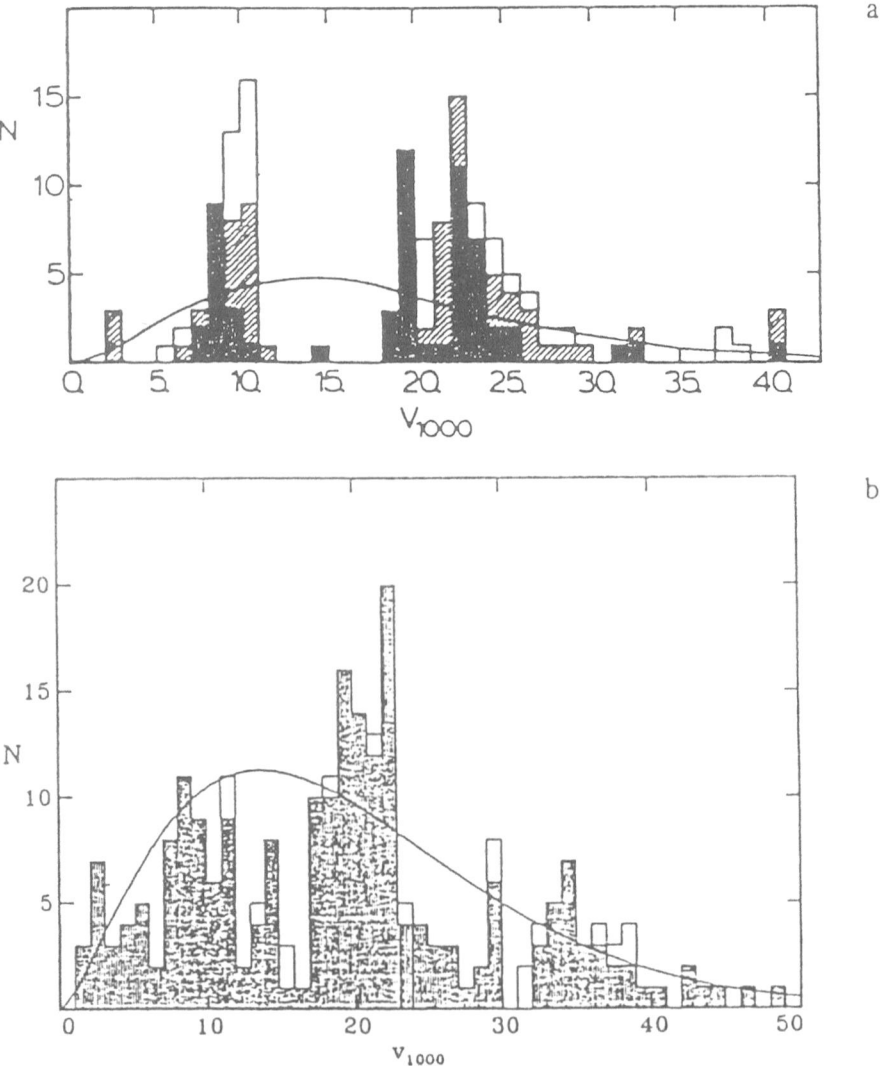

Fig. 9.   The number of galaxies in the recession velocity range of 1000 km s^{-1} versus the recession velocity, summarized by Rood (1988). Data from different fields of view are distinguished by open, hatched and filled histograms. The upper (a) and lower (b) figures are based on 133 and 231 galaxies, respectively. The curves represent the distributions expected in a homogeneous universe.

Since the positive value of the correlation function implies the density higher than the average value, there should be regions with lower densities. In fact, such regions have been found and are called voids. The redshift distribution of 133 galaxies in the Boötes region showed a large, clear gap at about $z \simeq 0.05$, as shown in Figure 9(a) (Rood, 1988). As the number of galaxies is increased to 231, the gap is not conspicuous but is still seen (Figure 9(b)). The study of voids will be extended to more distant regions with use of large telescopes. It seems likely that dwarf galaxies cannot fill a void, but it is an interesting question whether or not the distributions are the same for all types of glaxies and clouds. The question will be answered if the observation of faint objects is attainable with large telescopes.

References

Arnet, W. D., Branch, D., and Wheeler, J. C.: 1985, *Nature* **314**, 337.

Bahcall, N. A.: 1988, *Ann. Rev. Astron. Astrophys.* **26**, 631.

Calzetti, D., Giavalisco, M., and Ruffini, R.: 1988, *Astron. Astrophys.* **198**, 1.

Chokshi, A. and Wright, E. L.: 1988, *Astrophys. J.* **333**, 491.

de Lapparent, V., Geller, M., and Huchra, J.: 1986, *Astrophys. J.* **302**, L1.

Frogel, J. A., Persson, S. E., Aaronson, M., and Matthews, K.: 1978, *Astrophys. J.* **220**, 75.

Grasdalen, G. L.: 1980, in G. O. Abell and P. J. E. Peebles (eds.), 'Objects of High Red Shift', *IAU Symp.* **92**, 269.

Heisler, J. and Ostriker, J. P.: 1988, *Astrophys. J.* **332**, 543.

Jarvis, J. E. and Tyson, J. A.: 1981, *Astron. J.* **86**, 476.

Kristian, J., Sandage, A., and Westphal, J. A.: 1978, *Astrophys. J.* **221**, 383.

Krolik, J. H.: 1988, in 'Active Galactic Nuclei', *IAU Symp.* **134**.

Lebofsky, M. J.: 1981, *Astrophys. J.* **245**, L59.

Le Fèvre, O. and Hammer, F.: 1988, *Astrophys. J.* **333**, L37.

Lily, S. J. and Longair, M. S.: 1984, *Monthly Notices Roy. Astron. Soc.* **211**, 833.

Oort, J. H.: 1983, *Ann. Rev. Astron. Astrophys.* **21**, 373.

Peterson, B. A., Ellis, R. S., Kibblewhite, E. J., Bridgeland, M. T., Hooley, T., and Horne, D.: 1976, *Astrophys. J.* **233**, L109.

Rood, H. J.: 1988, *Ann. Rev. Astron. Astrophys.* **26**, 245.

Sandage, N. A.: 1988, *Ann. Rev. Astron. Astrophys.* **26**, 561.

Shanks, T., Stevenson, P. R. F., Fong, R., and MacGillivray, H. T.: 1984, *Monthly Notices Roy. Astron. Soc.* **206**, 767.

Wooseley, S. E., Taam, R. E., and Weaver, T. A.: 1986, *Astrophys. J.* **301**, 601.

Yoshii, Y. and Takahara, F.: 1988, *Astrophys. J.* **326**, 1.

HOW THE SPACE ASTRONOMY HAS EXTENDED THE HORIZON OF ASTRONOMY*

MINORU ODA

The Institute of Physical and Chemical Research, Wako-shi, Saitama, Japan

(Received 21 February, 1989))

Abstract. On the basis of my experience in the X-ray astronomy, and on some typical highlights of multi-wavelength observations, I emphasize the importance of collaboration between space astronomy and ground-based astronomy.

Space astronomy or astronomy from space acquired its citizenship in astronomy when the optical counterpart of the first discovered X-ray star Sco X-1 was identified by the Tokyo Astronomical Observatory and the Palomar Observatory in June 1966 (Oda, 1968). Let me describe some personal accounts of the early days of the X-ray astronomy.

After the discovery of the celestial X-ray source by Giacconi *et al.* (1962), the basic question was whether the source was nebular or star-like. We needed some kind of new techniques to measure or to set an upper limit on the angular size of the source. In 1964, Giacconi offered a space in his instrument bay of a sounding rocket. In this space was installed a gadget called the modulation collimator (Bradt *et al.*, 1968). The location, angular size and angular structure of the X-ray source may be worked out from modulation of the X-ray flux through the modulation collimator. With a rocket-borne modulation collimator we could place an interesting upper limit of 10', i.e., the source is star-like (Oda *et al.*, 1965).

It was predicted that, if X-rays are generated by thermal emission (free-free emission) of thin hot plasma from the star-like object, it could be an optical source of 12.5th mag with an optical spectrum of UV excess. I asked the following question of Prof. Strömgren who was then visiting professor at MIT: How precisely must we locate an object to interest optical astronomers in undertaking a search for the optical counterpart? His answer was 10'.

Then, Gursky and I designed a rocket experiment to measure the position of Sco X-1 with this precision. The MIT/ASE experiment from White Sands showed that the source is smaller than 10" and led to a location within two small error boxes with a total area of 4 square arc min (Gursky *et al.*, 1966). Tokyo and Palomar Observatories were then prepared to search for the optical counterpart. Jugaku and Osawa of the Tokyo Astronomical Observatory obtained two-colour double-imaged photographs of the sky region and found an unusually blue star, which was proven to be in one of the error

---

* Paper presented at the Symposium on the JNLT and Related Engineering Developments, Tokyo, November 29–December 2, 1988.

boxes. Sandage *et al.* (1966) at Palomar Observatory made comprehensive photometric observations of the star.

In Figure 1 showed are the optical spectra of Sco X-1; indeed the spectra exhibit H and He emission lines which are the characteristic features of the emission from the thin hot plasma. The two spectra indicate a clear contrast of the powers of 1.88 m and 5.1 m telescopes.

Okayama

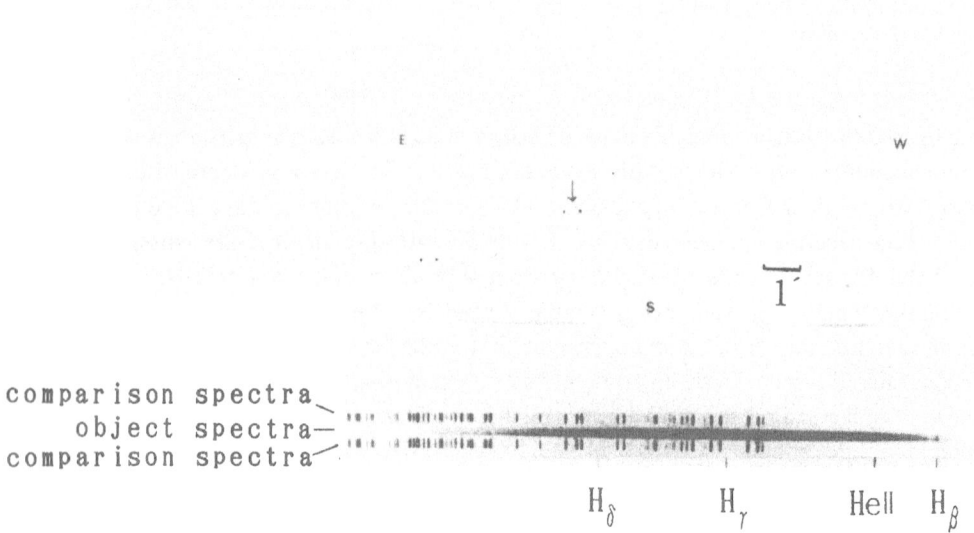

comparison spectra
object spectra
comparison spectra

H_δ H_γ HeII H_β

Palomar

(1966 July)

$16^d\ 5^h21^m-\ 8^h21^m$UT

$17^d\ 5^h10^m-\ 8^h27^m$UT

$18^d\ 4^h18^m-\ 7^h18^m$UT

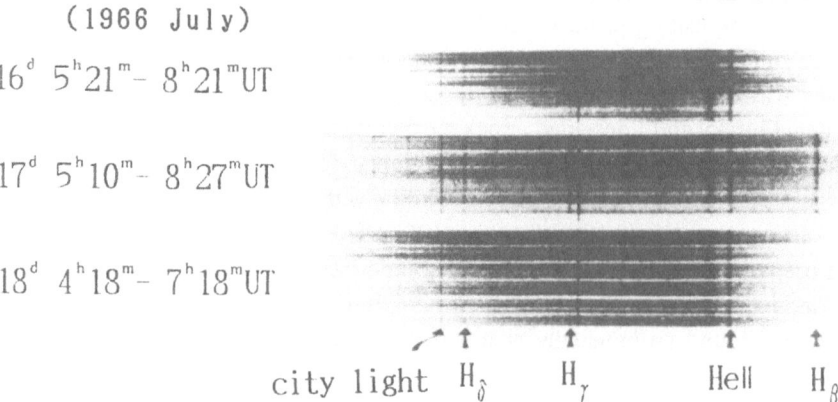

city light H_δ H_γ HeII H_β

Fig. 1.   Optical spectra of Sco X-1, taken at the Okayama Astrophysical Observatory (*upper panel*) and Palomar Observatory (*lower panel*).

Figure 2 is a chronological diagram to show the X-ray missions and their achievements since then. In the diagram represented are the major missions including UHURU, *Einstein* Observatory, a series of Japanese missions and highlights of findings on physics

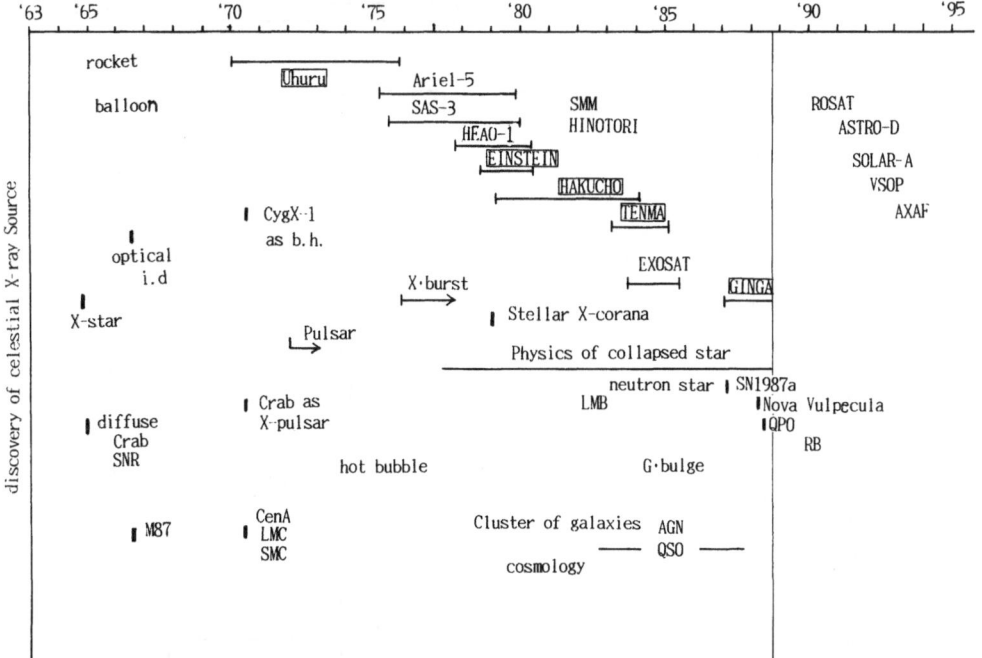

Fig. 2.   Chronological diagram of X-ray missions and discovery of important celestial objects.

of collapsed stars, SN 1987a, QSO, and other extra galactic sources. The diagram is brief, but it may indicate a typical example of the history of the progress of science being alternate succession of unification by theorists and diversification by experimentalists. Space astronomy together with ground-based astronomy has promoted science in diverse directions.

Figure 3 represents some typical space missions on a diagram of the wavelength range versus astrophysical objects; they include the missions for solar observations, several modest missions for hard X-ray band covering physics of collapsed objects and galactic bulge and AGN, and the *Einstein* Observatory which was the first and only satellite with the reflective X-ray telescope on board with high sensitivity for a soft X-ray band.

Figure 4 exhibits the missions on the diagram of sensitivity versus angular resolution for position and the field of view and the diagram of sensitivity versus energy coverage. These diagrams are prepared only to present a bird's eye view of astronomy from space.

In what follows I shall present a few typical highlights of multi-wavelength observations by which diverse directions in astronomy were opened, new astrophysical objects were added to our repertory and deeper insights in various phenomena were achieved.

(1) Bursting radiations from a neutron star were utilized as a search light to explore

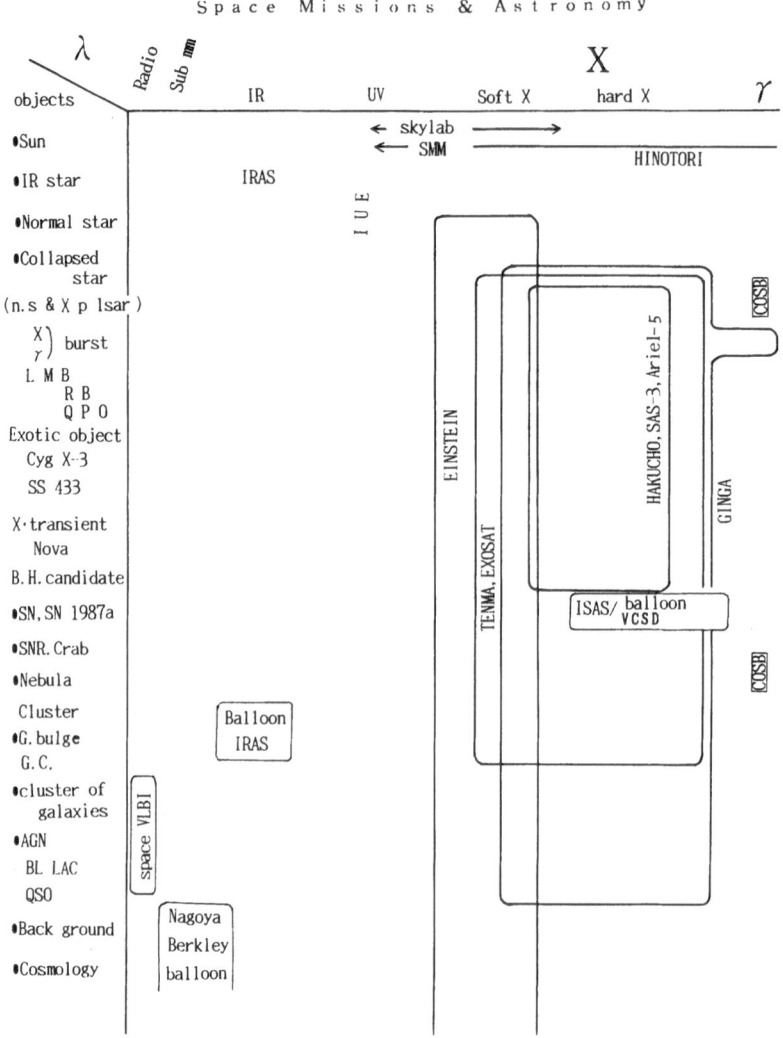

Fig. 3.   Space Missions and Astronomy. The spectral domain and celestial objects.

the environment of the neutron star by means of X-ray/optical simultaneous observations: The cloud in the vicinity of the neutron star is irradiated by the X-ray burst from the neutron star and emits the optical burst. About ten cases of simultaneous X-ray/ optical observations were obtained (Pedersen *et al.*, 1982; Matsuoka *et al.*, 1984). The delay and smearing (broadening) in time of the optical burst with respect to the X-ray burst led to knowledge on the distance to and the size of the cloud. Further observations for more cases of the bursts with larger optical telescopes will provide a deeper understanding in the environment around the neutron star.

(2) Numerous galactic objects, which have been observed over multiwavelength bands, promoted physics of the collapsed stellar objects (see, e.g., the books of *Recent*

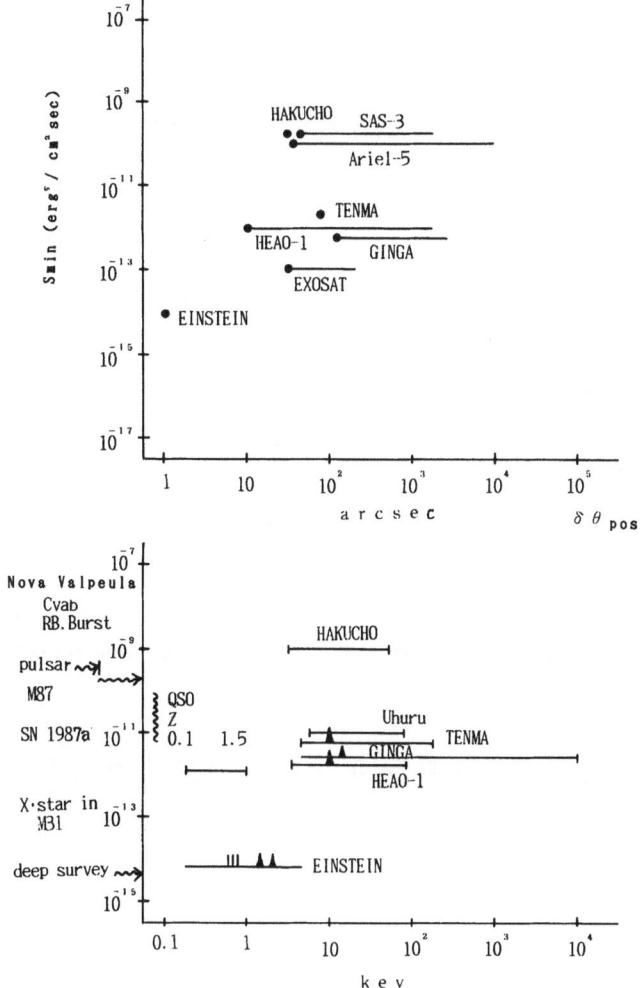

Fig. 4.   Sensitivity versus angular size. The filled circle denotes the angular resolution and the right-hand end of the bar denotes the field of view (*upper panel*). Sensitivity versus energy coverage, along with the typical energy levels for some X-ray sources (*lower panel*).

Multi-Frequency Observation and Theoretical Development, edited by H. Drechsel, Y. Kondo, and J. Rahe, D. Reidel Publ. Co., Dordrecht, Holland, 1986; and *Multi-wavelength Astrophysics*, edited by F. Cordova, Cambridge University Press, Cambridge, 1988. References are found therein.)

(3) Figure 5 summarizes the temporal variations of neutrino, optical, X-ray flux from SN 1987a until mid 1988. Of course, SN 1987a continues to be still the hot issue.

(4) Balloon observations of Crab Nebula had been performed from Palestine, Texas, by the collaboration of ISAS/UCSD in early 70's. Figure 6 illustrates the results together with that of *Einstein* Observatory: Now we know that the X-ray structure of

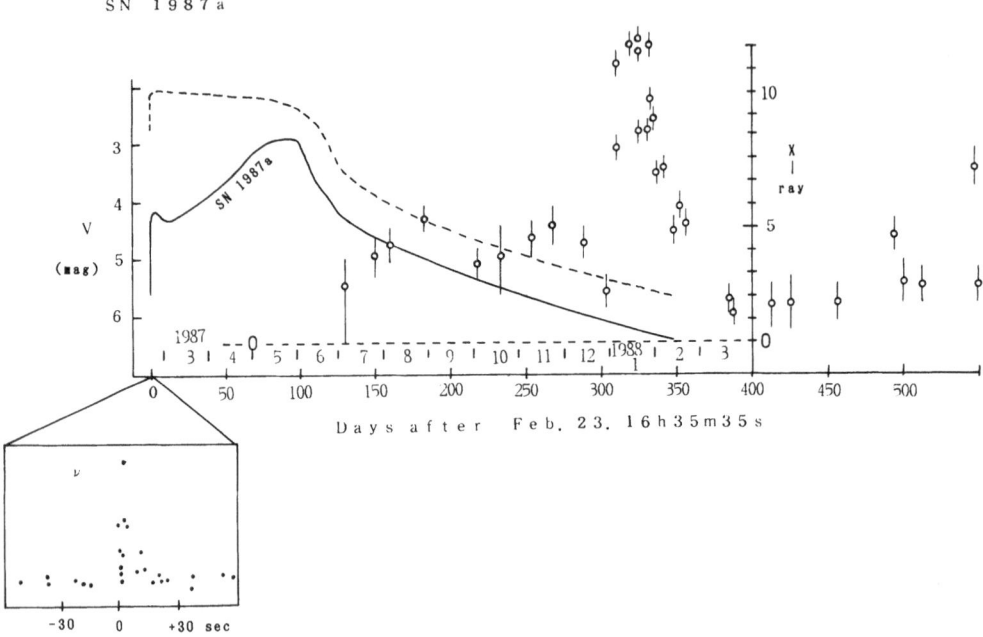

Fig. 5.   Temporally variations of Neutrino (ν), Optical (V), X-ray fluxes from SN 1987a.

Crab nebula

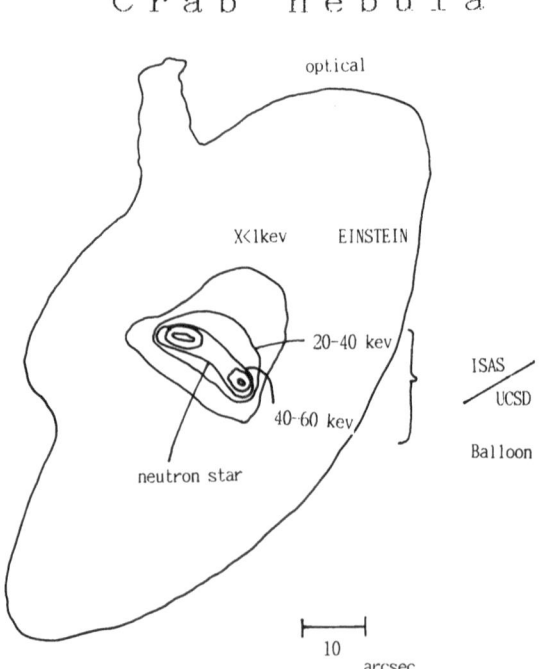

Fig. 6.   X-ray Structure of the Crab Nebula.

the Crab may be represented by a torus or a doughnut and a cavity around the neutron star (Pelling *et al.*, 1987).

(5) X-ray observations of clusters of galaxies with the *Einstein* Observatory compared with optical observations gave clues to understand the evolution and dynamics of the clusters of galaxies (see, e.g., Forman and Jones, 1982).

Quoted above are only a few examples, but I think these cases already indicate how fruitful future we may expect by space observations together with optical observations of high quality.

References

Bradt, H., Garmire, G., Oda, M., Spada, G., Sreekantan, B. V., Gorenstein, P., and Gursky, H.: 1968, *Space Sci. Rev.* **8**, 471.

Forman, W. and Jones, C.: 1982, *Ann. Rev. Astron. Astrophys.* **20**, 547.

Giacconi, R., Gursky, H., Paolini, P., and Rossi, B.: 1962, *Phys. Rev. Letters* **9**, 439.

Gursky, H., Giacconi, R., Gorenstein, P., Waters, J. R., Oda, M., Bradt, H., Garmire, G., and Sreekantan, B. V.: 1966, *Astrophys. J.* **146**, 310.

Matsuoka, M. *et al.*: 1984, *Phys. Rev.* **283**, 774.

Oda, M.: 1968, *Space Sci. Rev.* **8**, 507.

Oda, M., Clark, G., Garmire, G., Wada, M., Giacconi, R., Gursky, H., and Waters, J. R.: 1965, *Nature* **205**, 554.

Pedersen, H., Lub, J., Inoue, H., Koyama, K., Makishima, K., Matsuoka, M. Mitsuda, K., Murakami, T., Oda, M., Ogawara, Y., Ohashi, T., Shibazaki, N., Tanaka, Y., Hayakawa, S., Kunieda, H., Makino, F., Masai, K., Nagase, F., Tawara, Y., Miyamoto, S., Tsunemi, H., Yamashita, K., Kondo, I., Jernigan, J. G., Van Paradijs, J., Beardsley, A., Cominsky, L., Doty, J., and Lewin, W. H. G.: 1982, *Astrophys. J.* **263**, 325.

Pelling, R. M., Paciesas, W. S., Peterson, L. E., Makishima, K., Oda, M., Ogawara, Y., and Miyamoto, S.: 1987, *Astrophys. J.* **319**, 416.

Sandage, A. R., Osmer, P., Giacconi, R., Gorenstein, P., Gursky, H., Waters, J. R., Bradt, H., Garmire, G., Sreekantan, B. V., Oda, M., Osawa, K., and Jugaku, J.: 1966, *Astrophys. J.* **146**, 316.

RADIO ASTRONOMY AND VERY LARGE TELESCOPES*

NORIO KAIFU

Nobeyama Radio Observatory, National Astronomical Observatory, Nobeyama, Minamimaki, Japan

(Received 30 March, 1989)

Abstract. The current achievements of the observational abilities of radio astronomy is briefly reviewed putting emphasis on the imaging capability. The new projects in radio astronomy are discussed in connection with the new generation of optical/IR telescope projects.

The imaging with higher resolution and higher quality has been a central target of the developments of the astronomical instruments for hundreds of years. The recent remarkable situation is that the radio, IR, and X-ray images are attaining the comparable (sometimes exceeding) qualities to those of the images with the optical telescopes, and such improvements in various wavelength regions provided us the multi-dimensional aspects of the distant objects which lead us to the direct understanding of the substantial of the phenomena in the Universe.

1. Imaging with Single Dish Radiotelescopes

The single dish radio observations have achieved $10''$ spatial resolutions in the mm and sub-mm wavelength regions. The homologous deformation technique, as well as the various developments on the telescope design, measurements and adjustment, provided a jump of the surface accuracy of large paraboloid dishes. Very recently the JCMT (JAC, Hawaii) 15-m telescope has achieved 8 arc sec resolution at 350 µm wavelength. The discovery of small protostellar disks, structure of the bipolar molecular flows, and numbers of new results in interstellar clouds, star forming regions, late-type stars, the galactic center and external galaxies were produced by recent high-resolution observations with these large mm and sub-mm wave telescopes.

The main high-resolution single dish telescopes in the mm and sub-mm wave regions are:

Nobeyama	45-m telescope	$15''$ resolution at 115 GHz,
Pico Velta	30-m telescope	$10''$ resolution at 230 GHz,
Hawaii (JCMT)	15-m telescope	$8''$ resolution at 860 GHz.

However, the further improvements seem not to be rosy. One of the difficulties is the pointing accuracy problem of large dishes. Instead, the higher capability of the imaging

---

* Paper presented at the Symposium on the JNLT and Related Engineering Developments, Tokyo, November 29–December 1988.

Astrophysics and Space Science **160**: 27–29, 1989.
© 1989 *Kluwer Academic Publishers.*

by applying the sensitive multi-receiver system or multi-elements heterodyne detector receiver (10–20 elements so far) are being developed intensively in some radio observatories and it will add considerable power to the mm and sub mm single dish telescopes by combining with the larger spectrometers and improved imaging software systems.

2. Imaging with Super Synthesis Telescopes

The cm and mm wave supersynthesis arrays obtain the 'radio pictures' with 5″–0.1″ resolution, comparable to those of optical/near IR images (e.g., VLA, 0.1″ at 15 GHz, MERLIN, 0.35″ at 1.6 GHz). The mm wave arrays can provide velocity resolved high spatial resolution images of cold/warm molecular gas. Although the objects are more or less limited because of relatively poor sensitivity, the mm-wave supersynthesis arrays (listed below) are producing numbers of exciting results especially on the external galaxies.

Hat Creek	6-m × 3 (planned to be 6-m × 6),
Owens Valley	10-m × 3,
Nobeyama	10-m × 5,
Plateau de Buel	15-m × 4 (under adjustments).

The new powerful mm-wave imaging instruments which aim to the 1″ or higher resolution, very quick and high-quality images with various molecular lines and continuum emission are proposed in Japan (Nobeyama) and U.S.A. (NRAO).

Nobeyama plan	10-m × 30,
	40–230 GHz, baseline $\fallingdotseq$ 600 m ;
NRAO plan	7.5-m × 40,
	100–300 GHz, baseline $\fallingdotseq$ 500 m .

These instruments can be called as 'mm-wave VLA', but with very high abilities of spectroscopic imaging and will produce enormous amount of velocity resolved pictures. The Nobeyama plan puts emphasis on the high quality imaging of cold and general medium in the galaxies and in our Galaxy, and the NRAO plan aims at rather warm and compact sub-mm objects.

3. Imaging with VLBI

The extremely high-resolution images (milli-arc sec or higher) can be obtained by the VLBI observations for limited objects with very high brightness like quasars and maser spots. Recent successful experiments of the mm-wave VLBI (Japan–Sweden–U.S.A.) resulted the first mm-wave maps of 3C84 and 3C273 with the resolution of 50 μ arc sec. The high quality VLBI maps will become to be available soon by new large instruments in U.S.A. (VLBA, 25-m × 10) and in Australia (AT, 25-m × 6).

A considerable progress of the VLBI imaging quality as well as the resolution will be achieved by the space VLBI. The baseline between the orbit antenna and the ground-based antenna rotates rapidly and the rich Fourier components can be used for the image construction. The undergoing first space VLBI projects are:

RADIOASTRON   USSR, 1994–95?
                20-m orbit antenna, 0.37, 1.6, 5, 22 GHz ,

VSOP   JAPAN, 1994–95?
                10-m orbit antenna, 1.6, 5, 22 GHz .

High-quality VLBI images from above-mentioned newly coming instruments should be compared with the Space Telescope images or with the optical/IR interferometers which are also expected to be available in the near future.

4. Spectroscopy, sub-mm and IR Regions

It should be emphasized that the high spectral resolution mm wave observations combined with the high spatial resolution images have provided extremely rich information on the structure and dynamics of the molecular clouds and related various phenomena. The obvious next targets are the sub-mm wave and IR regions, and the developments of the techniques and instruments for the velocity-resolved imaging in these regions are essentially important.

The JNLT could be one of the most powerful instruments for the near IR and intermediate IR (1–30 μm) with its design suitable for high resolution (0.2″–0.3″) and IR-oriented optics. The developments of new and excellent IR detector (and spectrometer) systems for JNLT should be exciting job.

The sub-mm wave capability of JNLT is also high. The sensitivity of JNLT for compact sources at 350 μm (860 GHz) is comparable to that of JCMT 15-m sub-mm telescope, and the beam will be very well shaped with HPFW of 11.5″.

Finally I wish to stress again the importance of the efforts to develop novel instruments which make it possible or much easier to observe the new world.

SECOND SESSION

LARGE TELESCOPE PROJECTS

STRUCTURE OF THE KECK TELESCOPE – AN OVERVIEW*

STEFAN J. MEDWADOWSKI

Consulting Engineer, San Francisco, U.S.A.

(Received 18 January, 1989)

Abstract. The structure of the Keck Telescope is briefly described. The design required an innovative approach made necessary by the revolutionary nature of the segmented primary mirror, by the very stringent weight and cost limitations, and by observational and operational needs. Analysis of a progressively more detailed computer model predicts that all design objectives will be met, as shown in a summary of performance characteristics. The paper is illustrated with a number of drawings.

1. Introduction

The W. M. Keck Telescope, currently (December 1988) in fabrication, is a 1:1.75 altazimuth telescope with a 10 m (equivalent) mirror. On completion of installation near the peak of Mauna Kea on the island of Hawaii, it will be the world's largest optical telescope. As is well known, its mirror is of a novel, segmented type, and consists of thirty six elements, hexagonal in the projected plan. Each segment is only 75 mm thick, so that the total weight of the mirror is approximately 14.4 tons. Because of their thinness, each segment is supported in the normal direction at 36 points using a whiffle tree-type device. In the radial direction, each segment is supported on a single radial post.

Requisite stiffness of the primary is provided through a combination of structure stiffness and electronic stiffness. Motions in the plane of the mirror and rotation about optical axis are resisted by the stiffness of the structure itself. Motions normal to, and rotations out of the plane of the mirror are resisted electronically, by sensing and sending to a computer the actual position of the segments, calculating required corrections, and finally by executing appropriate actuator motions.

The use of a novel technology was imperative. The moving parts of the 5 m Hale telescope weigh approximately 550 tons. If the same technology were used, the weight of the moving parts of the 10 m Keck telescope would have been on the order of several thousand tons, placing extravagant demands on the mechanical system, and making construction cost prohibitively large.

From the point of view of the designer of the structure this novel design of the mirror presented a particular challenge. In an early report (Medwadowski, 1981a), we wrote "The structure of the telescope constitutes a platform which supports the primary mirror, the secondary mirrors, and the many sophisticated instruments which are used during observing. In this sense, the structure has no intrinsic value of its own. However, the demands placed on its performance, particularly on its stiffness, are exceedingly

---

* Paper presented at the Symposium on the JNLT and Related Engineering Developments, Tokyo, November 29–December 2, 1988.

stringent. As an example, the deflection to span ratio of the mirror cell is approximately 1 to 10 800, while in a more typical structure the same ratio is approximately 1 to 360, or some thirty times larger. One may expect that, to meet such stringent requirements, the structure of the telescope has to be as innovative in its design as the segmented mirror itself, and that was the spirit in which we approached the task. Furthermore, we were not unmindful of the aesthetic as well as the operational and economic attributes of a weight-efficient design."

The writer first became involved with the University of California Ten-Meter Telescope in the Fall of 1979 at the invitation of Jerry Nelson, the initiator of the project and the moving force behind it. At that time, the first concepts of the structure were developed. A fairly detailed report covering conceptual design, including estimates of cost, and construction time considerations, was completed in December 1980 (Medwadowski, 1981a). In the ensuing years we performed a number of investigations aimed at refining and optimizing the structure, and at studying specific aspects of its response (Medwadowski 1981b, 1982a, b, c, 1984, 1985). In 1985 California Institute of Technology joined the effort. The project became the Keck Telescope, operated by CARA, with Gerald Smith in charge as project manager. Rapid progress ensued. Preliminary design was completed in 1985, and final design, started in December 1985, was completed in early 1987 (Medwadowski, 1987a). All final analyses, drawings, weight and balance control tables, as well as specifications for the structure were prepared in the writer's offices in San Francisco during this period.

The methodology of design was to devise a geometry of the structure or a given sub-assembly based on past experience, on anticipated performance requirements, and on ease and economy of fabrication, transportation and erection. Next, this geometry, as well as its topology and mass distribution, were optimized using formal and informal techniques. Throughout, we strived to transfer forces as directly and as efficiently as possible. This led to the conclusion that, as much as possible, axial force transfer mechanism should be used, since its stiffness to structure weight ratio is greatly superior to the force transfer mechanism associated with bending. For this reason, sub-assembly geometries were designed to be those of a truss, with elements intersecting at centers of nodes. It was recognized that this approach leads also to superior dynamic behaviour. Also, particularly careful attention was paid throughout the design to the detailing of joints and connections for ease of fabrication and assembly, bearing in mind the need for shop assembly, tests and disassembly, followed by crating, shipping to the site, and then the final assembly at the observatory site. For this reason essentially all shop joints were designed welded, and all field joints bolted using high tension bolts.

Throughout the design of the structure, a period of approximately 7 years, we worked very closely with Jerry Nelson, currently Principal Scientist of the Keck Telescope Project at CARA, and with Hans Boesgaard, currently the Telescope Structure Manager for CARA, who joined the project in 1984 to produce preliminary mechanical design (final mechanical design was performed by TIW Systems, Inc. of Sunnyvale, California, the effort being headed by Joe Steffey). A detailed description of the Keck Telescope and Observatory project was given by Nelson et al. (1985). In a paper parallel to this contribution, Nelson (1988) presents the current status of the project as a whole. Here we describe briefly the structure of the Keck telescope itself. Its fabrication in Taragona, Spain, has been nearly completed (cf. Figures 5 and 6), and we anticipate that it will be shipped for final assembly at the site of the Observatory near the peak of Mauna Kea in early 1989.

2. General Description of the Structure

The structure consists of two principal parts: the *tube* which rotates on elevation axis, and the *yoke* which rotates in azimuth on azimuth journal, and which supports the tube. The tube is allowed to rotate through an angle of slightly beyond zenith and slightly below horizon. The yoke rotates in azimuth through a total angle of approximately 229°. The overall structure is shown in Figure 1. Additional illustrations can be found in a previous publication (Medwadowski, 1987b).

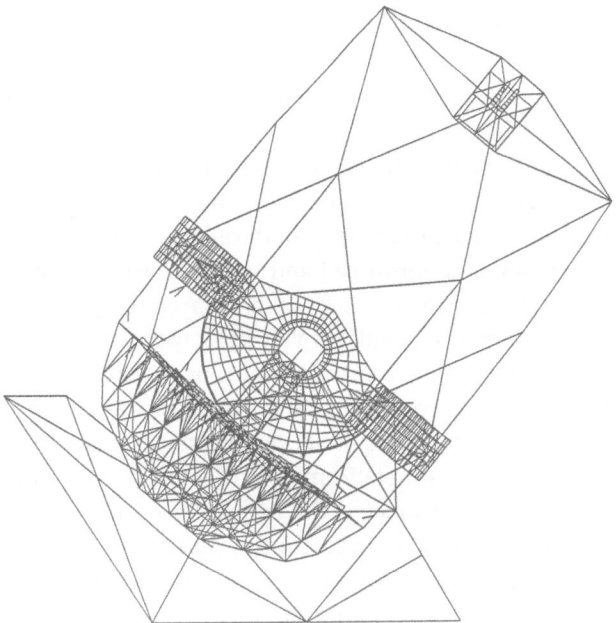

Fig. 1.   Keck Telescope computer model – side elevation.

The tube, in the large, is a hollow non-prismatic tube-like space frame; its hexagonal cross-section derives from the shape of the primary. The tube consists of the following principal components.

Upper assembly, which consists of the secondary socket, spiders, and upper ring. The spiders are pre-stressed for two reasons. The first is to prevent compression in the spiders under gravity loads. The second is to provide satisfactory torsional frequency of the secondary socket, necessary because the spiders intersect at the optical axis to reduce light loss.

Upper tube, which is a spatial equivalent of the classical Serrurier truss. Its geometry was first proposed by Lubliner (1981). Later, extensive optimization studies disclosed that more efficient geometry exists; it was not used because of operational considerations, i.e., the ease of primary segment replacement.

Elevation ring (cf. Figure 2), a hexagonal in plan non-prismatic box girder 0.76 m wide, from 2 m to 4 m deep, with 1.8 m Nasmyth openings, and with Nasmyth stubs

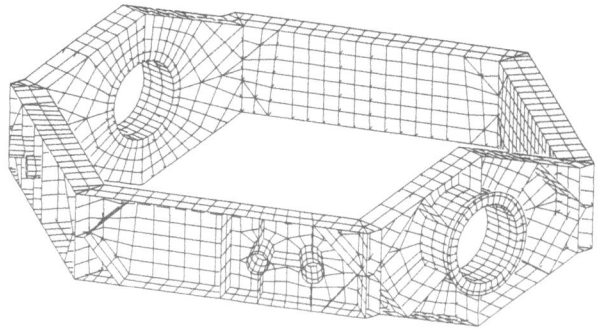

Fig. 2.   Computer model of elevation ring with outer web removed.

supporting elevation journals and longitudinal support journals. Attached to the elevation ring are drive disks 4 m radius, one at each side to enhance smooth operation, reduce tangent arm forces and the tendency for the tube to twist under unbalanced loads.

Lower tube, which supports mirror cell and mirror cover assembly.

Mirror cell with tertiary tower: Mirror cell is a three-layered, non-prismatic space frame (Medwadowski 1982a, 1985), very stiff yet very light, with the weight of the mirror cell of the Keck Telescope being approximately 8 tons, or roughly only 40% of the weight of the primary with its support system (PMSS). Top layer was designed to fit the layout of segment actuators, and is curved to fit the curve of the primary. Middle layer is level, and supprots a permanent service platform. It is of interest to note that the mirror cell structure is point symmetric. Symmetry with respect to *x*- or *y*-axis was abandoned in order to match the point symmetry of the primary mirror support system. Elements of the mirror cell are either pipes or built-up sections designed for ease of assembly and ease of interfacing to the remainder of the structure. Essentially all field connections, including connections of the tertiary tower, are designed to be bolted. Tertiary tower serves to receive and support instrument modules at forward Cassegrain, and other positions. It is a nonprismatic hollow hexagonal space frame, with joints welded to ensure stiffness of the cross-section. Tertiary tower is designed to be fabricated as a single unit which is later bolted to the mirror cell.

The *yoke* is also a space frame structure (Medwadowski, 1982b, c). Its principal components are the *main triangles* and the *transverse structure* which, together, support the tube. Their geometry, topology and mass distribution were optimized for lowest image motions associated with given mass of the structure. In addition, the yoke supports a number of operational platforms, including two Nasmyth platforms with the capacity of 10 tons each, the Cassegrain platform used to install Cassegrain modules into the tertiary tower, and several access and equipment platforms with access stairs.

Fabrication of the structure, tube and yoke, was specified to tolerances readily attainable in normal structural steel work. All dimensions shown on the drawings refer to ambient temperature of 2 °C.

Supports: The tube is supported by the yoke with the aid of hydrostatic elevation bearings. These are spaced 12.2 m apart, and are capable of transferring forces parallel to the z-axis (vertical), and to the y-axis (horizontal axis perpendicular to the elevation axis). Restraint in the direction of the x-axis (along the elevation axis) is provided by hydrostatic bearings at each end of the elevation axis. Finally, unintended rotation of the tube about the elevation axis is controlled by tangent arms, part of the elevation drive

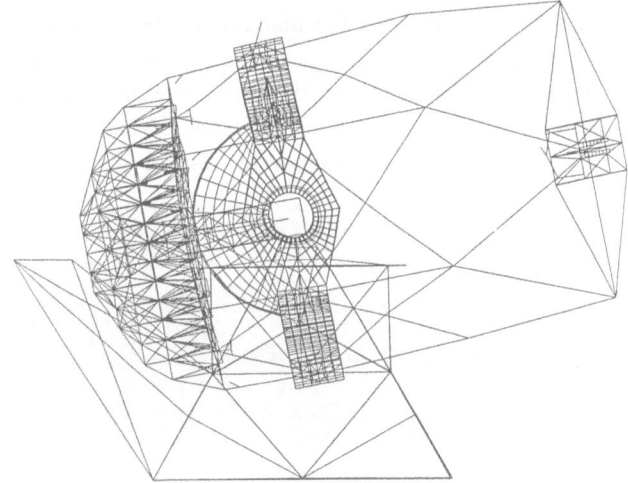

Fig. 3. Dynamic response – tube rotation about elevation axis f = 5.12 Hz.

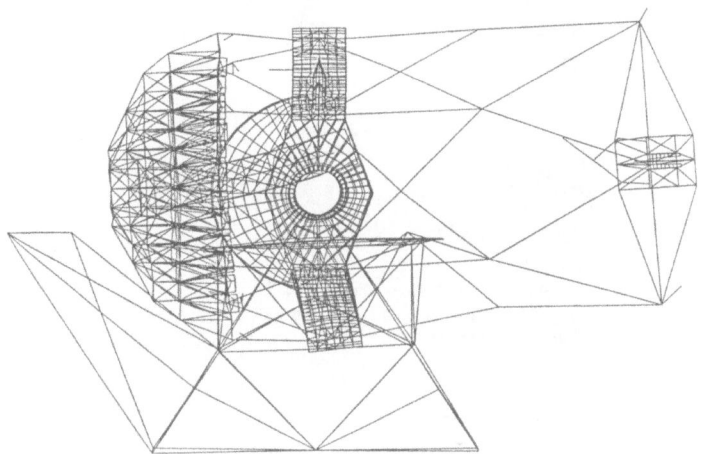

Fig. 4. Dynamic response – tube rotation about elevation axis f = 9.90 Hz.

system, which are attached to the yoke and to the rim of the drive disks 4 m in radius, located at each end of the elevation axis. We note that the tube with all instruments is balanced about the elevation axis; provision is made for fine tuning of the balance by supplying two adjustable and movable loads at the front of the elevation beam.

Hard stops, using shock absorbers as required, are provided to prevent over-motion due to static or dynamic (seismic) excitations, both at tube and at yoke levels.

To provide for a secure tie-down of the structure with one or more instrument modules removed (i.e., when the tube is no longer balanced on the elevation axis), there is a provision for a positive static restraint of the tube, pointing to horizon, by the yoke structure.

The yoke is supported at four points, at the base of the two main triangles, in such a way that the hydrostatic azimuth bearings are allowed to travel only along the perimeter of the azimuth journal circle. The diameter of the latter is equal to the diagonal of the square of the base of the main triangles with the side of 12.2 m. The height of the yoke was determined from the condition that the angle of the main triangle be 60°, shown acceptable in an optimization study and desirable because of ease of fabrication.

Interfaces: The tube and the yoke support a number of mirrors and instrument

Fig. 5.   Fabrication in progress-tube on temporary supports, side view (photo courtesy J. Steffey). For colour reproduction of this figure see colour section.

modules, and are interfaced to several mechanical devices required for the operation of the telescope. These are described briefly in the following.

Each of the 36 primary mirror segments is supported on a subcell, fabricated to very close tolerances, and connected to the structure of the mirror cell in such a way that fine adjustment of final segment location is possible. Conceptual design of the sub-cell was prepared by the writer to optimize geometry and mass distribution to meet stringent stiffness, weight limitation, and flexibility of attachment requirements (Medwadowski, 1984); detailed design was performed at Lawrence Berkeley Laboratory under the direction of Andy Dubois. In order to facilitate sub-cell interface to the telescope structure, the top layer of the mirror cell (see below) was specified to fabrication tolerances more stringent than those of the structure as a whole.

One of the requirements placed on the design of the structure was ease of exchange of instrument packages at the secondary and at the tertiary tower, with the target time

Fig. 6. Fabrication in progress-tube on temporary supports, front view (photo courtesy J. Steffey). For colour reproduction of this figure see colour section.

of exchange being not more than 20 min. For this reason, instrument packages at specific locations were designed to have identical mass and location of center of gravity. Appropriate rails were designed into the secondary socket and tertiary tower to facilitate the exchange.

3. Analysis

Throughout the design period, from conceptual through final design, we analyzed the structure of the telescope and its various sub-assemblies and elements by using progressively more refined models. Paper and pencil and well as computer techniques were used.

Computer models themselves became progressively more refined the closer we were to the final analysis. Final performance characteristics were established on the basis of a finite element based computer model which consisted of a total of 7529 elements (4178 plates, 2845 beams, and 146 others) and 4525 nodes; the structure had 26153 degrees of freedom. It is of interest to note that the static run time (on a VAX 8650) was approximately 20.8 hr, and the dynamic run time, with 24 frequencies, was over 50 hours; elapsed times were 3 to 5 times longer. The final element-based software was ALGOR SUPERSAP. An idea of the complexity of the final model can be gained by examining the elevation ring shown in Figure 2. Hard disk space required was in excess of 1 gigabyte. Effects of various loads were evaluated with the tube pointing to $0°$, $20°$, $40°$, $60°$, $80°$, and $90°$ measured from zenith. Loads included gravity; wind at secondary, primary, Nasmyth platform and Cassegrain platform; the effect of unevenness in the azimuth journal; and seismic actions. In addition, various sub-assemblies, elements and joints were analyzed in detail from the point of view of static response, dynamic behaviour, and stability (Medwadowski, 1987a).

4. Performance Characteristics

In the following we summarize the performance characteristics which we reported at the CARA Telescope Structure Critical Design Review (CARA 1987), and in our Summary Report (Medwadowski, 1987a).

Physical characteristics: These are summarized in Tables I and II.

Image motions: Performance characteristics of the structure under gravity and wind loads, for tube pointing to $40°$ measured from zenith, are summarized in Table III.

Dynamic characteristics: A partial listing of dynamic characteristics, based on a half-model of the structure, is given in the following.

– Global modes:

(1) Rotation of tube about elevation axis: first mode: 5.12 Hz; second mode: 9.90 Hz (refer to Figures 3 and 4).

(2) Rotation of structure about azimuth-tube to horizon: 8.33 Hz.

– Local modes:

(1) Torsion of secondary socket about optical axis: 4.13 Hz.

TABLE I

Keck telescope weights and moments of inertia

Element	Weight (kgf)	Millions of (kgf s^2 cm^{-1}) cm^2 (mass × length^2 units)		
		I_x	I_y	I_z
Tube to zenith	109000	39.4	43.8	24.3
Tube to horizon	109000	39.4	24.3	43.8
Yoke	161000	–	–	104.5
Total to zenith	270000	–	–	128.8
Total to horizon	270000	–	–	148.3

TABLE II

Summary of Keck tube weight

Tube element	Weight (kgf)
Secondary socket	2957
Spiders, upper ring and upper tube	16724
Elevation ring	25439
Nasmyth stubs	6465
Disk drives	5323
Lower tube and mirror cover supports	4055
Mirror cell	13212
Tertiary tower	3250
Primary mirror & its support system	21053
Instruments and baffles	9495
Miscellaneous items	1270
Total, tube on elevation bearings	109244

(2) Pendulum motion of lower assembly (partial model): x-direction: 10.85 Hz; y-direction: 11.40 Hz.

(3) Translation along x-axis: first mode: 4.04 Hz; second mode: 7.76 Hz.

5. Summary and Conclusions

This paper contains a brief description of the structure of the Keck Telescope, currently in fabrication. The structure represents an innovative departure from the traditional telescope structures used as recently as in the design of the CFH telescope on Mauna Kea. As a result, its weight to mirror diameter ratio has been dramatically reduced compared to other telescopes. Its lowest observationally meaningful natural frequency is approximately 5.12 Hz, which is associated primarily with the rotation of the tube about elevation axis. Analysis of an idealized computer model predicts that despite decenter, net tilts and image motions will all be within acceptable limits. It is expected that the Keck telescope will soon take its place in the forefront of the many instruments which wll expand the limits and contribute to the understanding of the visible universe.

S. J. MEDWADOWSKI

TABLE III

Performance characteristics at secondary, $f/15$

Quantity	Angle from zenith				Average, hor.-zenith	Objective
	0°	40°	60°	90°		
Gravity loads						
Despace (cm)	0.00344	0.00259	0.00163	− 0.000105	0.00177	<0.002
Decenter (cm)	− 0.00272	− 0.00763	− 0.00891	− 0.00877	− 0.00303	<0.02
Net x-tilt (arc sec)	− 0.383	− 4.92	− 4.58	− 3.06	0.385	< 15
Im. motions (arc sec)	9.89	11.32	12.40	14.38	2.245	
Wind at secondary, 50 kgf, x-direction						
Im. motions (arc sec)	0.124	0.110	0.099	0.090		<0.25 @ 15 kgf
Wind at secondary, 50 kgf, y-direction						
Im. motions (arc sec)	0.243	0.231	0.220	0.200		0.25 @ 15 kgf
Wind at secondary, 50 kgf, z-direction						
Despace (cm)	0.000692	0.000692	0.000692	0.000692		<0.005
Im. motions (arc sec)	0.001	0.022	0.030	0.036		<0.25 @ 15 kgf

Notes:
(1) All quantities are listed as calculated.
(2) In all cases, image motions listed are those calculated due to motions and tilts in the tube.

Acknowledgements

Several individuals in the writer's office contributed at various times to the design of the structure, most particularly Dave McCormick, Jose Sandoval and Peter Wrona. Work on the design of the structure of the Keck Telescope was perhaps the happiest experience of the writer's professional career, due primarily to the opportunity of working with many first class scientists and engineers, including Hans Boesgaard, Andy Dubois, Bill Irace, Terry Mast and Joe Steffey. Above all, the inspiration resulting from working closely for a number of years with Jerry Nelson has been much appreciated.

References

CARA: 1987, *Telescope Structure Critical Design Review*, pp. SJM/87.
Lubliner, J.: 1981, *Preliminary Design of the Tube for the TMT*, UC TMT Report No. 50.
Medwadowski, S. J.: 1981a, *Conceptual Design of the Structure of the UC Ten Meter Telescope, Final Report, Phase II*, UC TMT Report No. 59.
Medwadowski, S. J.: 1981b, *UC TMT Pier Rotation due to Wind Action on the Observatory Dome*, UC TMT Report No. 53 (revised 1984).
Medwadowski, S. J.: 1982a, *An Investigation of the UC TMT Mirror Cell Structure*, UC TMT Report No. 70.
Medwadowski, S. J.: 1982b, *An Investigation of the UC TMT Yoke Structure*, UC TMT Report No. 71.
Medwadowski, S. J.: 1982c, *An Alternate Geometry for the UC TMT Yoke Structure*, UC TMT Report No. 72.
Medwadowski, S. J.: 1984, *Subcell Structure for the Ten Meter Telescope*, UC TMT Technical Note No. 121.

Medwadowski, S. J.: 1985, *Optimized Mirror Cell for the Ten Meter Telescope*, UC TMT Technical Note No. 137.

Medwadowski, S. J.: 1987a, *W. M. Keck Telescope – Summary of Results of Analyses*, Report submitted to CARA March 27, 1987.

Medwadowski, S. J.: 1987b, *Space Frames in Architecture and Science*, Invited General Lecture presented at the IASS Osaka Symposium on Membrane Structures and Space Frames 1986, *Bulletin of the IASS*, Vol. 28, No. 1, pp. 7–12.

Nelson, J. E.: 1988, *W. M. Keck Telescope and Observatory – Current Status* (in press).

Nelson, J. E., Mast, T. S. and Faber, S. M.: 1985, *The Design of the Keck Observatory and Telescope (Ten Meter Telescope)*, Keck Observatory Report No. 90.

CONCEPT AND STATUS OF THE VLT PROJECT*

D. ENARD

European Southern Observatory, Garching bei München, F.R.G.

(Received 1 February, 1989)

Abstract. The ESO Very Large Telescope project consists of an array of four 8-m telescopes. This concept presents a very great flexibility and covers a broad range of scientific applications. The most far reaching one is the co-phasing of the four telescopes which will open the way to very high resolution imaging.

The funding has been granted in December 1987. The first telescope is scheduled for 1995. The whole array is expected to be completed by the year 2000.

1. General Concept

The ESO VLT concept was presented for the first time at the IAU Colloquium No. 79 of April 1984.

The concept presented at that time only intended to illustrate a number of conceptual ideas:

(a) The trade offs between technical, scientific and financial considerations resulted in an array concept on four 8-m telescopes.

(b) Considering the topography and orography of existing sites in Chile, the 4 telescopes were optimally arranged in a linear configuration.

(c) A somewhat provocative idea was to operate the telescopes in quasi open air. This idea was deliberately proposed as a research goal with the intention to solve 2 major problems, the minimization of building costs and the elimination of dome seeing. Wind buffetting effects were, of course, expected, and two strategies were envisaged: a reduction of the wind speed using permeable screens or/and the introduction of active compensation for tracking.

(d) One of the main goal of a very large telescope is the increase of the spatial resolution. Aperture synthesis or interferometry is therefore a most important observing mode. Smaller auxiliary telescopes were thought to be necessary for the development and fulltime use of the interferometer and were already at that stage 'integrated' into the concept.

These conceptual ideas remarkably survived 4 years of analysis and after a number of iterations concretized into a more realistic design which is the present basis for the VLT construction. This latest concept is presented here below and is illustrated by Figures 1 and 2. Figure 1 shows a model built in 1986 in which inflatable domes are supported by a common platform. Figure 2 shows a later design in which the telescopes have truly independent enclosures.

* Paper presented at the Symposium on the JNLT and Related Engineering Developments, Tokyo, November 29–December 2, 1988.

D. ENARD

Fig. 1.   1986 model built after preliminary engineering studies.

Fig. 2.   Latest concept of the VLT building. Only the top of the tube is exposed to wind. The primary mirror is embedded into the shelter. Openings allow a ventilation of the bottom part according to prevailing wind conditions.

2. 8-m Unit Telescopes

The architecture of the unit telescopes is driven by two main considerations:

(i) The telescopes have a quasi-fixed configuration, in opposition to the previous generation which could be adapted to perform a number of different tasks generally at the expense of operational efficiency. In the case of the VLT, flexibility comes from the concept which, with its 4 telescopes and 17 various foci, allows to perform most tasks without major change of the telescope configuration. This is also made necessary to comply with the requirements of remote control and flexible scheduling.

(ii) Wherever possible heavy and passive technologies are replaced by control techniques. This simply follows the modern trend in all technical areas: computer controlled devices tend to replace passive mechanical systems. This approach generally leads to lower construction costs and better performance but increases development costs. This may not necessarily pay off for a prototype but the VLT, which has 4 identical units, justifies the use of advanced technologies.

2.1. ALT–AZ MOUNT

The telescopes have a conventional alt-azimuthal mount with the elevation axis above the primary. This has been found to provide a light structure and minimises the building size. The possibility to have the elevation axis behind the primary as radio-telescopes would be preferable only if the primary mirror were very light, had a very small F/ratio and if it were much larger than 8 m so that enough space would be left for instrumentation.

2.2. OPTICAL SCHEME

The primary mirror has an aperture of $F/1.8$. The optical combination is of Ritchey–Chrétien type with two $F/15$ Nasmyth foci and an $F/13$ Cassegrain focus (Figure 3). The combination is optimized for the Nasmyth focus. For the Cassegrain, the secondary mirror is moved to change the position of the focal plane and the primary mirror deformed to correct the spherical aberration thus introduced. There is a Coudé focus in the telescope base. The telescope beam is picked up by a removable mirror or prism before the Nasmyth focal plane and relayed to the Coudé focus by a train of mirrors (Figure 4). High efficiency multi-dielectric coatings are mandatory in order to maintain a good optical throughput. Such coatings are limited in bandpass to less than one octave which implies that several sets of mirrors need to be used to cover the full spectral range.

2.3. ACTIVE PRIMARY AND SECONDARY MIRRORS

Each primary mirror consists of a thin meniscus made of Zerodur. The thickness is 175 mm. Such a mirror is very flexible and a correct figure can only be obtained and maintained through the use of an active support. The present analysis calls for about 150 supports in order to maintain the high-frequency ripple below 25 nm r.m.s.

The support system combines a passive and an active system. The passive support

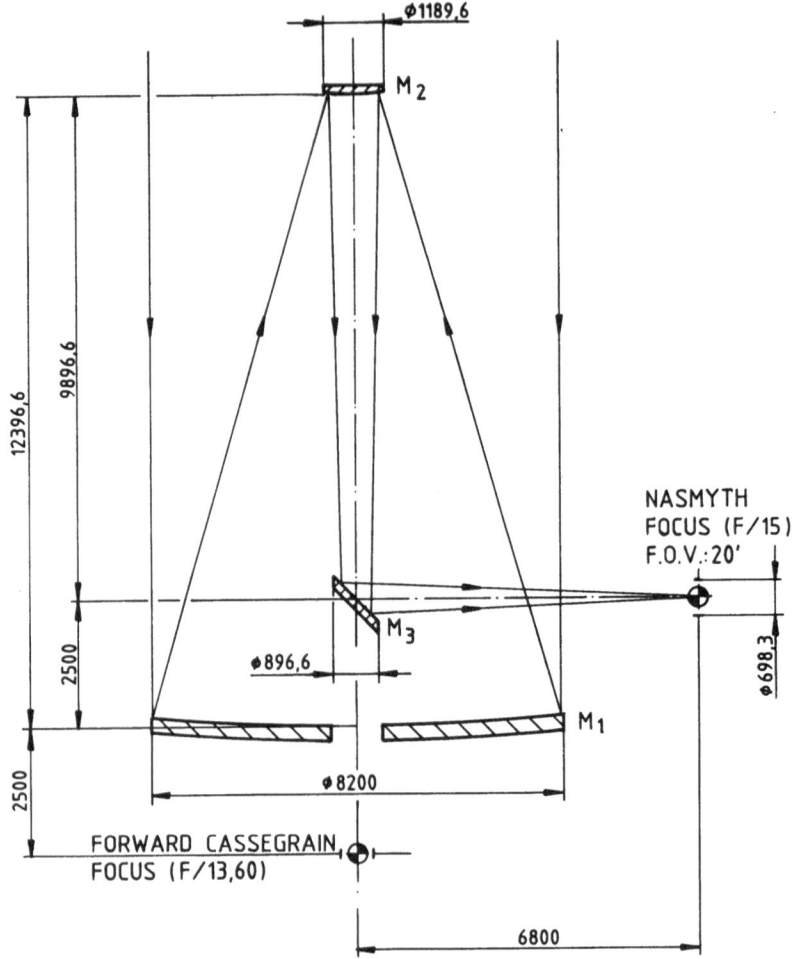

Fig. 3.   Optics scheme of the 8 m unit telescopes.

system consists of an hydraulic wiffle tree. Each support is made of a twin chamber in order to compensate the hydrostatic oil pressure when tilting the telescope. Supports are connected together in three independent 120° sectors containing 50 supports each.

Each actuator of the active system is directly attached to the backside of the passive support and consists of a spring compressed by a nut and lead screw system driven by a D.C. motor. A force sensing device is located between the active and passive stages.

Because the telescope may be operated under relatively strong winds, and because of the tough tracking requirements for I.R. imaging at the limit of diffraction, it is necessary to introduce an active compensation of tracking errors. The best approach is to use the secondary mirror which is the smallest optical element (diameter 1200 mm). A concept of an active secondary mirror is under study and aims at correcting small tracking errors up to a frequency of 20 Hz. Another goal of the study is to design the

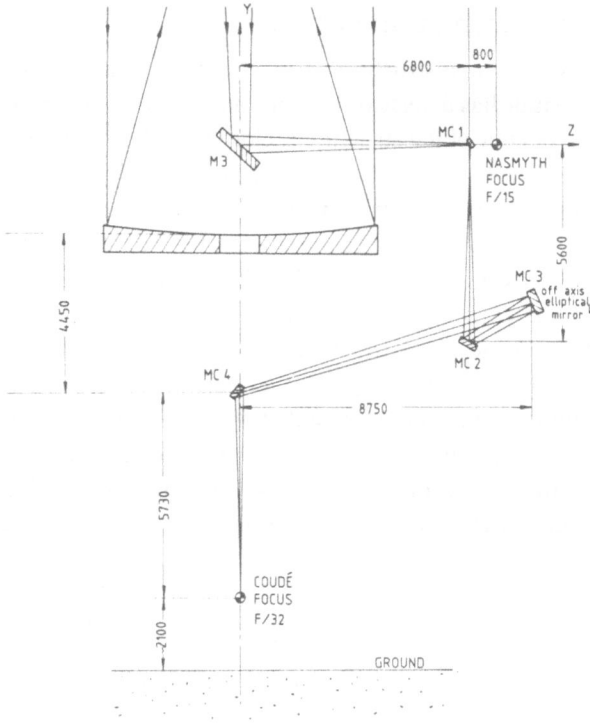

Fig. 4. Coudé beam.

secondary mirror to work as a chopper for the I.R. The design goal is to reach 5 Hz with an amplitude of 1 arc min on the sky. Such a secondary would need to be extremely light. A total mass of about 130 kg for a lightweight Zerodur mirror and its support is considered. Lighter mirrors could be obtained with beryllium or carbon fibers, but there is not yet any evidence that such technologies could provide an adequate performance and stability for a mirror of that size.

2.4. NASMYTH MIRROR

The Nasmyth mirror is designed to cover an unvignetted field of 20 arc min. Its size is 0.9 × 1.3 m. The mirror will rotate around the optical axis is order to feed the 2 Nasmyth foci. In addition, another rotation will position the mirror parallel to the optical axis for observation at the Cassegrain focus. The F.O.V. at Cassegrain will then be limited to about 15 arc min and the Nasmyth mirror will slightly protrude out of the central obstruction which is otherwise only set by the secondary mirror. The VLT is not envisaged to have conventional baffles. Because a re-imaging is anyway necessary to match the detector pixel size, it is thought that the baffling is better done inside the instrument. The central obstruction will be about 2% instead of about 10% with baffles. It is important to keep central obstruction small in view of the availability of new glasses which makes possible the development of spectrographs and cameras without any central obstruction.

2.5. MECHANICAL STRUCTURE AND DRIVES

Figure 5 shows the telescope structure. It is a framework with an optimized stiffness to mass ration. The tube has a mass of 110 tons and the total moving mass is 2530 tons. The lowest eigenfrequency of the structure is close to 10 Hz. Two solutions are investigated for the drives. The baseline solution is a conventional gear wheel and pinion with 2 pairs of motors working in opposition for backlash compensation.

A solution based on directly coupled motors has been analysed and seems extremely attractive from the performance point of view. The elevation axis would use 2 motors, 2.4 m in diameter and each with a peak consumption of about 6 kW (mean would be 10 to 50% of peak depending on the wind torque). The azimuth axis would use four quasi-linear motors 3.5 m long arranged on a circle of 10 m diameter. The total peak power for the azimuth motors would be only 6.4 kW owing to their large diameter. All motors would have to be water-cooled. Despite its potential advantages, it remains to be established whether the cost of a direct drive solution is acceptable and whether the problems of heat dissipation, and of electromagnetic interferences produced by the high power switching amplifiers can be given adequate solutions.

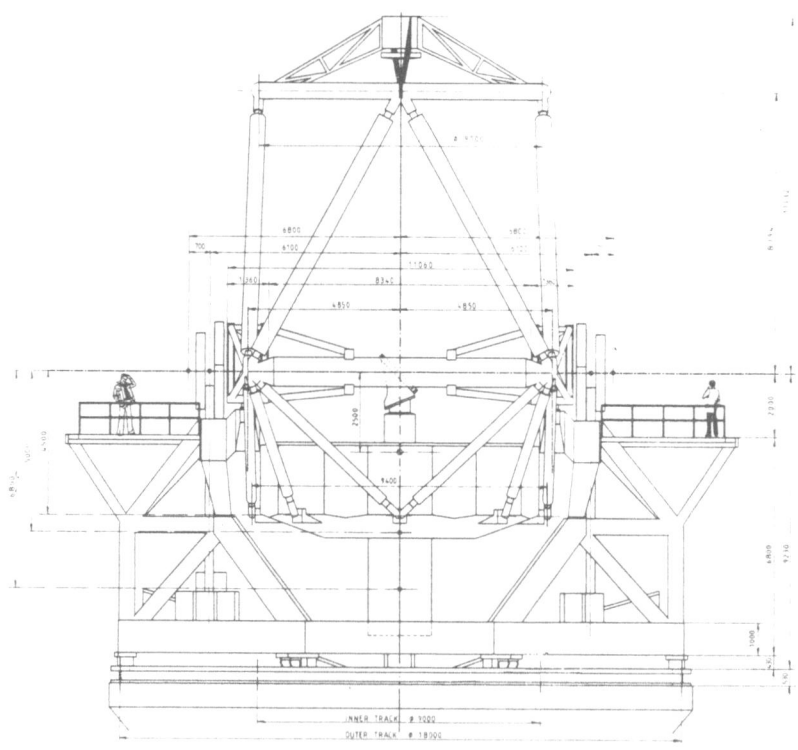

Fig. 5.   Unit telescope mechanical structure.

2.6. Telescope enclosure

The telescope enclosures are envisaged to be based on the inflatable dome concept. Alternative designs such as a rotating building are also considered though they are expected to be more expensive. A 15 m diameter prototype (half the diameter needed for the VLT) has been built and is experimented at La Silla (Figure 6). It consists of double wall fabric lenticular ribs which are supported by rigid hoops articulated on 2 opposite shafts located at the basis of the dome. The dome is thus consisting of 2 quarters of a sphere which can be independently maneuvred. When closed, the dome as well as the volume contained between the 2 fabric walls are inflated thus making a rather stiff surface. This technology is derived from that used in inflatable radomes.

The closing mechanism is based on cables for the present prototype. It is expected to use hydraulic actuators for the final VLT dome.

3. Beam Combination/Interferometry

Recombining the beams of the 4 telescopes is essential in order to take advantage of the full collecting power of the VLT.

There are 2 combined modes of the VLT.

3.1. The incoherent combination

The 4 beams are combined in the center of the array without any correction for pathlength differences. This combined focus will be mainly used for spectroscopy in the visible and near-I.R.

3.2. Interferometric combination

The pathlength differences between the telescope beams must be compensated. The beams will come out of the 8 m telescopes perpendicular to the array direction and will be combined in a long building parallel to the array. This building can be seen on Figure 2. It will contain delay lines as well as various correction devices needed to phase the 4 beams and the detection.

It is clear that interferometry with large dishes make sense only if the pupil itself can be phased, i.e., the seeing corrected. It appears possible within the VLT construction time frame to apply adaptive optics techniques at least down to near-I.R. wavelengths even though extension to the visible may require more time. Adaptive optics is of great significance for large telescopes. It may provide not only gains in efficiency but may ultimately solve the problem of instrumentation matching by reducing the image size. A demonstration prototype of an adaptive system is being built and is expected to be tested in 1989 on a 1.5 m telescope.

It was realized already at the start of the project that combining the 8 m telescopes would not be realistic without the prior experience with telescopes of an intermediate size. These telescopes could also be coupled with one or several 8 m telescopes once the interferometric techniques are fully developed. This smaller interferometer is likely to consist of 2 to 4 mobile telescopes of about 2 m diameter.

D. ENARD

Fig. 6.   Demonstration prototype of an inflatable dome. This prototype is being tested at La Silla Observatory.

4. Instrumentation

For each 8 m telescope the following foci are available:
- 2 Nasmyth foci: $F/15$ and 20 arc min FOV.
- 1 Cassegrain focus: $F/13$ and 15 arc min FOV.
- 1 Coudé focus: 0.5 to 2 arc min FOV.

The combined focus may feed several large instruments with a FOV less than 1 arc min.

The instrumentation will be conceived as quasi-fixed packages combining a number of functions and observing modes.

5. Site

The site for the VLT has not yet been definitely selected. Two sites are under consideration: a relatively large summit in the immediate surroundings of La Silla called Cerro Vizcachas and Cerro Paranal, a 2650 m high summit close to the northern coast of Chile at about 150 km south of Antofagasta.

The two sites are being monitored with identical instrumentation. Three years of meteorologic data are already available, which show the superiority of Paranal for atmospheric water vapour content and the number of clear nights. Negative aspects compared with La Silla are the very limited space available at the summit and the wind which is on average 20% stronger and less stable in direction. For these reasons another summit located in the immediate surroundings of Paranal may be preferred. This summit called La Montura offers more space and seems to be less exposed to strong winds. Seeing data based on differential image motion are being recorded but do not extend yet over a sufficiently long period to be significant.

6. Present Status

A contract with Schott has been signed early September, 1988, for the procurement of the primary mirror blanks. The contract for the polishing is expected to be awarded by mid-1989. A technology alternative to glass is being developed in case of difficulties with the production of the Zerodur mirrors. It is based on aluminium and is being tested with 1.8 m mirrors.

The conceptual design of the entire telescope should be definitely frozen and the site selected by mid-1990 and the first telescope erected in 1995.

KODAIRA – Did you consider a possibility to have a second set of the secondary mirror?
ENARD – No. We intended to cover all observing modes with one mirror unit. This unit will provide fine active tracking for IR imaging with arrays at the limit of diffraction and chopping for broad IR observing. The $F/15$ aperture is thought to be acceptable for work in the visible and IR.

Recent Progress on the Columbus Project and the Mirror Laboratory Program *

J. R. P. Angel, J. M. Hill, H. M. Martin and P. A. Strittmatter
Steward Observatory, University of Arizona

(Received 15 March, 1989)

Abstract

This paper summarizes recent advances on the Columbus Project telescope and in the University of Arizona Mirror Lab. The Columbus telescope structure has been re-optimized to allow rapid changes between foci, while still maintaining high rigidity. Room has been made to translate secondary and tertiary spiders out of the light path to the center. A bill allowing construction of the telescope on Mt. Graham, Arizona, has been passed by Congress and signed into law. Two alternative enclosure designs, one with a co-rotating building and a second which opens like a flower, are being explored.

A common baseline design for the 8 m honeycomb mirrors for both the Columbus and Magellan telescopes has been developed. It has stiffness comparable to that of the Palomar 200 inch mirror. The Mirror Lab has successfully cast two 3.5 m honeycomb blanks and expects to begin casting at the 6.5 and 8 m scale at the end of 1990. Interferometric tests of the Vatican f/1 1.8 m borosilicate honeycomb mirror show good stability of figure with the air jet ventilation system. A 60 cm stressed lap has been completed, and will be used to parabolize this mirror which is now polished as an f/1 sphere. Plans for a polishing facility to house two 8 m machines and a test tower are complete, with construction starting in April 1989.

1. Introduction

The Columbus Project has as its goal the construction of a binocular telescope with two 8 m mirrors on a common mount. A unique design has evolved which takes maximum advantage of the binocular configuration, and of lightweight, rigid and short focus primary mirrors being made by the University of Arizona Mirror Laboratory. The Ohio State University ("Ohio"), the University of Arizona ("Arizona") and the Italian Astronomical Community represented by the Arcetri Observatory ("Italy") have combined forces to build and operate the telescope. The project, and the preparations to cast and polish 8 m borosilicate honeycomb mirrors were recently described in a number of papers at the 1988 ESO Conference on Very Large Telescopes and their Instrumentation. In this paper we will give an overview that emphasizes developments not already reported. The material was presented at the Japanese National Large Telescope conference by J.R.P. Angel in two separate talks.

* Paper presented at the Symposium on the JNLT and Related Engineering Developments, Tokyo, November 29–December 2, 1988.

Astrophysics and Space Science **160**: 55–70, 1989.
© 1989 *Kluwer Academic Publishers.*

2.   The Columbus Project

2.1 Optical Design

The telescope is based on the principle of multiple co-mounted telescopes that has proven so successful in the Multiple Mirror Telescope (MMT). Even though the Columbus project calls for only two mirrors, it nevertheless has many features in common with the MMT. A single 11.3 m single dish would have the same collecting area, but would be inconveniently large to manufacture in a single piece. With the two 8 m primaries, the simplicity of large monolithic axisymmetric optics and a single mount is maintained.

Interferometry has become an important use of the MMT and is especially suited to co-mounted telescopes. Beam combination is made with only two reflections following the Cassegrain telescopes, yielding an interferometric focus of great power. No relative motion of the mirrors is required for any telescope orientation, and in practice stable fringes are easily obtained. More important, the combined focus has a wide coherent field (6 arc minutes at F/33). This means that if the equal path length condition for interferometry is met for any one star in the field, it is met for all stars. Thus a field star can be used to stabilize atmospheric wavefront errors, while a program star is observed. This property will be very important in the infrared, where the isoplanatic patch is large and there is a good probability of finding bright field stars. The design calls for a 6 m separation between primaries, permitting interferometry at baselines up to 22 m.

A feature of the individual telescope design is the use of primary mirrors of focal ratio f/1.2. Such fast mirrors have not been used before, because of the limitations in polishing technology, but have significant optical and mechanical advantages. The same fast primary will be used in the Magellan telescope, and it is anticipated that the secondary mirrors will be identical. Dual chopping foci at f/15 with 71 cm secondaries, and fast Cassegrain foci at f/5 with affordable and manageable 2 m secondaries, are planned. Recently there has been considerable interest in the fast Cassegrain foci used with no correcting elements, except for the detector window for very clean imaging at the optical confusion limit. A large Tektronix chip sees a field of 5 arc minutes and sampling at 0.15 arc sec. Recent calculations by Epps (1989) show that extremely well corrected fields of 40 arc minutes are obtainable at the fast Cassegrain foci with correctors having only spherical elements.

2.2 Mechanical Design

Rigidity is important in any telescope, and especially in one to be used as an interferometer. The binocular configuration of the telescope has allowed the project to develop a telescope mount of extraordinary rigidity (Figure 1). This involves twin wheels that are attached to and pass under the optical support structure. These wheels turn on hydrostatic pads to give the elevation motion, and the load path goes directly to a circular track for azimuthal motion. The indirect load paths of a conventional fork mounting are thus completely circumvented.

The binocular configuration offers the possibility of storing optics and instruments on board the telescope in the central gap, ready for rapid interchange. This feature is fully exploited in the Columbus design. The wide field Cassegrain secondary mirrors and the tertiary mirrors will both be mounted on trolleys, so they can be driven out of the light path to the central region. A recent optimization of the structure has been made to permit these motions (Majorana et al. 1988). The lowest resonant frequency remains at 10 Hz for the entire telescope, whose mass including instruments is 450 tons. To our knowledge, this is the highest projected frequency of any of the planned new large telescopes, despite the fact that this telescope will have the largest collecting area. The structural design that takes the mirror weight directly downwards also leaves the ends of the structure open. This has allowed us to design for the removal or insertion of new optics or even whole spiders from the sides of the telescope.

Figure 1. Elevation structure model for the Columbus telescope. The two wheels are supported by hydrostatic pads on an azimuth platform.

2.3 Enclosure Design

It is interesting that the envelope traced out by the telescope motions is very closely spherical, with a diameter of 13.5 m. Thus no more space is required in the enclosure than is needed for single 8 m telescopes of more conventional focal ratio. The extraordinary compactness and high stiffness derive from the short focal length of the primaries in combination with the Davison mount.

The type of enclosure to be used for the telescope is still under debate. A design for a co-rotating building along the lines of the MMT enclosure has been worked out in some detail by Gallieni et al. (1988). This offers good wind protection, and an automated method for introducing the aluminizing "bell jar" and alternate secondary spiders. These are lifted to the primary mirror level by a portable elevator brought up to the side of the telescope. They then pass through building wings either directly on to the telescope by rails, or into an elevator in the wing that lifts secondaries to rails on the top of the telescope. The concept is illustrated in Figure 2a.

An alternative that allows completely open air operation of the telescope when desirable is also under consideration (Davison and Angel, 1988). In this concept, shown in Figure 2b, the non-rotating building is made like a closed flower, with petals that fold back to reveal the telescope. Each petal has a top triangular section that folds out and down against the wall section. In high wind, the independently actuated walls would be kept up as high as possible, but in low wind they too would be folded out to allow free flow of air across and even underneath the telescope. In this design off-telescope handling of the aluminizing tank and spiders is accomplished with a single telescoping boom crane, operating from a fixed point just outside the petal hinges.

The choice between the different designs will center on quantitative estimates of enclosure-induced seeing and cost, as well as practical engineering details.

2.4   The Site

The telescope is to be built on Mt. Graham in the Pinaleno range in southern Arizona. These mountains share the same clear weather as Kitt Peak, but at 3200 m, are considerably higher. A bill has recently passed through both houses of the United States Congress to assure the availability of a 150 acre site for the observatory (Figure 3). The bill was signed by President Reagan on November 18, 1988.

The scientific research area defined by the bill is on the western edge of the highest region of the Pinalenos. Site tests show that this region, on the windward side, has the best seeing. The median image size derived from image motion of Polaris, measured near ground level, is 0.85 arc seconds full width at half maximum. The area includes four primary peaks of 3205 m, 3192 m (Emerald Peak), 3150 m and 3140 m elevation. Detailed studies to find which of these is best for the Columbus Project are now under way.

The binocular telescope and a 10 m telescope for submillimeter astronomy, a collaborative venture between the Max Planck Radio Astronomy group and Steward Observatory, will be the first large facilities of the Mt. Graham International Observatory. These will join the 1.8 m Vatican Advanced Technology Telescope. Four additional telescopes, including an interferometer, are envisaged, after definition and approval by the U.S. Forest Service. The site has extended plateau-like regions appropriate for interferometry with baselines up to 1 km.

2.5   The Columbus Partners

The project has now reached the end of the first phase telescope concept development. This has been marked by the publication of a Phase I report, edited by Richard Kron and available from the Project Office at Steward observatory. Phase II, construction, requires the participating institutions to commit the necessary funding. The University of Chicago, which has played a critical role in Phase I, is unable at this time to commit to Phase II. The remaining Phase I partners, Ohio, Arizona and Italy, expect to be able to commit in the near future, but will leave open the opportunity for participation by a fourth partner. The total project cost is estimated to be $60M.

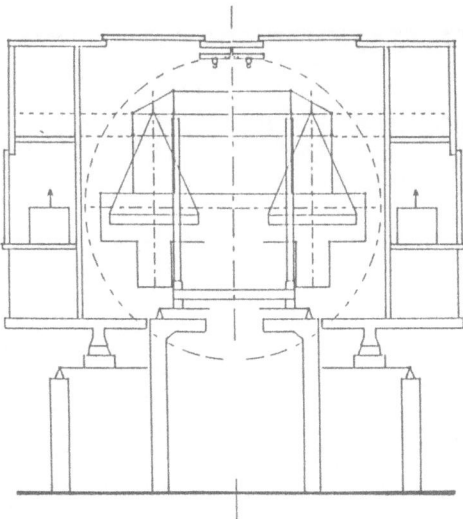

Figure 2a. MMT style co-rotating building. The shutters open by moving out to the side. The secondary and tertiary spiders can translate to the center and also outward on rails, onto elevators in the side wings. Access for realuminizing is by an external elevator that reaches the side wing doors at the elevation axis level.

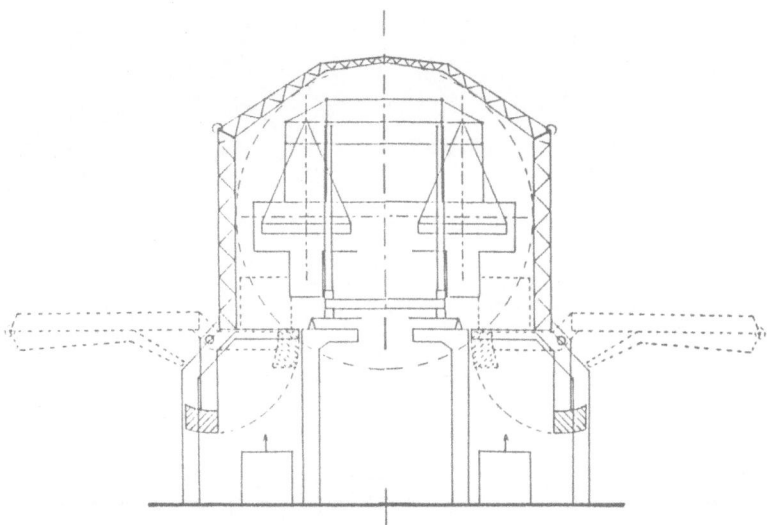

Figure 2b. Flower type building. Hinged triangular roof sections rotate up and around, to be against the side walls. These walls can in turn be folded out independently, according to the direction of observation, and the strength and direction of the wind. Access for realuminizing and spider exchange is by a hydraulic boom crane, permanently installed on the outside of the enclosure at the observing floor level.

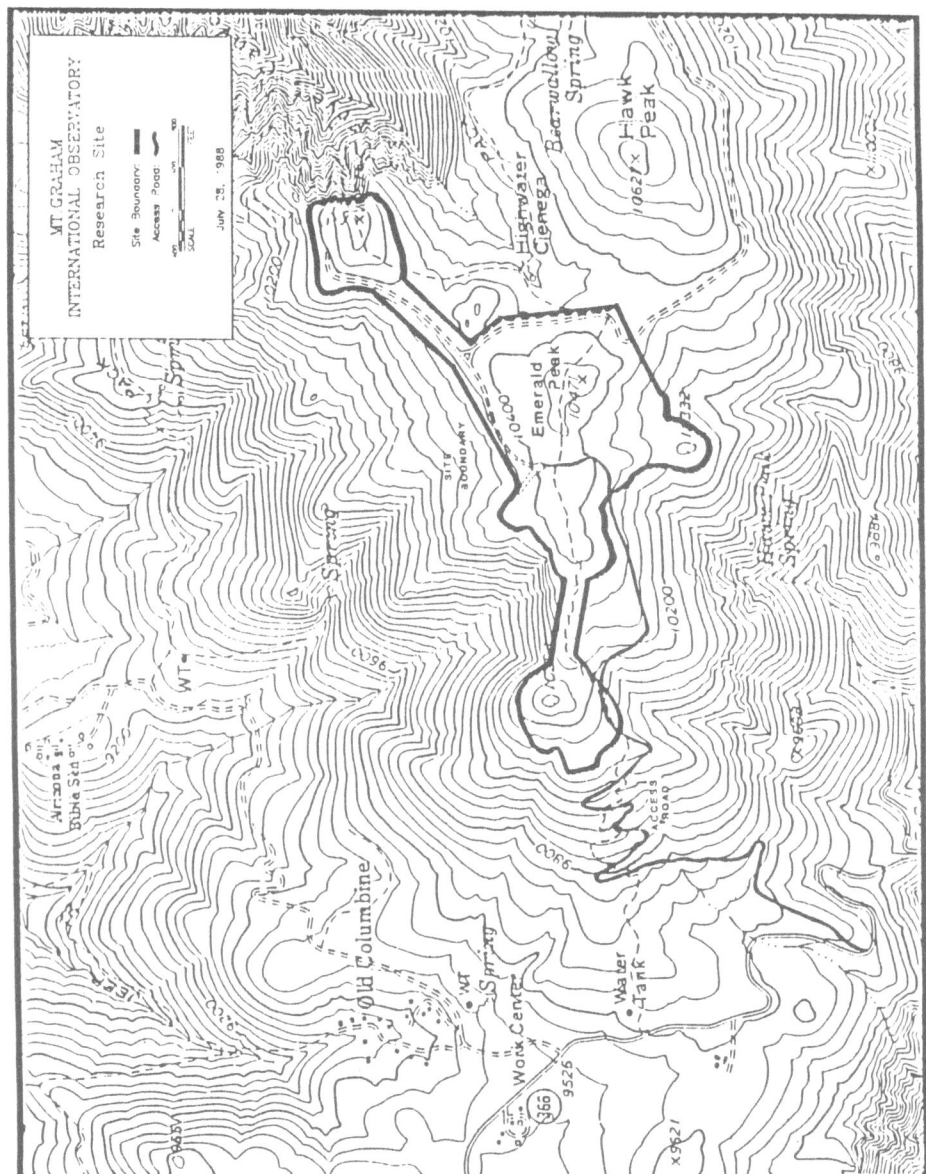

Figure 3. This map shows the biological and astronomical research area which has been established
in the Emerald Peak area of the Pinaleno Mountains. As many as seven telescopes will be
constructed on 24 acres within the 150 acre site.

3. The Mirror Lab Program

3.1 Mirror Substrate Design

The ideal 8 m mirror substrate would be rigid, lightweight and dimensionally stable over many years. It should not distort when the ambient air temperature changes, and should quickly equilibrate to air temperature to avoid mirror seeing. Honeycomb sandwich mirrors of glass are best able to meet these mechanical and thermal goals.

The baseline design adopted for the mirrors for the Columbus and Magellan telescopes is given Table 1. In arriving at this design we have used as a consistent

Table 1. Baseline 8 m Honeycomb Mirror Design

shape:	plano-concave
focal ratio:	f/1.2
diameter:	8 m
facesheet thickness:	25 mm
rib thickness:	12 mm
outer edge thickness:	.84 m
inner edge thickness:	.43 m
cell shape:	hexagonal
cell spacing:	193 mm
mass:	14000 Kg

yardstick that errors from any one source should give an image of no more than 0.06 arc second FWHM. Allocation for errors on different scales is made by defining an equivalent wavefront structure function, with the parameter r_o = 180 cm at 500 nm wavelength. In the case of small scale wavefront errors with amplitude much less than the wavelength of light, which cause scattering, we have set a criterion that the Strehl ratio from any one source should be > 90%, i.e. no more than 10% scattering from the central Airy disc.

A configuration of 186 axial support points for this baseline mirror has been designed. The array has six-fold symmetry, and the solution requires only 20 different force values. The computed surface deflection under gravity is 14 nm rms, and ray tracing gives 90% of the reflected energy in 0.14 arc seconds. In Figure 4 we give the structure function for a wavefront reflected from this surface.   It meets the specification structure function corresponding to seeing of 0.06 arc seconds.

The lateral support system must support all the weight of the mirror when it is horizon pointing. We have adopted a unique solution found by Ballio and Parodi (1988) which requires the application of forces at only the outer perimeters of the two facesheets. The trick is to apply forces at the edges of the faceplates that squeeze the sides of the mirror as well as pushing or pulling up against gravity. This is like the force that would be felt by the mirror floating in a liquid. The best solution is found by allowing complete freedom in the direction and strength of the edge forces on each facesheet. When these are optimized, the rms surface error is 22 nm, and the ray deviations show 90% of the total energy in a cone of 0.25 arc seconds. The reflected wavefront structure function is shown in Figure 4. Although the distortion slightly exceeds the r_o = 180 cm specification for lateral support, this error occurs only when horizon pointing.

3.2 Bending Stiffness

The stiffness advantage of the honeycomb mirrors is best illustrated by making a comparison of different designs. If there are random support force errors, or unbalanced wind forces, the most likely distortion is astigmatic bending, against which the mirror has least resistance. The stiffness of different mirrors to this bending can be compared directly, with the aid of analytical expression derived by Nelson, Lubliner and Mast (1982). We take the loading case in which four forces F act at the outer perimeter, two pushing upwards at the N and S positions, and two downwards at the E and W.

The surface deformation is given by

$$ z(r,\theta) = \frac{24\ F\ \cos 2\theta}{\pi\ E\ t^3} \ \frac{1-\nu^2}{3+\nu} \left[\frac{r^2}{1-\nu} - \frac{r^4}{6a^2} \right] $$

and the image blur (full diameter) is given by

$$ \Delta\theta = \frac{96}{\pi}\ \frac{Fa}{Et^3}\ \frac{1-\nu^2}{3+\nu} \left[\frac{2}{1-\nu} - \frac{2}{3} \right] $$

Here each of the individual forces is F, a is the radius and t the thickness. E and ν are the Young's modulus and Poisson's ratio of the glass. These expressions are strictly for flat, thin discs, however, for astigmatic bending the effects of shear and shell stiffness are very small. Thus the results are quite accurate for both thick mirrors and meniscus shapes. Parodi (1989) has made finite element calculations for the 8 m honeycomb design, loaded in the same way. The results are in remarkably good agreement with the above expressions, provided an effective Young's modulus is used, equal to the glass modulus reduced by the light-weighing factor, ff (mirror mass/mass of solid mirror of same external dimensions). In Table 2 are shown the dimensions and mass of various existing and proposed telescope mirrors, and the size of astigmatic image under a loading in which each of the 4 forces is 0.1% of the mirror weight.

Our expectation that the 8 m honeycomb mirrors will be stiff enough to allow basically passive support is borne out by the comparison with smaller mirrors. Their bending under gravity is only a little larger than that of the Palomar 200" mirror, and about 1.6 times that of the individual Keck telescope segments. Both are supported by completely passive mechanical systems. The 22 ton VLT design and a ULE meniscus of 8.0 x 0.2 m show an order of magnitude larger distortion than the baseline

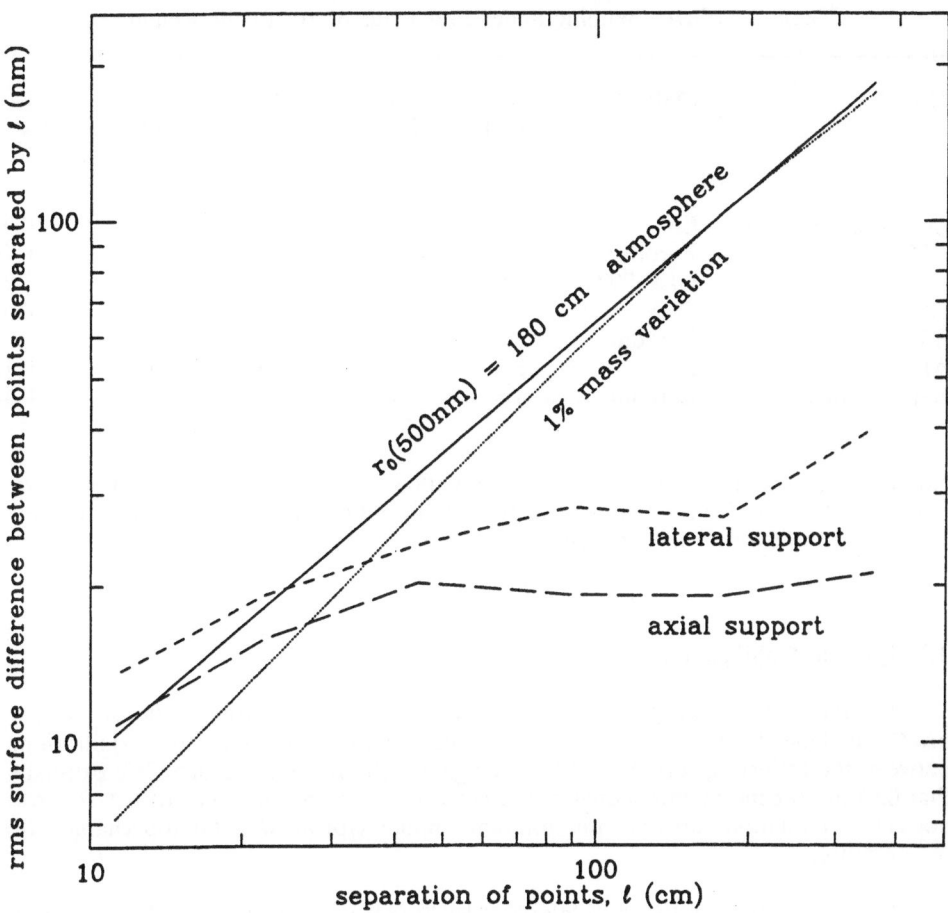

Figure 4. This figure shows the surface structure functions for several 8 m mirror support cases. The solid line represents the structure function corresponding to an $r_o = 180$ cm atmosphere scaled to the equivalent reflective surface. This is typical of the support specifications of the primary mirror. No correction for diffraction has been made as the finite element grids do not include spacings smaller than 10 cm. The long-dashed line shows the structure function from the axial support pattern. As can also be seen from the surface isocontours, the axial support errors are dominated by small-scale distortions. The short-dashed line shows the structure function for lateral support when pointing at the horizon and the dotted line shows the effects of a radial temperature gradient of 0.2°C.

14 ton Columbus/Magellan design. The contrast is seen also in existing mirrors. The 3.5 m E6 honeycomb castings for the ARC and WIN telescope are three times lighter than the 3.6 m ESO NNT Zerodur meniscus, and yet are twice as stiff under support force errors.

Table 2. Mirror bending under four loads of 10^{-3} of the weight

Mirror	Material	d (m)	t (m)	ff (rms)	M (metric tons)	$\Delta\theta$ (arc seconds)
Columbus Baseline	E6	8.0	0.63	0.207	14.0	0.64
ESO VLT	Zerodur	8.2	0.175	1	22.6	6.74
-	ULE 7971	8.0	0.20	1	21.3	5.23
Palomar	CGW 7160	5.0	0.35	1	15.6	0.55
ESO NTT	Zerodur	3.6	0.24	1	5.5	0.28
ARC	E6	3.5	0.39	0.246	1.9	0.13
Keck Segment	Zerodur	1.8	0.075	1	0.5	0.40

Note: Mirror masses were computed with an allowance for center holes. An effective thickness of 0.35 m for the Palomar mirror is believed to represent both the mass and stiffness of the 0.6 m ribbed structure.

3.3 Thermal Stabilization

The faceplate thickness of 25 mm for the honeycomb mirrors is chosen to give low thermal inertia. Internal ventilation can then be used to force thermal equilibrium between the reflecting surface and the changing night air temperature. We estimate that 0.06 arc second mirror seeing would correspond to an imbalance of 0.2° (mirror hotter). Ventilation can hold this tolerance under typical mountaintop changes of temperature.

Thermal stability of figure in telescope mirrors is typically achieved by the use of zero expansion materials. These are necessary because internal temperature gradients are inevitable in thick glass sections. In the honeycomb structure no section is thicker than 25 mm, and internal gradients will be avoided by a ventilation system. It is then advantageous to use borosilicate glass, which is very favorable from the standpoint of manufacture and cost.

The large mirrors will be ventilated by jets of air that blow into every honeycomb cell from below. The air enters and exits each cell through the 90 mm hole in the backplate. Lab experiments with single cells show that uniformity to 0.1°C is realized under changes in ambient temperature typical of the mountaintop environment (Cheng and Angel 1988). Recently a jet system has been put into operation to ventilate the 80 cells of the 1.8 m f/1 mirror for polishing and testing. This is incorporated into the polishing machine, and can be run during polishing, while the mirror rotates. Preliminary results with a non-optimized system already show a large improvement in the figure stability. Surface contour maps obtained 30 minutes after polishing and again after all night ventilation differ by only $\lambda/8$. With this stability, figuring to near diffraction limited quality has already been achieved (Figure 6). At this stage we expect that stabilization to 0.1° peak-to-valley will be possible for the 8 m mirrors, thus eliminating thermal figure distortion to the 0.06 arc second level.

3.4 Honeycomb Mirror Casting

The Mirror Lab is currently casting a series of honeycomb blanks of 3.5 m diameter. These are being used to complete refinements to the technique before 6.5 m and 8 m blanks are cast. The first 3.5 m trial was cast in April 1988 (Figure 5a), and the second in December 1988 (Figure 5b). Both are to be used in new 3.5 m astronomical telescopes, the ARC consortium telescope in New Mexico (Washington, Chicago, Princeton and New Mexico) and the WIN telescope in Arizona (Wisconsin and NOAO). A third trial blank is now being made and, if successful, will be used in an Air Force telescope. All three blanks have the same size hexagonal cells (20 cm) rib thickness (1 cm) and faceplate thickness (2.5 cm) as will be used in the 8 m mirrors. The quality of the two blanks cast so far is excellent. Bubble content is very low, and stress levels are 10± 10 nm/cm, characteristics of a fine anneal. The rib geometry is good, with about 1% variations in mass distribution on the spatial scale of the axial supports.

The 3.5 m mirrors are being cast in a furnace of 6 m internal diameter. On completion of this series of trial castings, the furnace will be expanded to an internal diameter of 9.5 m, with additional refractory panels. The turntable and power slip rings are already sized for the larger furnace. It is planned that the full size furnace should be commissioned in the spring of 1990, and the first large casting, at 6.5 m, will be made by the end of 1990. If this is successful, subsequent castings will be at 8 m diameter. The present interval between castings is 7 months. We expect this will grow to nearly a year for the first 8 m casting, but the goal is for subsequent trials to follow at 8 month intervals. The first successful 8 m blank is currently planned for the Columbus Project, and the second for the Magellan project.

3.5 Polishing of Fast Paraboloids

As preparation for the challenging task of polishing a number of 8 m f/1.2 primary mirrors, we have developed a polishing technique specifically geared to fast aspherics. An actively stressed lap changes its shape continuously as it moves across the mirror surface, maintaining an accurate fit to that surface at all times. A relatively large and rigid lap can be used, provided that the forces required to bend it can be produced with sufficient accuracy. This lap can travel over the entire mirror surface rapidly and in arbitrary patterns, giving the optician great flexibility in figuring the mirror and the advantage that slow variations of the wear coefficient will be averaged over the surface.

A stressed lap has been built for use on the 1.8 m f/1 primary for the Vatican Advanced Technology Telescope. The principles and general design of this lap are described by Martin, Angel and Cheng (1988). Briefly, the lap consists of an aluminum disc 750 mm in diameter and 25 mm thick. Twelve moment generating actuators are mounted around its perimeter and connected in such a way that all the required moments can be generated internally. These actuators produce forces of up to 3.5 kN in order to bend the lap by 1 mm at the edge. Initial tests indicate that these forces can be produced in a closed-loop fashion to an accuracy of 3.5 N, the design goal.

A continuous distribution of edge moments can produce all the required distortions to very high accuracy, as shown by Lubliner and Nelson (1980). With a

Figure 5a.   First 3.5 m trial casting, now being polished for the ARC telescope.
For colour reproduction of this figure see colour section.

Figure 5b.   Second 3.5 m trial casting, to be polished for the WIN telescope.
For colour reproduction of this figure see colour section.

finite number of actuators, errors are introduced near the edge of the lap. These can be reduced to any desired level by not allowing an outer ring of the lap to contact the mirror. Finite element studies by Walter Siegmund have determined that twelve actuators produce distortions accurate to 0.3% over the inner 600 mm diameter area of the lap. This inner circle will be the area of contact with the mirror.

The stressed lap for the 1.8 m f/1 mirror is currently being calibrated and tested. In the meantime the 1.8 m mirror is being polished as a sphere with radius of curvature equal to 3.6 m. The figuring is done with a rigid passive lap 600 mm in diameter so as to mimic the figuring of the paraboloid with the stressed lap. During this exercise, computer controlled polishing techniques are being developed, as well as computer optimization of polishing strokes. These methods have provided good control of the mirror figure. At the time of writing, the figure is accurate to within a surface error of 0.07 waves rms, 0.49 waves peak-to-valley, as measured with a phase-shifting interferometer at 633 nm (Figure 6).

The mirror was not generated prior to loose-abrasive grinding, and contained approximately 50 μm of surface astigmatism peak-to-valley. Using the computer-controlled polishing algorithm this has been reduced to 60 nm at the time of writing.

The 1.8 m mirror will be finished as a sphere to a specification producing 0.125 arc second FWHM images. It will then be generated as a paraboloid using the Large Optical Generator at the University of Arizona Optical Sciences Center, which has generated aspheric surfaces accurate to 3 μm rms over an area nearly as large as the 1.8 m mirror. All loose-abrasive grinding and polishing will be done with the stressed lap. The surface will be tested interferometrically from the center curvature, using a two-element null lens to correct the approximately 0.5 mm asphericity of the wavefront. Both transmissive and reflective null lenses with spherical surfaces have been designed.

3.6 New Polishing Facility

The University of Arizona is constructing a new polishing facility to provide all necessary equipment and support for the finishing of 8 m mirrors. The detailed architectural design is complete and the construction bid package is being prepared at the time of writing. The building plan is shown in Figure 7. It consists of a high-bay area measuring 18 m x 36 m, for two polishing machines and a test tower, auxiliary lab space for optics and lap development, and approximately 200 m² of offices. The test tower is enclosed in a double-walled steeple 30 m tall, providing excellent insulation. Two 8 m machines will be required in order for the generating, grinding and polishing operations to keep up with the 8-month schedule for casting 8 m blanks. The first of these machines will be a modification of the existing 8 m capacity Large Optical Generator. It will be capable of all finishing work from generating through polishing. In the long-range plan, we expect to use this machine for generating and loose-abrasive grinding. The second machine will not be a generator, and will be used exclusively for polishing and, if necessary, some loose-abrasive grinding.

The new test tower will be used for all mirrors up to 8 m diameter. It has been designed to minimize vibration, which can make optical testing over long path lengths difficult or impossible. The tower is a steel structure rigidly connected to a ribbed

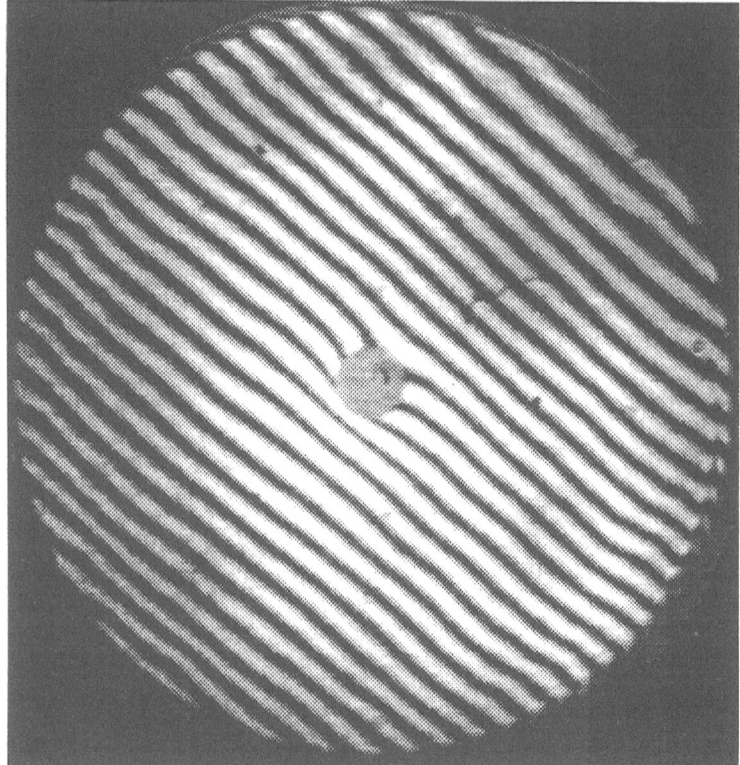

Figure 6. Interferogram obtained during the polishing of the 1.8 m mirror as a sphere. Deviation from straight fringes by one dark-dark separation corresponds to a wavefront error of one wavelength (633 nm). After removal of tilt and focus, the residual surface error is 0.07 waves rms, 0.49 waves peak-to-valley.

concrete base on which the mirror sits. It contains five platforms for test equipment located at heights appropriate for center-of-curvature testing of mirrors with focal lengths between 5 m and 12 m. The platforms are offset to avoid any obstruction of the light paths due to lower platforms or test equipment. All large structural members are designed to provide sufficient clearance for testing of an 8 m f/1.8 mirror.

The tower is designed to have a lowest resonant frequency of approximately 10 Hz, and will be supported on airspring isolators with resonant frequency of 1.2 Hz. Any floor vibrations capable of exciting the tower are reduced in amplitude by a factor of about 70. Along with fast phase-shifting interferometers under development, the new test tower will reduce vibration to levels that should allow testing to $\lambda/100$ accuracy.

The test tower will be shared between the two machines. Mirrors will be moved between the machines and test station on air casters built into the polishing cells. From the time after the back of the mirror is generated to the time it leaves the shop, the mirror should not have to be taken off its cell. The lift provided by the air casters will serve to place the mirror and cell on and off the polishing turntables and defining

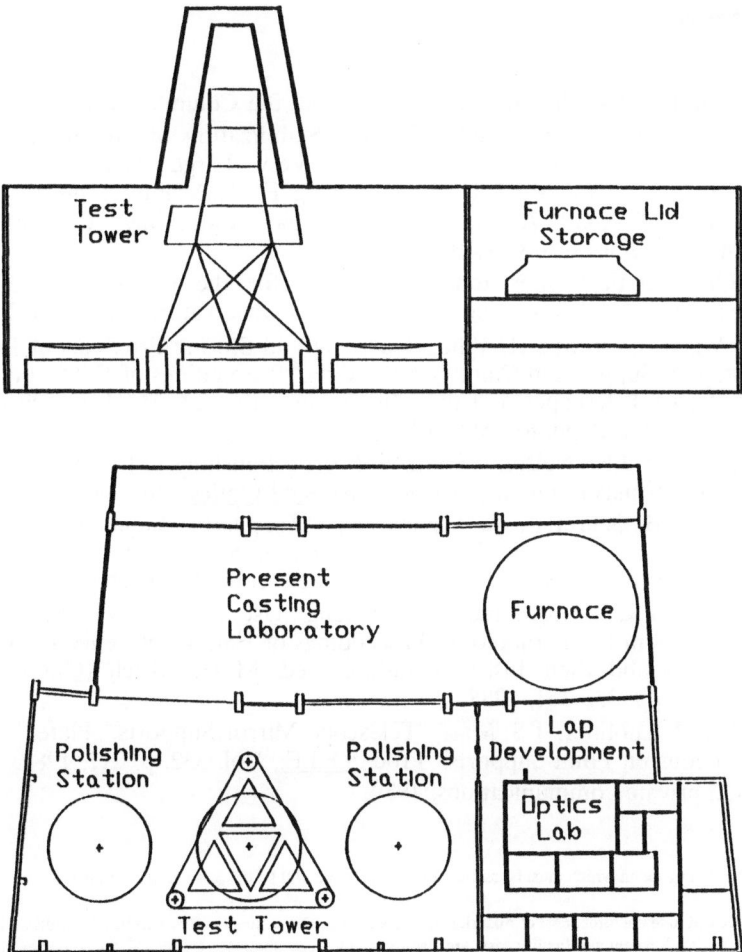

Figure 7. Elevation and plan views of the new polishing facility that will be built adjacent to the Steward Observatory Mirror Lab. The new laboratory will house two 8 m capacity polishing machines and a rigid, vibration-isolated test tower capable of testing mirrors up to 12 m focal length.

points at the test station. The mobile cell will contain the multi-point mirror support system and the ventilation system to accelerate thermal equilibration for testing.

4. Summary and Acknowledgements

Progress on the Columbus Project and in the Mirror Lab program during the period March - December 1988 has been reviewed We wish to acknowledge the contributions of many colleagues, including members of the Columbus Council and Scientific Advisory Committee. We are especially indebted to Dr. Piero Salinari, head of the Italian Project Office. We also wish to acknowledge support of the Mirror Development Program by the National Science Foundation (through NOAO), the University of Arizona, the Columbus Project and by the U.S. Air Force.

5. References

Ballio, G. and G. Parodi, "BCV Report #112 to the Columbus Project," 1988.

Cheng, A.Y.S. and J.R.P. Angel, "Thermal Stabilization of Honeycomb Mirrors," Proceedings' of ESO Conference on Very Large Telescopes and their Instrumentation, ed. M.-H. Ulrich (Garching, ESO), Volume I, pp. 467-477, 1988.

Davison, W. and J.R.P. Angel, "Enclosures that Fold Away," Columbus Project Technical Memo, UA-89-1, 1988.

Epps, H.W., "Optical Subsystem Alternatives for the Columbus and Magellan Telescopes," Mt. Wilson and Las Campanas Observatories, January 27, 1989.

Gallieni, W., R. Gatti and R. Villa, "Handling and Storing Problems of Instruments and Optics Supports in Columbus Building," Proceedings of ESO Conference on Very Large Telescopes and their Instrumentation, ed. M.-H. Ulrich (Garching, ESO), Volume II pp. 889-895, 1988.

Lubliner, J. and J.E. Nelson, "Stressed Mirror Polishing. 1: A Technique for Producing Nonaxisymmetric Mirrors," Applied Optics, Vol. 19, p. 2332, 1980.

Majorana, C., B. Schrefler, A. Vigilante, A. Zaupa, F. Zaupa and G. Zavarise, "Feasibility Study of the Elevation Structure, Static and Dynamic Analysis," Columbus Project 2 x 8 m F 1.2 Optical Telescope, SIGE Report, 1988.

Martin, H. M., J.R.P. Angel, and A.Y.S. Cheng, "Use of an Actively Stressed Lap to Polish a 1.8 m f/1 Paraboloid," Proceedings of ESO Conference on Very Large Telescopes and their Instrumentation, , ed. M.-H. Ulrich (Garching, ESO), Volume I, pp. 353-361, 1988.

Nelson, J.E., J. Lubliner, T.S. Mast, "Telescope Mirror Supports: Plate Deflections on Point Supports," Proc. S.P.I.E., Vol. 332, p. 212, 1982.

Parodi, G., private communications, 1989

MACK – This is a question which must be asked. How much, what will be the price of an 8 m blank?

ANGEL – We estimate the direct cost of materials and labor for repeat castings at 8 m will be $2.6 million. If past and future research, development and facility costs are also included, these amount to another $2.5 million per blank, averaged over the first few 8 m blanks.

NARAI – Can you explain your null-lens system?

ANGEL – Epps has described a two-element Offner-type lens for null testing an f/1 paraboloid of 6.5 m diameter (MMT Technical Report No. 18, 1986). The field element is 195 mm in diameter and the relay element 277 mm. All four surfaces are spherical. The rms wavefront arror is 38 nm rms.

KODAIRA – How accurately have you to control the circulating air temperature?

ANGEL – We plan to control the air supply to 0.03°C. This is to ensure the figure distortion is no more than 0.06 arc seconds (0.1°C peak-to-v alley in the glass).

BECKSTETTE – Do you expect the stressed lap to be used for the final polishing?

ANGEL – Yes.

THE U.S. NATIONAL 8 M TELESCOPE PROJECT*

L. D. BARR

National Optical Astronomy Observatories†, Tucson, U.S.A.

(Received 2 December, 1988)

Abstract. The National Optical Astronomy Observatories (NOAO) are planning to build two 8 m telescopes, one for Mauna Kea, Hawaii, the other for a site in Chile. Optical configurations, primary mirror systems, and the telescope mounting are discussed. A new optical testing method is outlined. The system imaging goal is 0.25″ FWHM. Construction could begin in the early 1990's.

1. Introduction

Modern large telescope facilities consist of the telescope, usually with more than one well-instrumented focal position, housed in a protective enclosure, operated from remote consoles through computers, and preferably located on a dark, dry site where most of the nights are clear. This description presently applies to the 4 m telescopes now operated by NOAO at Kitt Peak and Cerro Tololo and it will apply to the 8 m telescopes we propose to build on the summit ridge of Mauna Kea and on Cerro Pachon in Chile near Cerro Tololo. One of the 8 m telescopes is illustrated in Figure 1.

In the 20-year period since the NOAO 4 m telescopes were designed, many factors that influence the telescope design have changed. We now recognize that the free atmosphere, undisturbed by interaction with the ground, delivers far better optical performance than was heretofore commonly believed. Moreover, as a direct result of our own site survey efforts, confirmed by a number of independent tests, we have identified at least one site – Mauna Kea – where the free atmosphere performance can be routinely realized (Merrill and Forbes, 1987). We anticipate that conditions at Cerro Pachon will also frequently approach these same high performance levels. Recent progress in identifying and minimizing the principal contributions to image degradation not associated with the free atmosphere – localized thermal pollution (dome and mirror seeing), tracking errors, and residual optics errors – have shown that our collective observing experience has generally been limited, even at the best of sites, by factors not directly related to atmospheric seeing.

Over the past twenty years, structural analysis techniques have improved to the point where telescope system performance can be accurately predicted and designs can be truly optimized for best performance. Recent technological breakthroughs within a number of key areas related to telescope construction and design lead us to believe that an 8 m telescope can be constructed to take full advantage of the best performance that

* Paper presented at the Symposium on the JNLT and Related Engineering Developments, Tokyo, November 29–December 2, 1988.
† Operated by the Association of Universities for Research in Astronomy Inc., under contract with the National Science Foundation.

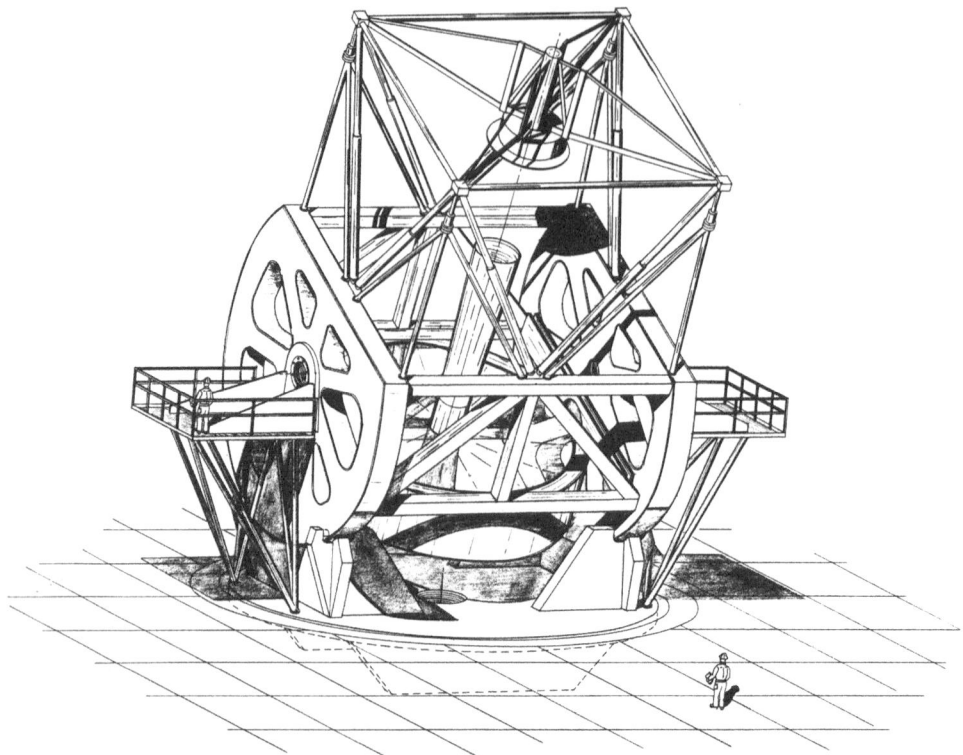

Fig. 1.   The 8 m telescope proposed for construction on Mauna Kea and a site in Chile. The design allows air to flow through the structure while retaining high stiffness properties. The top end is capable of extension to serve different optical configurations.

the atmosphere can deliver. For example, the testing of large optics is now done in a few minutes to accuracies impossible 20 years ago. Microprocessor technology has brought about fundamentally different approaches to engineering design and enables electronic coordination of widely separated functions that formerly were impossible or else obtained coordination through awkward, mechanical hoop-ups. Because of these and many other recent gains in technology, we believe that a doubling of the telescope size is possible along with the achievement of better performance relative to earlier telescopes. Toward achieving this goal, we have adopted the design guidelines listed below.

(1) The telescope and all of its subsystems will be designed to preserve the median image quality at Mauna Kea. The NOAO site tests, supported by other test data, have shown that median upper atmosphere seeing (i.e., ≥ 10 m above the ground) is below 0.5″ FWHM. To avoid degrading this natural performance by more than 10%, the telescope facility must be designed for 0.25″ FWHM performance.

(2) Man-made seeing effects will be minimized by avoiding heat transfer to the air in and around the telescope and by allowing air to pass naturally through the telescope enclosure to keep it flushed. Hot or cold surfaces exposed to the air must be avoided.

The telescope will be positioned well above the naturally occurring ground layer thermal turbulence.

(3) The primary mirror will be circular and monolithic to simplify polishing and support. Guided by the present work on mirror blank manufacture, we have chosen 8 m as the largest size likely to be available in the early 1990's. Blank of this diameter are expected to be available from Corning Glass Works, Schott Glaswerke, and the University of Arizona (Steward Observatory Mirror Laboratory).

(4) The primary focal ratio will be set to the lowest value consistent with the ability to design simple auxiliary optics and to polish the mirror with refined conventional methods. The chosen value of $f/1.8$ is a substantial gain over the fastest large telescope (the 4.2 m Herschel Telescope at $f/2.5$) produced to date, but still enabled us to design virtually all of the optical correctors with easy-to-make spherical/flat surfaces. Similar f/ratios have been chosen for the European Southern Observatory's Very Large Telescope (ESO–VLT) and the Japanese National Large Telescope (JNLT) which tends to confirm our own conclusions on the feasibility of polishing an $f/1.8$ mirror.

(5) Computer technology will be integrated into the design of telescope subsystems. 'Smart' instruments are familiar to most astronomers, but 'smart' telescope subsystems are rare. Major gains in performance can be obtained, sometimes at lower cost by integrating a computer to control positioning devices and force actuators instead of relying strictly on relatively cumbersome mechanical techniques.

A brief description of the telescope follows next. We plan to submit a proposal for construction of two telescopes to the U.S. National Science Foundation in early 1989.

2. Optical Configurations

The 8 m telescopes have been designed for 'true' Cassegrain operation which means that the primary mirror (a paraboloid) will not incorporate corrections specific to any one secondary. We can, therefore, obtain an equivalent degree of correction for images produced by more than one secondary. The price paid for this is slightly greater difficulty in making prime focus corrections. We have specified three different secondary mirrors and have retained the future option to use the prime focus. Optical correctors and/or atmospheric dispersion compensators (ADCs) have been designed for all but the IR configuration which does not require them. General properties of the basic optical configurations discussed below are given in Table I. In general, image quality $\leq 0.25''$ FWHM has been maintained over all of the fields for the wavelength span indicated except prime focus which is slightly worse. Imaging outside of the design wavelength span is also possible at some expense of quality and throughput.

2.1. THE $F/7$ 'WIDE-FIELD' CASSEGRAIN ($F/6.6$ UNCORRECTED)

A 45' corrected field of view (FOV) has been designed for use from 0.33 to 1.5 µm wavelength. A relatively large optical corrector is required which can be removed for other operational modes. $F/7$ was chosen for good performance with fibers (Barden, 1987). The focal surface is curved concentric with the exit pupil.

L. D. BARR

TABLE I

The basic optical configurations and related characteristics planned for the NOAO 8 m telescopes

OPTICAL MODE	FIELD OF VIEW (Arcmin)	FINAL FOCAL RATIO	SECONDARY MIRROR DIAMETER (m)	IMAGE SCALE (μm/π)	
Prime Focus	14	f/2		77	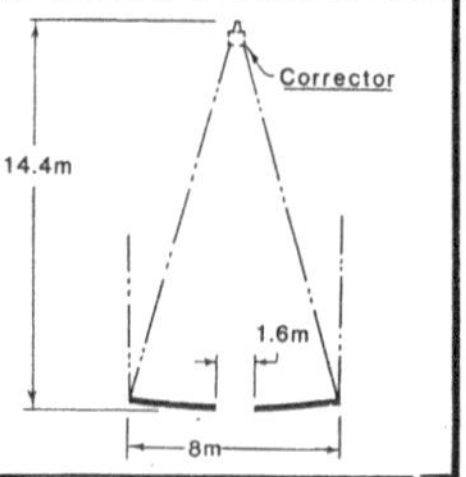
Cassegrain (only)	45	f/7	2.4	270	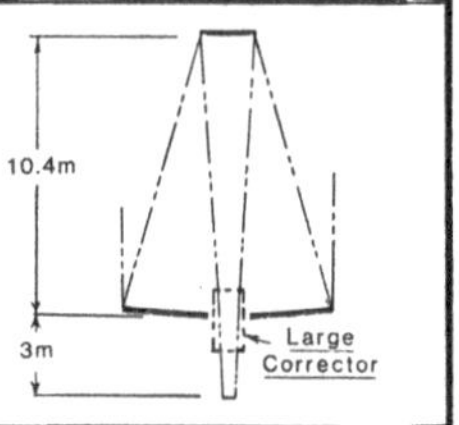
Cassegrain or Nasmyth (w.≠3 mirror)	6	f/7	2.4	270	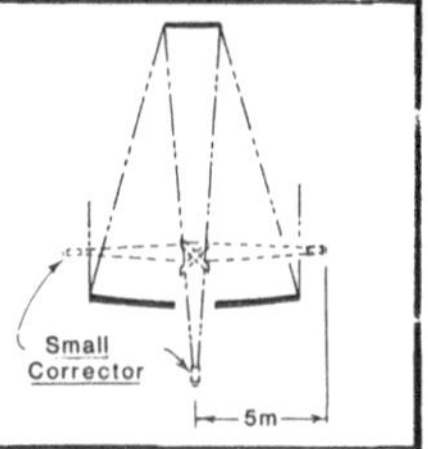
Cassegrain or Nasmyth	2	f/35	0.5	1350	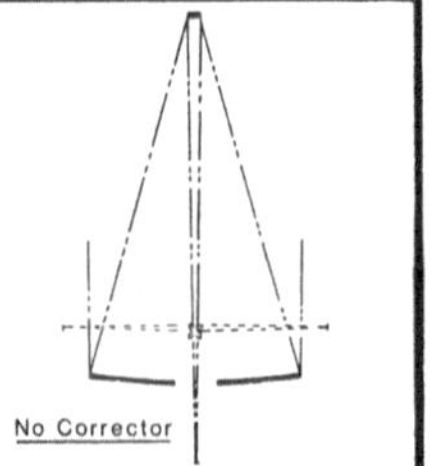
Nasmyth (only)	6	f/12	1.4	465	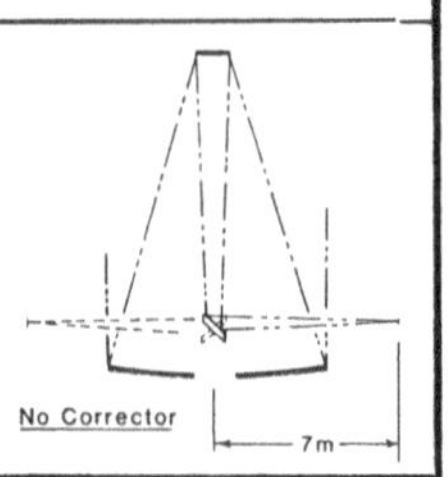

Instruments under consideration are:
- Optical imager with option of CCDs or photographic film.
- Moderate-resolution spectrograph with the options of 100 fiber feeds (possibly mounted away from the telescope on a stable platform) or long slit (to be mounted at Cassegrain focus); $R = 300–500$.

2.2. THE $F/7$ 'SMALL-FIELD' CASSEGRAIN OR NASMYTH

In order to obtain a flat field and the best possible imaging over a more restricted FOV (i.e., 6′), we have designed a compact corrector with fewer elements, but covering the same design wavelength range from 0.33 to 1.5 μm used for the wide-field corrector. The final f/ratio is 6.91, although we still call it $f/7$ for convenience. The small corrector size will permit easy relocation to/from the Nasmyth foci if desired. The Nasmyth mode uses a No. 3 flat mirror mounted to the corrector support housing in the primary central hole. Light can be fed directly into Nasmyth instruments placed near the inner platform edge or optically relayed further away on the platform.
Instruments under consideration:
- Optical imager (Cassegrain).
- Long-slit moderate-resolution spectrograph (Cassegrain).
- High-resolution optical spectrograph (Nasmyth); $R = 30\,000$, fiber-fed multiple-object spectrograph capability over limited spectral range.

2.3. THE $F/12$ 'EXTENDED BEAM' NASMYTH

An $f/12$ secondary was designed to obtain good images over 6′ FOV without using a corrector. The focal distance was extended an additional 2 m (i.e., to 5 m beyond the primary vertex before folding by the No. 3 mirror) to avoid the need for an optical relay. Design wavelength range from 0.33 to 1.5 μm.
Instruments under consideration:
- High-resolution beam-fed echelle spectrograph; $R = 10\,000–120\,000$.
- High-resolution infrared echelle spectrograph plus adaptive optics; $R = 100\,000$.

2.5. THE PRIME FOCUS OPTION

Although we do not plan to use the prime focus initially, the option of using it has been explored. A 14′ FOV is possible with corrected image quality of about 0.33′ FWHM for a wavelength range from 0.36 to 1.0 μm.
Instrument possibilities:
- Direct imaging with the option of CCDs or photographic film. Prisms.

2.6. ADAPTIVE OPTICS

The potential for adaptive optics correction of wavefront aberrations in the future is very great and we have explored how this could be done. The system tentatively proposed would use an $f/65$ adaptive secondary and a Shack–Hartmann wavefront sensor. Its optical characteristics include:
- Imaging down to the diffraction limit of the telescope.

- Image scale: 2.52 mm arc sec^{-1}.
- Wavelength coverage: infrared wavelengths ≥ 1.6 μm.
- Final f-ratio ($f/65$) chosen to give critical sampling at 1.6 μm for a detector with pixel spacing 50 μm ($= 0.02''$).
- Field of view: 2.5$''$ for 128×128 detector array.

Instrument available:
- Infrared array detector with option of use with filters or prism.

3. The Primary Mirror

Two primary mirror blank options have been investigated; a borosilicate glass honey-comb, and a low-expansion glass meniscus. Meeting the imaging requirement with either option requires success with significantly different engineering designs for the finished primary mirror assembly. This complicates our choice, but we draw great encouragement from our investigation because the manufacture of 8 m mirrors is clearly now within our grasp. Table II summarizes the most important technical characteristics of the two

TABLE II

Comparison summary of two 8 m mirror options

Characteristic	Honeycomb Style	Meniscus Style
Source for raw blank	Steward Observatory Mirror Laboratory	Corning Glass Co. or Schott Glaswerke
Raw blank material	Borosilicate glass, possibly Ohara E-6	Corning "ULE"™ or Schott "Zerodur"™
Est. weight of raw blank	16,800 kg	20,430 kg
Coefficient of thermal expansion (CTE)	$\sim 3 \times 10^{-6}/°C$	$\sim 0 \times 10^{-9}/°C$
Expected variation in CTE	$1\text{-}2 \times 10^{-8}/°C$	$1\text{-}2 \times 10^{-8}\cdot°C$
Polishing & Testing		
Polishability	Good	Good
Polishing method planned	Conventional lap methods. Special care needed to avoid rib structure print-through onto front surface.	Conventional lap methods. Uniform pressure support needed to cope with blank flexibility.
Principal testing method	Interferometric methods. Need high spatial resolution to measure rib print-through. Need to compensate or control thermal effects on figure.	Interferometric methods

Table II (continued)

Characteristic	Honeycomb Style	Meniscus Style
Thermal Control		
Principal reason for thermal control	Warpage of mirror due to internal temperature gradients	Avoidance of mirror seeing
Thermal control technique planned	Radiation to/from the mirror cell plus controlled temperature air circulation in each honeycomb cell (~ 1500 cells)	Liquid circulation through ~ 200 internal cooling channels
Tolerable ΔT in the blank to meet spec. of 0.05 arcsec FWHM	~ ±0.04°C depending on temperature distribution	More than 10°C
Estimated ability to control ΔT	~ ±0.05°C in any given cell. ~ ±0.25°C overall	~ ±0.07°C overall
Estimated ability to control front surface temperature	Within 0.3°C of ambient air temperature	Within 0.1°C of ambient air temperature
Support		
Mirror passive support	Levers and interconnected hydraulic pistons	Levers and interconnected hydraulic pistons
No. of supports to meet spec. of 0.05 arcsec FWHM	135 axial acting at rib junctions. 135 lateral acting in backplane cell holes	162 combined axial and lateral acting on common attachment points
Active control of support forces	Stepper motors and springs acting on 135 axial supports	Stepper motors and springs acting on 162 axial supports
Estimated precision requirement for support forces to meet spec. (active or passive)	5×10^{-3}	1×10^{-3}
Principal reason for active force control	To correct for thermal warpage of mirror	To keep passive support forces at correct values

primary mirror systems as we perceive them now (November, 1988). Future developments may change our perceptions.

Borosilicate honeycomb mirror castings are under development at the University of Arizona's Steward Observatory Mirror Laboratory and a low-expansion glass meniscus may be obtained from commercial glass-makers. To date, a 3.5 m-diameter honeycomb and a 4.1 m-diameter meniscus have been produced successfully. Projected costs to complete the first 8 m-dameter blank are roughly equal for either option, but the cost for succeeding blanks is expected to favour the borosilicate honeycomb, principally because of lower raw material costs. Offsetting this advantage are the engineering studies at NOAO that show that polishing and thermal control will be easier for a meniscus mirror than for a borosilicate honeycomb principally because of its low-expansion property. Recent analysis at NOAO also shows that the effects of supports for a meniscus mirror can be more accurately calculated than those for a honeycomb which should result in less empirical calibration and field adjustment. A future choice between options will be made based upon manufacturing feasibility and delivery schedules when funding is available.

The technical issues of concern for the mirror blanks have centered principally on the relatively large coefficient of thermal expansion for borosilicate glass (CTE $\sim 3 \times 10^{-6}/^\circ C$) and the relatively low bending stiffness of a meniscus blank with a diameter-to-thickness ratio of 40 : 1. The high CTE value necessitates a thermal control system capable of maintaining temperature throughout the mirror within a narrow band of $\pm 0.04 \,^\circ C$ (Pearson *et al.*, 1986; Pearson and Stepp, 1987). Large temperature differences will cause thermal distortion of the mirror blank which, in turn, causes image enlargement progressively larger than the 0.05″ FWHM error budget presently allowed. The meniscus mirror, because of its near-zero CTE value does not have this problem, but it still requires low-precision thermal control for its front surface to remain near the ambient air temperature and, thus, avoid mirror seeing. The meniscus mirror requires a more accurate support involving about 20% more supporting mechanisms than the borosilicate honeycomb.

Satisfactory support systems have been designed for both mirror blank options within an error budget allowance of 0.05″ FWHM. Both will have the ability to control the forces applied to the mirror as a function of zenith angle. Force sensors and actuators will be built into each support mechanism. Each mechanism can also operate passively in the event of control system failure or when corrections are not required. Corrections for residual polishing error, calibrated gravitational effects, and non-uniform mirror blank properties can be made with these controlled supports. If the mirror surface or

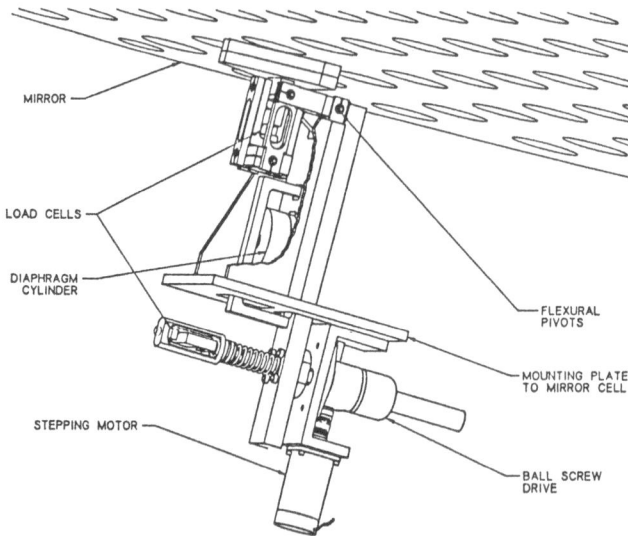

Fig. 2.   The prototype axial support mechanism for the 3.5 m borosilicate mirror. Similar mechanisms are planned for the 8 m mirror.

the image is monitored (e.g., by Shack–Hartmann techniques), it will be possible to make low-frequency corrections for thermal distortion and other unpredictable changes in the mirror surface. Figure 2 illustrates one axial support design under consideration

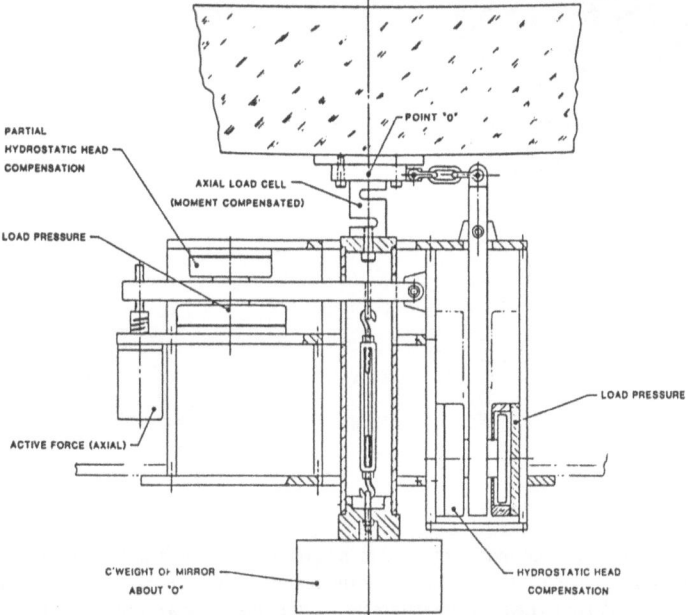

Fig. 3.   The axial and lateral support mechanism that will support the 8 m mirror at 162 locations. The mirror weight pressurizes the 'load pressure' pistons. Opposing pistons offset hydrostatic pressure at non-zero zenith pointing angles. The active force device corrects for small errors in load detected by the load cell.

for the honeycomb mirror. Figure 3 illustrates a combined axial and lateral support for an 8 m meniscus. Figure 4 is an example of the type of correction we estimate is possible with an active support. The results of Figure 4 apply to an 8 m-diameter borosilicate honeycomb mirror model (described by Pearson *et al.*, 1988a) resting on 135 controlled force supports and affected by an assumed temperature distribution (maximum peak-to-valley of 0.7 °C) similar to an actual distribution measured in a 1.8 m honeycomb mirror at NOAO (Pearson *et al.*, 1988b). Modeling was done at NOAO using a thermo-elastic finite element program. The figure obtained after correction meets our specifications.

Polishing of the mirror figure within a residual error allowance of 0.05″ FWHM is regarded as feasible for either mirror blank. The recently acquired ability developed by C. Roddier *et al.*, at NOAO, to test the entire mirror surface on large optics to accuracies of less than 0.02 μm at spatial separations of less than 10 cm is one major reason for this optimism. Techniques for suppressing vibration and thermal turbulence in the optical testing setup have also been developed. An 'automated' testing procedure involving a CCD camera has shortened the time needed to obtain useful mirror surface information from hours (or even days) to only a few minutes. This gives us confidence that a highly accurate 8 m mirror can be polished in the same time formerly required

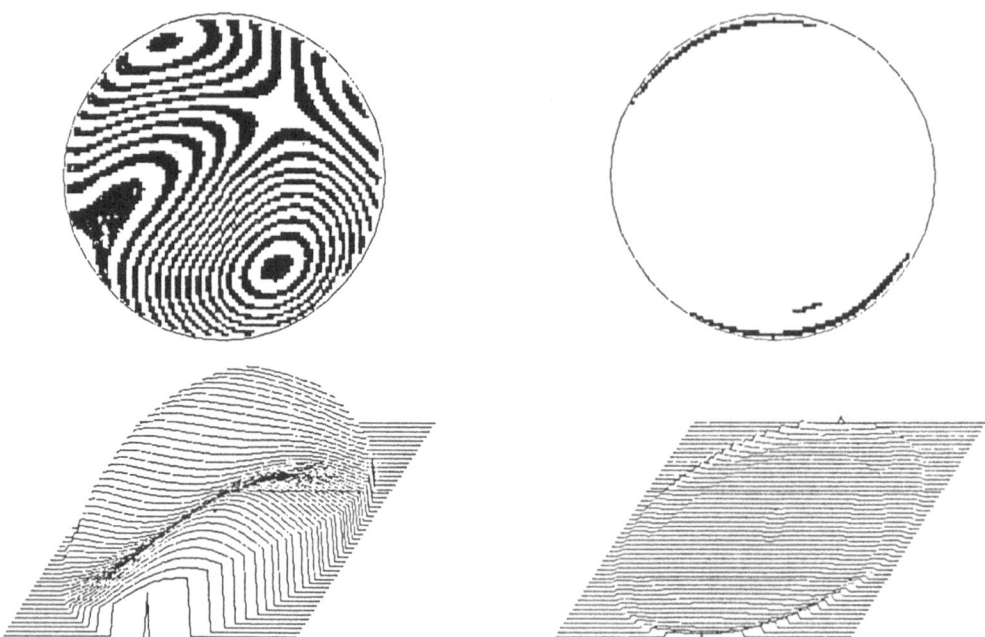

Fig. 4. Results of a computer model of an 8 m mirror affected by hypothetical internal temperature gradients (see text). The corrections possible with the planned active support were then applied. Contour interval is 0.18μ. $\lambda = 550$ nm. (*Left*) Figure before correction: P–V: 4.463λ; r.m.s.: $1 : 111\lambda$; image spot size: $0.52''$ FWHM. (*Right*) Figure after correction: P–V: 0.313λ; r.m.s.: 0.033λ; image spot size: $0.05''$ FWHM.

for a 4 m mirror with only one-fourth of the surface area. Figure 5 illustrates the new method tested at NOAO on a 1.8 m-diameter mirror. The Roddier data analysis method (Roddier and Roddier, 1987) was used.

For the 8 m telescopes the accuracy of the final mirror figure will be set by the test optics, especially the null lens. An all-spherical null lens has been designed by C. F. W. Harmer of NOAO with peak-to-valley residual wavefront error of less than 0.005 μm across the full aperture. By an iterative manufacturing process, we expect to obtain diffraction limited performance from this lens, a required condition for meeting the polishing specifications.

Thermal control of the primary mirror is needed for either mirror blank option. he borosilicate glass blank must be kept at a near-isothermal condition to avoid warpage. A combination of radiative heat exchange with the mirror cell and convective heat exchange with air blown actively into each honeycomb cell will be used to stabilize the blank temperature. The meniscus blank, because of its thicker faceplate (i.e., 20 cm total thickness vs 2.5 cm faceplate thickness for the honeycomb) requires thermal control to keep its front surface near the ambient air temperature and avoid mirror seeing effects on the image. A fluid circulation system, internal to the blank, has been designed for this purpose with the ability to maintain the front surface within ± 0.1 °C of ambient air temperature, which is more than adequate (Barr *et al.*, 1988).

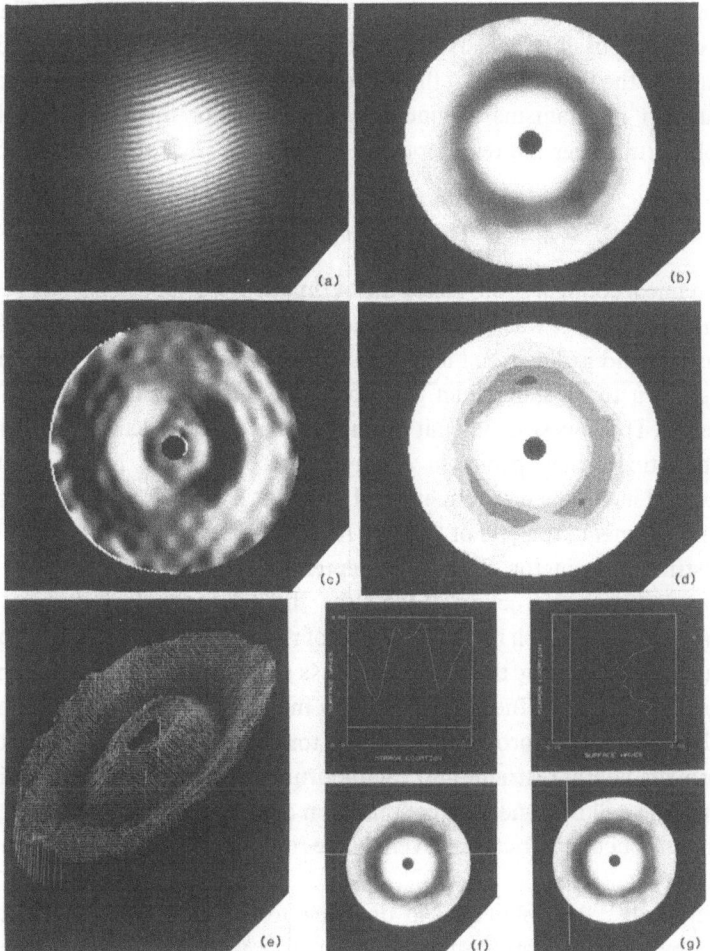

Fig. 5. Sequence of events in testing a 1.8 m diameter honeycomb mirror with a procedure developed at NOAO. Photos of computer monitor. (a) Scatterplate interferogram taken with a CCD camera having 512 × 512 pixels. (b) Phase map developed from (a) using intensity information and the Roddier analysis method. (c) Surface of the mirror developed from (b). Similar in appearance to a Foucaultgram taken with a knife-edge. (d) and (e) Surface contour maps developed from (c). (f) and (g) Surface profiles along the line cuts indicated. Maximum peak-to-valley shown is about 0.25 μm.

Aluminum has been specified as the initial primary mirror coating. A coating chamber with filament evaporators will be provided, but with structural provisions for future adaptation to sputter deposition. The possibility of long-lasting multi-layer coatings that provide higher reflectivity than aluminum is an option that may be feasible in the future and sputtering is one way they may be applied. At present, sputtering can be done 'linearly' with large panes of glass, but further work is needed to do as well with a rotating substrate.

Primary mirror removal for re-coating will be done by first detaching and removing the lower support cell containing the support hardware and, then lowering the mirror in a fixture that serves at the bottom part of the coating chamber. This two-step operation will be possible because of special detachment provisions in the support mechanisms and the fact that all of the mirror support will be done from its rear surface (i.e., no support mechanisms around the mirror rim or interior to the blank). An elevating platform under the telescope will be used for the removal operation.

4. The Telescope Mounting

A new mechanical configuration, called the 'Azimuth-Disc' mounting, has been adopted for the 8 m telescopes. It was designed in collaboration with the Magellan Project and L&F Industries and is illustrated in Figure 1. Structural analyses of the Azimuth-Disc mounting show it to be stiffer than the more traditional azimuth yoke mounting and lower in weight. The lowest significant resonance is 6.1 Hz, well above the critical region for wind buffeting which peaks in energy amplitude below 1.0 Hz. Total moving telescope weight including instruments, will be about 240 000 kg (527 000 lbs) (Concept Design Finite Element Analysis of 8 m Alt-Az Disk Telescope' for AURA (NOAO) by L&F Industries, Huntington Park, CA, September 1988).

The optical support structure (OSS) for the telescope is based on a pair of 12 m-diameter journals, each floated on a pair of radially-acting hydrostatic oil bearings, and truncated for supporting the secondary truss structure. Cross-beams connecting the journals provide lateral stiffness. The primary mirror cell attaches to the journals and the cross beams, thereby providing in-plane torsional stiffness. Caliper-style oil pad bearings provide lateral restraint and each journal is driven at its rim by friction roller drives. Encoding will be done by a combination of encoding tapes attached to the rims and incremental encoders friction driven by the rims. Incremental resolution of $\pm$ 0.02" is planned.

The azimuth disc, roughly equal in diameter to the altitude journals, is supported underneath by oil pad bearings positioned in line with the load reactions from the altitude axis bearings. This arrangement creates a nearly direct load path from the altitude bearings to the underlying foundation and does not bend the azimuth disc in any significant way. As a result, structural stiffness is effectively higher. The azimuth disc is also driven as its rim by roller drives. By driving the rims, the effective mechanical stiffness of the drives is enhanced.

In order to accommodate the three secondaries and prime focus options, an extendable top-end structure has been designed. Four screw jacks mounted in stationary corner trusses cause the secondary strut support frame to raise/lower to the desired positions. Figure 6 illustrates two of the possible configurations. Telescoping diagonal truss members with built-in adjustable mechanical stops are driven into a state of tension/compression by action of the jack screws. Load sensors and a subsystem control computer will be used to set desired stress levels in the diagonal trusses. After the stress levels are set, the screw jacks are turned off and the system operates passively under gravitational and thermal influences. Future experience may allow us to use this system to make flexure corrections although that is not initially planned.

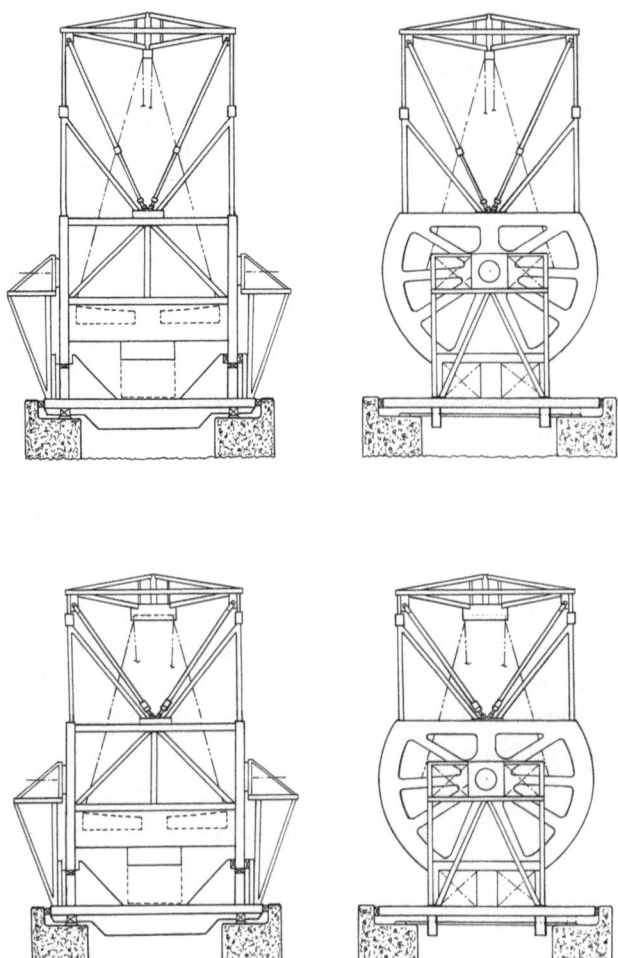

Fig. 6.   Front and side views of the 8 m telescope with the top end extended for IR operation (*upper views*) and retracted for wide-field Cassegrain operation (*lower views*). An intermediate position will serve the planned *f*/12 Nasmyth configuration.

Finite element analyses of the structure show that some of the required optics positioning tolerances can be achieved without active adjustment during operation while others require active control. Table III summarizes the 'zenith-to-horizon' situation for the *f*/7 wide-field Cassegrain mode which requires the most stringent secondary-to-primary position control. The means for controlling the secondary are discussed further on.

5. The Secondary Mirror Mounting and Alignment

The three secondary mirrors planned for the 8 m telescope (*f*/7, *f*/12, and *f*/35) will be mounted in individual cells capable of being taken on/off the telescope. Secondary

TABLE III

Secondary Position Parameter	Allowable Movement	Calculated Flexure
Decentration	±0.25mm	+0.18mm
Spacing Change Relative to Primary	±1mm	+2.84mm
Defocus	±.017mm	+2.84mm
Tilt	±15 arcsec	+31 arcsec

exchange will occur with the telescope pointed to horizon and will be done from a rising platform installed in the enclosure floor in front of the telescope.

All of the secondaries will be lightweight and will be made from low expansion glass. The $f/35$ secondary will be used as a chopping device for IR operation using a 2-axis design developed for the NOAO 2.1 m telescope. A contra-oscillating weight prevents most of the chopping accelerations from being transmitted to the telescope structure.

The $f/7$ and $f/12$ secondary cell assemblies will be controlled for tilt and focus motions by means of position actuators connecting the cells to the secondary support struts. Angular precision of 0.05″ is feasible and required because the secondaries will be actively controlled to maintain pointing precision of the telescope as measured by star-tracking guider probes in the focal planes.

6. The Telescope Enclosure

The principal functions of the telescope enclosure will be to protect the telescope from weather, prevent excessive interior warming during the daytime, and serve as a 'porous' wind barrier at night. Control rooms, workshops, and other activities requiring comfort control for people will be located in a separate, but connected, facility. A coating facility is also planned to be adjacent to the telescope enclosure. A description of the 8 m enclosure plans is provided in a companion paper presented by Wong at this conference.

7. Conclusion

A brief description of an 8 m telescope has been presented. In early 1989 NOAO plans to submit a proposal to the U.S. National Science Foundation to build two of these telescopes as national facilities on Mauna Kea, Hawaii, and a site in Chile on AURA property. Construction is anticipated to start in the early 1990's.

Acknowledgements

The design of the 8 m telescope described in this paper is the work of many people. Special credit goes to Steve Gunnels and L&F Industries for structural analysis. Earl

Pearson, Larry Stepp and Bill Keppel at NOAO have designed the mirror supports. Warren Davison (Steward Observatory) adapted the double-altitude journal concept to optical telescopes. Charles Harmer developed the optical designs. The list of contributors is long and is continuing to grow. I am especially grateful to Anne Beeler for manuscript support.

References

Barden, S.: 1987, 'Focal Ratio Degradation (FRD) Tests of Two 31.7-Meter Samples of Polymicro FLP 320385415 Fiber', *KPNO Fiber Optics Lab Reports* No. 1.

Barr, L. D., Beckers, J. M., Pearson, E. T., Hobbs, T. W., and Spangenberg-Jolley, J.: 1988, 'Reducing Mirror Seeing Problems in Meniscus Mirrors', *ESO Conference Proceedings on 'Very Large Telescopes and Their Instrumentation'*, Garching, F.R.G.

Merrill, K. M. and Forbes, F. F.: 1987, 'Comparison Study of Astronomical Site Quality of Mt. Graham and Mauna Kea', *NNTT Report* No. 10, NOAO, Tucson.

Pearson, E. and Stepp, L.: 1987, *SPIE Proc.* **748**, 215.

Pearson, E. T., Stepp, L., Wong, W.-Y., Fox, J., Morse, D., Richardson, J., and Eisenberg, S.: 1986, *SPIE Proc.* **628**, 91.

Pearson, E., Stepp, L., and Keppel, W.: 1988a, 'Support of 8-Meter Borosilicate Glass Mirrors', *Proc. ESO Conference on Very Large Telescopes and Their Instrumentation*.

Pearson, E., Stepp, L., and Fox, J.: 1988b, *Opt. Eng.* **27**, No. 2.

Roddier, C. and Roddier, F.: 1987, *Appl. Optics* **26**, No. 9.

ENARD – Some time ago, we have investigated linear actuators to replace the conventional gears for the elevation drives. We concluded that the stiffness of such actuators would not be good enough.
Have you investigated this matter for your movable top tube? Do you think that the actuators would not degrade the dynamic performance of the tube?

BARR – In our finite element analysis, we used spring constant values that are consistent with available screw actuators. The flexural results obtained appear to be satisfactory. Since the positioning of the top end is due to fixed mechanical stops built into the diagonal truss members, the accuracy of the screw actuators does not affect the positioning performance of the system.

HALL – Experience with infrared arrays now in use on a number of telescopes demonstrates that background limited performance can be achieved without a chopping secondary. Given the disadvantage of large throw chopping secondaries on 8 m class telescopes, could you comment on the inclusion of one for the NOAO 8 m?

BARR – We are continuing to plan for a chopping secondary for the needs of those astronomers who do not use array detectors. This situation could change in the time before actual construction begins. The issue, in my opinion, is still unresolved.

McLAREN – You referred to three Cassegrain mirror positions for the telescoping upper end. What are they to be used for?

BARR – There are only two Cassegrain mirrors planned, one at $f/7$ wide-field work, the other at $f/35$ for IR work. The third mirror, at $f/12$, is intended for Nasmyth usage only. I incorrectly mentioned this as a Cassegrain mirror.

THE PLANNING OF NOAO 8 M TELESCOPE ENCLOSURE AND FACILITIES*

WOON-YIN WONG and LAWRENCE D. BARR

National Optical Astronomy Observatories†, Tucson, U.S.A.

(Received 9 January, 1989)

Abstract. The National Optical Astronomy Observatories (NOAO) proposes to build 8 m telescopes at two locations. The Northern Hemisphere location is on Mauna Kea, Hawaii, and the Southern Hemisphere location is within the existing boundary of the Cerro Tololo Inter-American Observatory. This paper describes the sites under consideration and the facilities that support the operation of each telescope, namely, the telescope enclosure, the control facility, and the recoating facility.

1. Introduction

The main goal in the site development study and facility design for the NOAO 8 m telescopes is to keep the image degradation due to manmade seeing below 0.06″ r.m.s. Some yet-to-be-proven ideas are introduced. With lessons and experiences gained from existing facilities we are specifying the site facility design with the intention that the overall 0.25″ imaging goal can be achieved. We, therefore, propose to do the following: (1) thermally insulate the enclosure and avoid components of large thermal mass, to ensure that the enclosure temperature follows the ambient air temperature variation; (2) build a perforated enclosure to allow a larger amount of air flow; (3) keep all possible heat sources away from the telescope to minimize the thermal disturbance; and (4) provide a forced ventilation system for faster exchange of enclosed air.

2. Site Development

NOAO is proposing to build two 8 m telescopes, one in the Northern Hemisphere and one in the Southern Hemisphere. The northern site is Mauna Kea in Hawaii, and the southern site is within the boundary of CTIO in Chile, possibly at the high peak of Cerro Pachon.

Mauna Kea is the highest of the five large volcanoes that have coalesced to form the island of Hawaii. Some of the relevant features of the island are shown in Figure 1(a), including the positions and elevations of the volcanoes and the locations of major towns and principal roads.

Figure 1(b) shows the south slope of Mauna Kea from the Humuula Saddle to the summit. The Saddle Road connects Hilo and Waimea. From the Mauna Kea turnoff,

* Paper presented at the Symposium on the JNLT and Related Engineering Developments, Tokyo, November 29–December 2, 1988.
† Operated by the Association of Universities for Research in Astronomy Inc., under contract with the National Science Foundation.

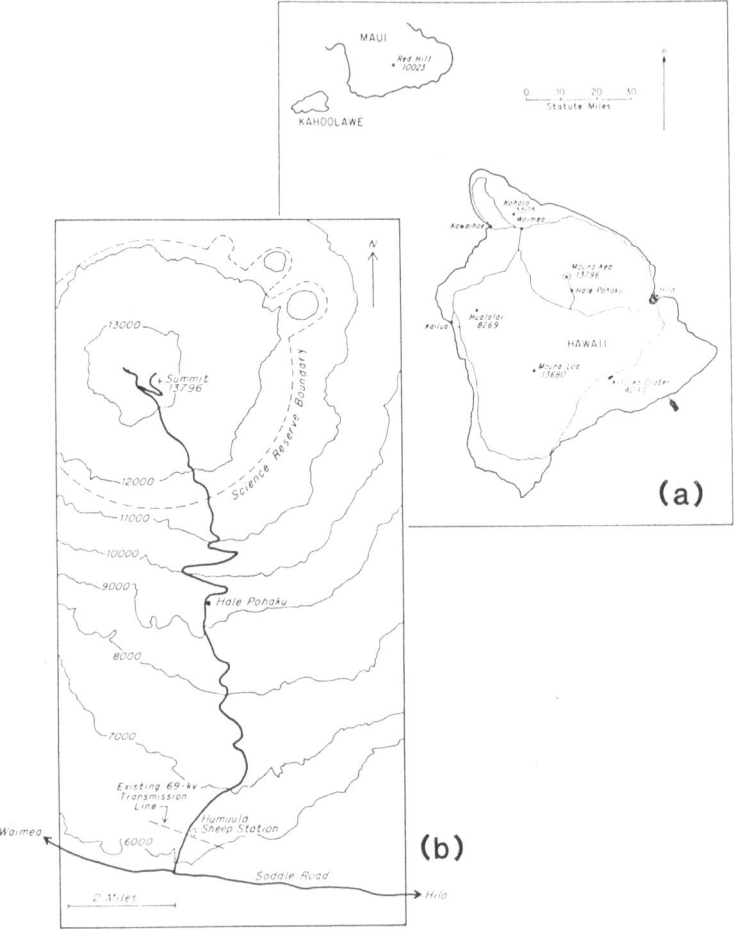

Fig. 1.   (a) The Island of Hawaii. To the northwest are Maui and the uninhabited island of Kahoolawe. The indicated altitudes are in units of feet above sea level. (b) The south slope of Mauna Kea. Hilo is 45 km to the east and Waimea is 51 km to the west. Hale Pohaku is a mid-level base. Above it is 22 km of unpaved road. All ground above 3660 m is designated as 'Science Reserve'.

it is 45 km to Hilo, and 51 km to Waimea. The distances from the turnoff to Hale Pohaku and to the proposed site are, respectively, 10 and 23 km. Hale Pohaku, originally a ranger station, serves as a mid-level base for the observatories and for the contractors working at the summit. The summit area, including all the ground above 3660 m, is designated as a 'Science Reserve' administered by the University of Hawaii, under a long-term lease from the State. The paved road up to Hale Pohaku is quite good. To negotiate the unpaved road beyond Hale Pohaku requires a 4-wheel-drive vehicle.

The nightly mean wind speed and the corresponding wind speed histogram for these data are shown in Figure 2, and the corresponding histogram for wind direction is given in Figure 3. The median wind speed is 7 m s^{-1}. Ninety-four percent of the measurements were at or below 18 m s^{-1}. The wind direction is ENE–ESE 37% of the time

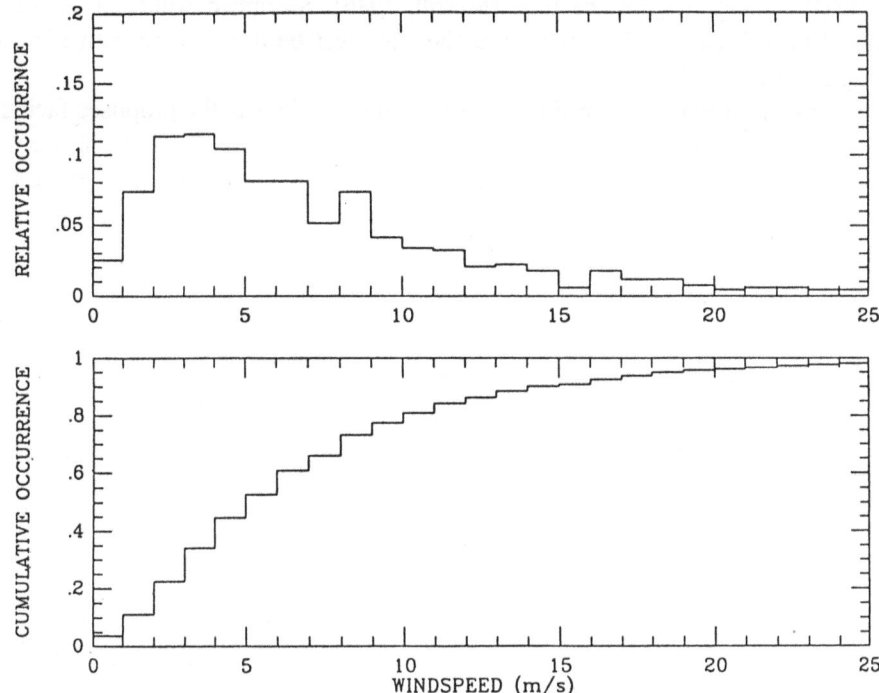

Fig. 2. Integrated occurrences of wind speed at NOAO test site on Mauna Kea. The medium wind speed was 7 m s^{-1}. Ninety-four percent of the measurements were at or below 18 m s^{-1}. Sustained wind in excess of 45 m s^{-1} have been noted. All buildings are designed to withstand 67 m s^{-1} wind.

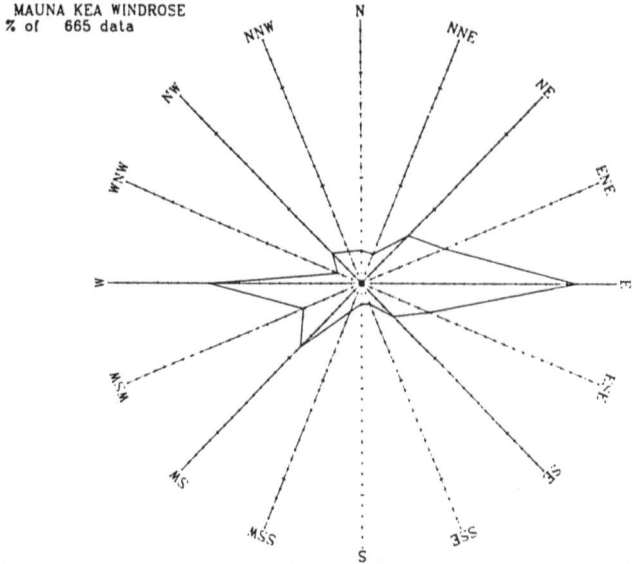

Fig. 3. Wind rose for the NOAO test site at Mauna Kea. Full scale in each direction is 25%. The wind direction is ENE–ESE 37% of the time and WNW–WSW 36% of the time.

and WNW–WSW 36% of the time (Merrill, 1986). Sustained winds in excess of 45 m s^{-1} have been noted. Buildings at the site must be designed to withstand ice loading and 67 m s^{-1} wind force.

The site plan, as illustrated in Figure 4, shows the locations of the proposed facilities relative to the existing UH 2.2 m and CFH 3.6 m telescopes. The site plan calls for four major items: (1) the telescope and its enclosure; (2) the control facility; (3) the 8 m

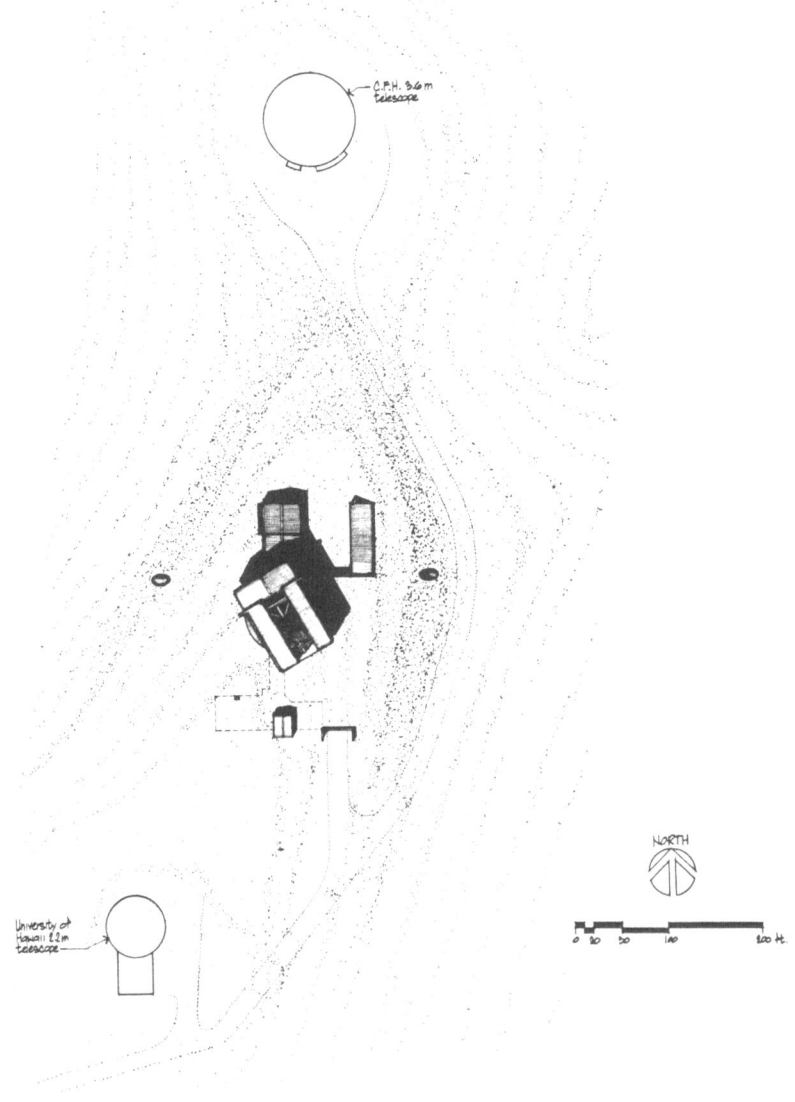

Fig. 4. The proposed site plan for Mauna Kea. The CFHT is to the north and the UH 2.2 m telescope is to the south. The rectangular corotating building option is shown here, between the control facility to its south and the recoating facility to its north. In this layout, the existing access road between the UH and CFHT facilities has been rerouted. The area to be leased for this project is approximately 6600 m².

recoating facility; and (4) a 3 m diameter tunnel, approximately 100 m long, for forced-air ventilation of the telescope enclosure.

Cerro Pachon is the highest peak within the boundary of CTIO. It is southeast of Cerro Tololo and rises 2725 m above sea level, about 500 m higher than Cerro Tololo. The inset in Figure 5 is a location map showing CTIO in relation to La Serena and the

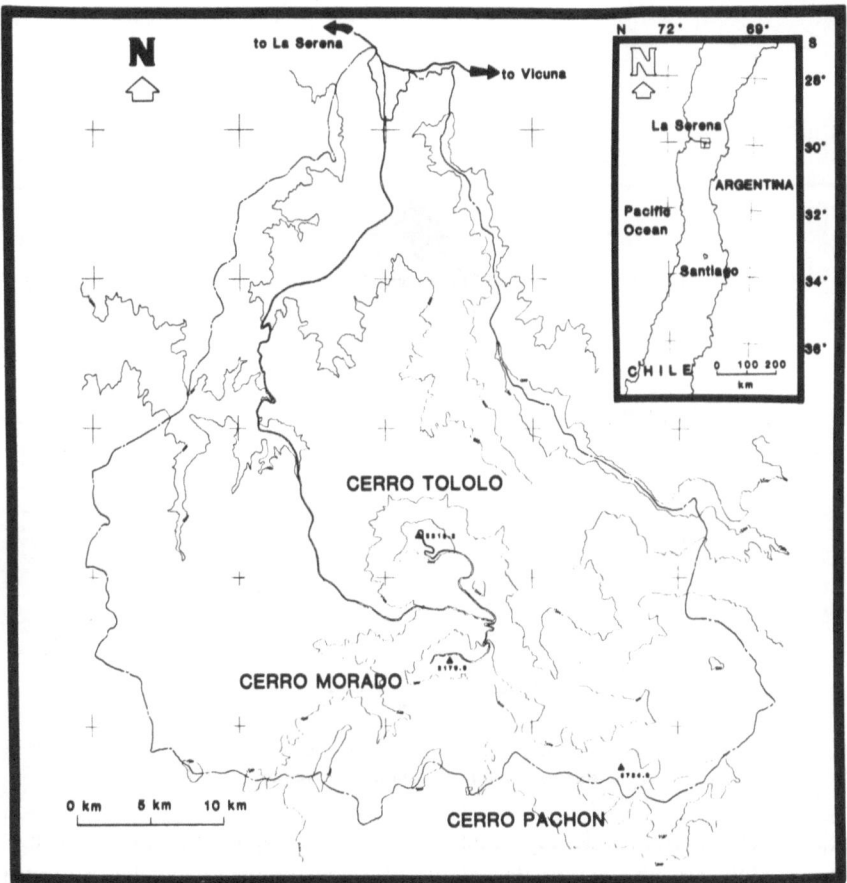

Fig. 5. Map of the three sites in Chile that are being tested. Cerro Pachon is approximately 500 m higher than the other two mountains. The dash-dot line is the boundary of the AURA property.

capital city of Santiago. The larger map shows the relative placement of Tololo, Morado, and Pachon, all of which are being tested as possible sites for an 8 m telescope. Preliminary site measurements indicate that the prevailing wind is from the NNE; about two-thirds of the time the wind comes from a direction within 45° of this. The mean wind speed is 5 m s^{-1}, with less wind in the summer months and more in the winter (NOAO Draft Report, 1988).

A site plan for Cerro Pachon is shown in Figure 6. The site plan is, to a large extent, the same as the one for Mauna Kea. Minor differences such as the arrangement of other

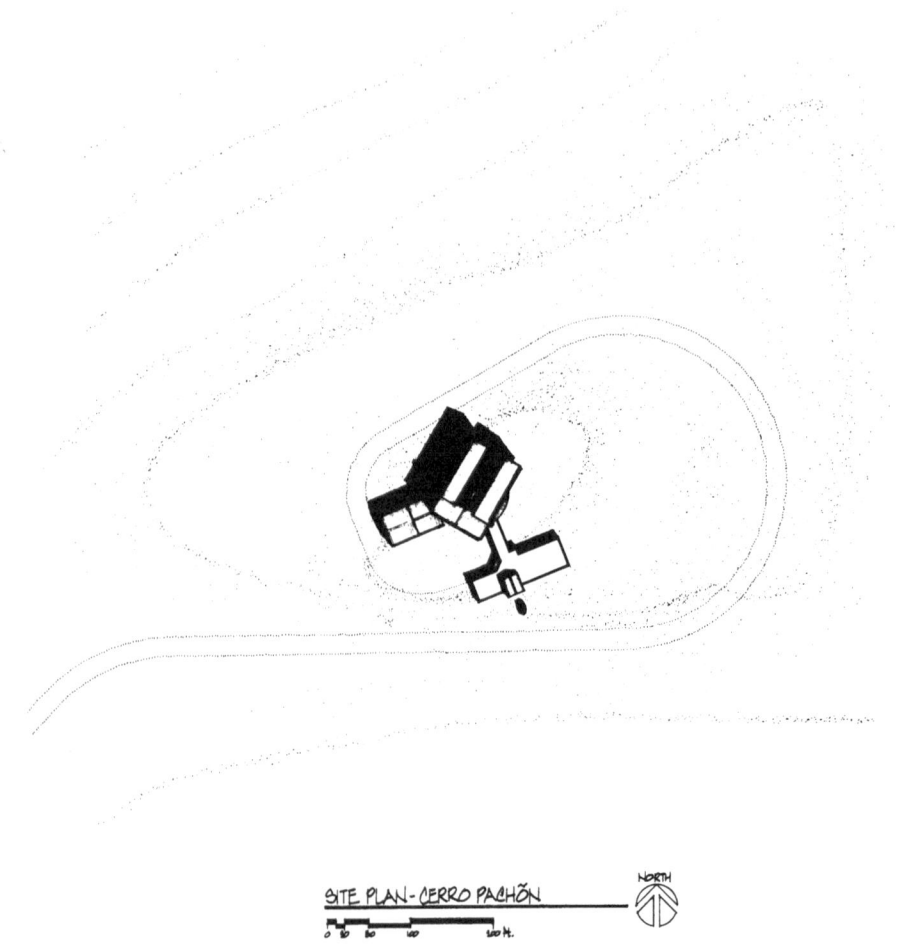

SITE PLAN- CERRO PACHÓN   NORTH

Fig. 6.   Preliminary site plan for Cerro Pachon, similar to the Mauna Kea plan except for building
arrangements due to different terrain and prevailing wind direction.

facilities in relation to the telescope enclosure, roadway alignment, size of parking area, and length of the air exhaust tunnel will have little effect on the facility descriptions given in the following sections.

Seeing-management measures are included in the plans. The control and recoating facilities are mostly underground; the parking area and the road are graded but not paved; and the air tunnel is aligned with the prevailing wind. It is suggested that the buildings be covered with the local cinder material wherever possible. Exposed walls, windows, and doors are to be insulated.

3. The Telescope Enclosure

Traditionally, telescope enclosures were made large compared to the telescope for three reasons. First was that the large air volume enclosed by a well-insulated shell provided a relatively stable air environment for the telescope, shielding it from the influences of outside temperature variation, particularly during the daytime when the outside temperatures are most likely to change rapidly. Second was the oversized dome shielded the telescope from the wind, reducing the image shake during observation. Third was that before the development of computer technology, equatorially mounted telescopes were technically more feasible to operate than altidue-azimuth telescopes. Because of their physical arrangement and the necessity to reach any part of the sky, equatorial telescopes normally require a large amount of space within the dome to allow for the telescope's motions in hour angle and declination. Provision of a large overhead crane further increases the dome size.

Figure 7 is a collection of scaled drawings of three existing 4 m class telescope buildings and two proposed 8 m telescope enclosures. The KPNO 4 m and the CFHT

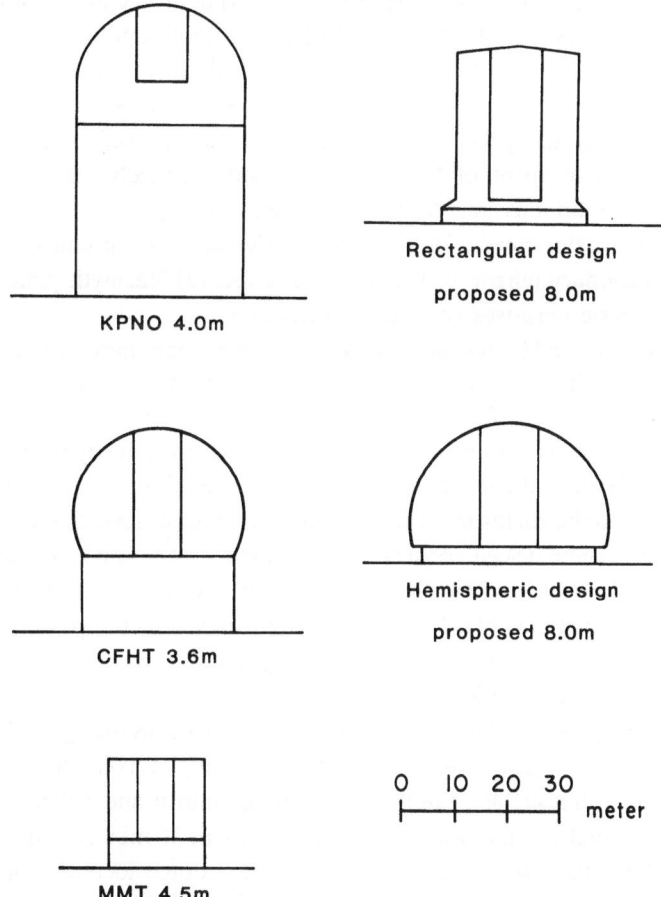

Fig. 7. Two enclosure design options for the 8 m telescope are shown on the right. Three existing domes for 4 m class telescope are shown in the left. In spite of the difference in telescope sizes, the dome sizes and the shutter widths are comparable. Hence, the construction cost for this 8 m telescope enclosure remains relatively unchanged since the construction of domes for the 4 m class telescopes.

3.6 m are both equatorially mounted telescopes, whereas the MMT 4.5 m is an alt-az mounted telescope. The difference in building size and cost between the former and the latter is substantial. The KPNO dome was built in 1970 and cost 5 million U.S. dollars; the MMT corotating dome was built in 1975 and cost 2 million U.S. dollars. The size and the shutter width of the KPNO 4 m dome and the proposed enclosures are comparable. This comparison indicates that building the future enclosures for the 8 m telescopes is a relatively straightforward engineering task. The challenge is not in how to build enclosures of this size, but in how to build them with minimal artificial seeing effects.

Figure 8 is an engineering drawing showing three views of one of the principal enclosure design options, and Figure 9 a detailed floor plan (for the Mauna Kea site). This preliminary enclosure design includes the following special features:

(a) *Advantage cranes and elevators for easy operation*: One of the most important functions of the enclosure, other than its prime function of protecting the telescope from the environment, is servicing the telescope to insure smooth and uninterrupted operation. For instance, the front end of the telescope needs frequent access for the exchange of secondary mirrors. Auxiliary equipment and other heavy components need to be installed. Periodically the primary mirror must to be removed and carried to the recoating facility. Movement of heavy objects, both vertically and horizontally, is important to the continuous functioning of the telescope.

There are four major needs for cranes and elevators: (1) primary mirror transportation, (2) secondary mirror exchange and storage; (3) Nasmyth platform accessibility, and (4) general purposes such as maintenance.

(1) Tentative plans call for a yearly recoating of the primary mirror. The primary and its supporting mechanism are to be transported as a unit, approximately 9 m × 9 m × 2 m, weighing about 50 tons. With the telescope at zenith, the primary mirror unit will be lowered to a trolley and then rolled through the opened shutter space onto a platform elevator, the mirror unit remaining at zenith throughout. The platform elevator, adjacent to the enclosure, will lower the mirror to the level of the vacuum tank. This plan imposes some design constraints on the enclosure. The load capacity of the floor must be sufficient to sustain the 50-ton mirror-and-cell unit plus the trolley. In event that the recoating facility is located elsewhere, an access ramp should have to be provided outside the enclosure, so that the trolley could move directly to a specially designed truck for transportation.

(2) One special feature of the optical support structure (OSS) design is the retractable or extensible top end to accommodate the wide-field, high-resolution optical spectrograph and infrared optics (Barr, 1989). The secondary mirror and cell combinations are to be designed so that changeover can be made in about 30 min. A dedicated elevator will serve as both the exchange platform and the storage location for the unused secondaries. A holding turret rigidly fixed to the elevator will hold the unused mirrors. The exchange procedure will begin with the horizon-pointing telescope lined up with the elevator and locked in place. At that point the elevator is raised from the storage level to the service level, and an empty holding turret is then aligned with the secondary mirror

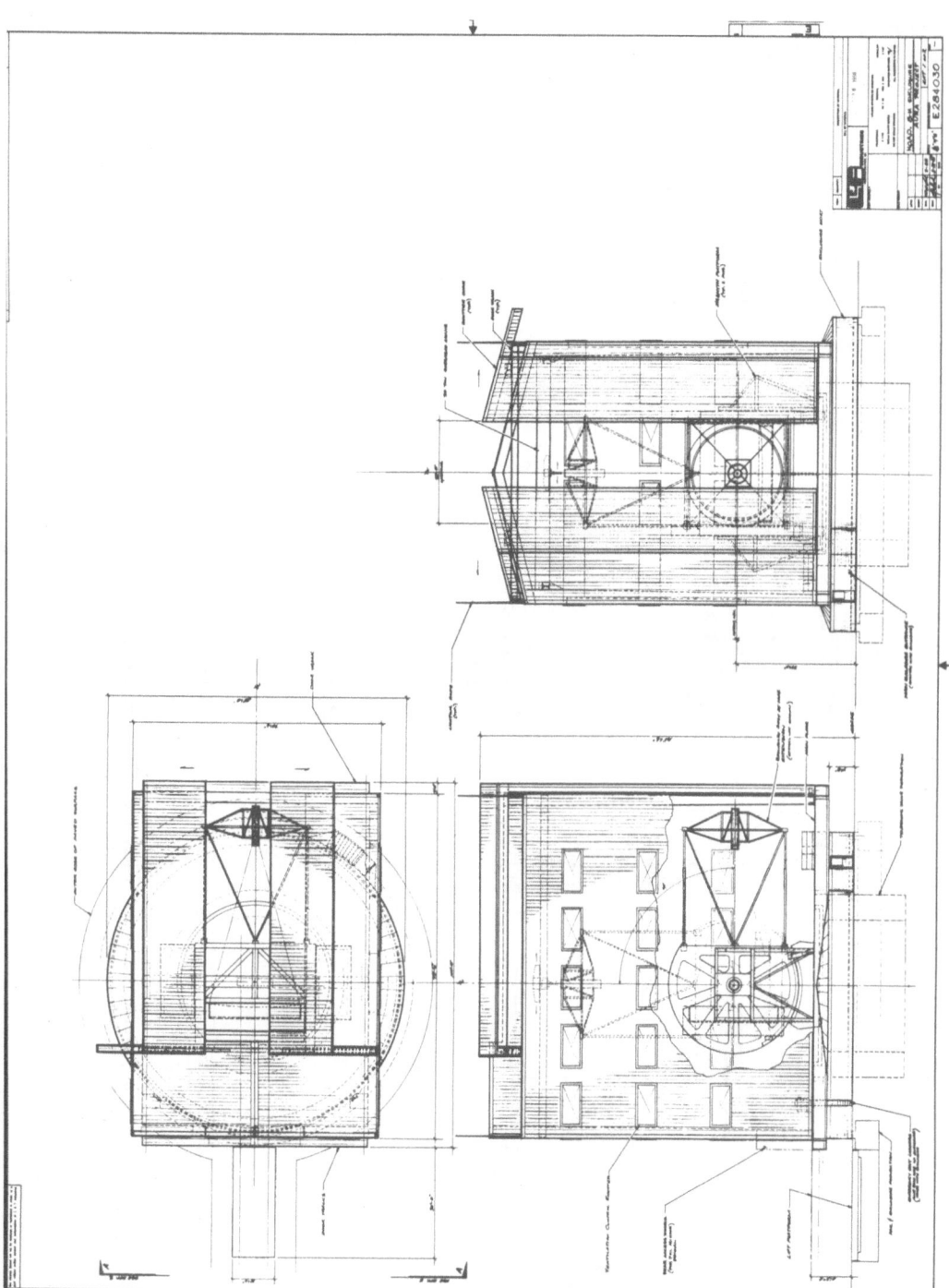

Fig. 8. An engineering drawing of the enclosure design by L&F Industries: a rectangular building that corotates with the telescope, similar to the Multiple Mirror Telescope and Apache Point Telescope enclosures. The special feature in this proposal is the large number of openings on the walls of the enclosure, offering the option of passive ventilation under mild wind conditions.

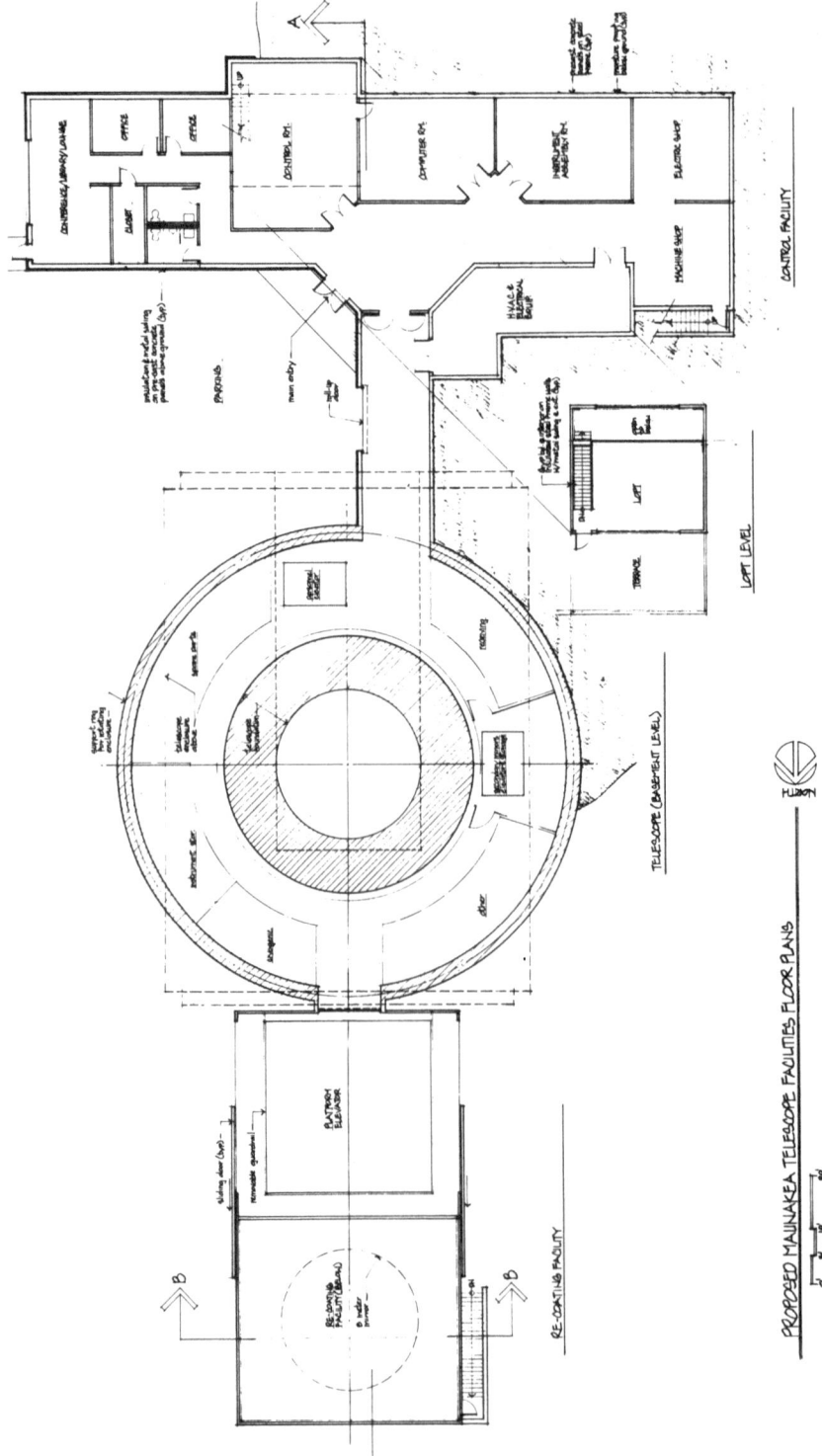

Fig. 9.   The proposed floor plan for the control facility, the storage area and the recoating facility. A corridor links the control facility to the telescope's storage area, but the environments of these two areas are separated by doors. The control facility is the only area where comfortable working environment is provided.

that is on the telescope. This secondary is transferred to the turret, the turret is rotated, and a new secondary is lined up and attached to the top end. The elevator is then ready to be lowered to the storage area, and the top end of the telescope is ready to be adjusted to the proper length for the newly installed secondary. The operation of this elevator is illustrated in Figure 10. During normal operation, the elevator and the unused secondary mirrors are kept in the stage area. The range of travel required for the elevator is estimated at 10 m. For more general use there will be an identical elevator in a different

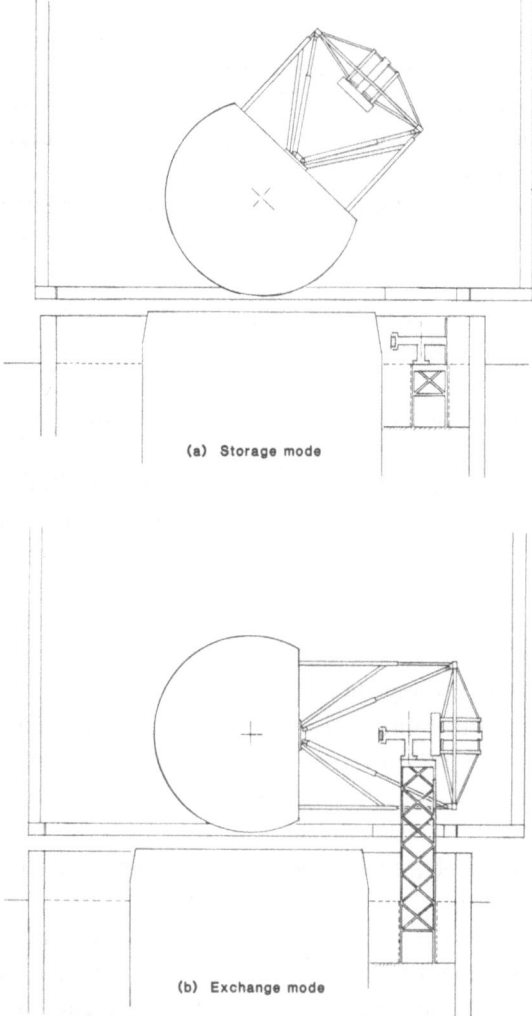

Fig. 10. Two highly stylized drawings showing the proposed scheme for storage and exchange of the secondary mirrors. (a) shows the storage position of the platform elevator, with the wide-field secondary on the telescope and the infrarecope secondary mirror in storage. (b) shows the position where the wide-field secondary mirror is being installed or removed. All secondary mirrors are to be exchanged with the telescope's optical support structure fully extended. After the secondary mirror and the light-baffle are installed, the telescope has to be adjusted to the proper length.

place that operates in the same way as the secondary mirror elevator. The function of this elevator will be to move equipment and people between the telescope floor and the storage floot, a distance of about 5 m.

(3) Two other elevators, one for each Nasmyth platform, will serve the purpose of linking the telescope's floor level and the Nasmyth platforms. The distance between these two levels is about 10 m. These elevators are fixed to, and supported by the enclosure structure.

(4) A 20-ton overhead crane will provide the means for general maintenance of the telescope. It will serve the telescope at both zenith and horizon-pointing positions. Additionally, two 5-ton jib cranes will be available over the Nasmyth platforms, for transportation of heavy equipment to and from all locations within the enclosed area.

(b) *Large equipment storage space*: The space between the telescope pier and the enclosure foundation can be used as storage area. With the exception that two sections of this area are occupied by the two platform elevators, and space for passage, the rest of the space is available for storage of non-heat-producing items such as instruments, cryogenic equipment, and spare parts. The preliminary study indicates that there will be about 130 m^2 of available space for this purpose. Of course this area is not to be air conditioned or heated. A suggested arrangement of this area is shown in Figure 9.

(c) *Detached control facility and recoating facility*: An arcade linking the control facility to the enclosure will offer sheltered access to the telescope. This arcade will also serve the purpose of housing all cable links between the control room and the telescope enclosure.

(d) *A fiber-fed laboratory*: Fiber-fed spectrographs, in both high-dispersion and multiple-object modes, will be an important part of the instrumentation on the 8 m telescopes. These will be located on the Nasmyth platforms; the constant-gravity environment offered by these platforms makes them suitable for laboratory instruments. At present it is considered acceptable that the fiber-fed laboratory be at ambient air temperature at night. Mechanical vibrations, however minute, will be isolated by three to four optical tables on the platform.

(e) *Concern with special safety requirement*: Hypoxia is a well-known physical disability at altitudes over 3000 m; mental capacity for any normal human being is not expected to be at its best on Mauna Kea. For this reason, special efforts must be made to design the equipment-moving facilities to be foolproof. Concern for human safety must be paramount. Experience gained by other observatories on Mauna Kea will be invaluable.

(f) *A perforated enclosure concept*: In recent years, the addition of forced-air ventilation to existing observatories has had some beneficial effects in reducing dome seeing. For example, new ventilation at the CFHT (Mauna Kea) and the AAT (Siding Spring) reduced the microthermal activity in the vicinity of the dome shutters (Abdel-Gawad, 1983; Gillingham, 1983). Active ventilation systems have now become standard items in the design of new telescope enclosures. Passive ventilation can also help to reduce dome seeing. Images at MMT (Mount Hopkins) are seen to improve when the wind speed reaches 5 to 13 m s^{-1}, high enough for good flushing of the air through the enclosure (Beckers, 1982). Louvers are already being used at the McGraw-Hill Telescope (Kitt Peak) and in the ARC Telescope building (Sacramento Peak) (Siegmund and Comfort, 1986). To use passive ventilation and yet be able to make the enclosure airtight in bad weather, we propose to use awning-type windows instead of louvers. The awnings will be made of the same material used for the enclosure walls. Thus the openings can be large and numerous, yet when they are closed the walls will act as well-insulated entities. When they are opened, the walls will be perforated and the enclosure will be ventilated naturally. One will also have the option of augmenting the ventilation by use of the forced-air system.

The sectional view of the enclosure (Figure 11), with the recoating facility on the left and the control facility on the right, illustrates the envisioned air flow pattern when the enclosure is fully opened and the force ventilation system is on.

4. Control Facility

At either site, the control facility houses the control electronics and computers, as well as small laboratories for observers to set up their observing equipment. Electronics and machine shops are available, but mainly for small-scale maintenance rather than full-fledged production work. The computer room is located between the control room and the instrument set-up room. The shops also are located next to set-up room to provide a logical working environment. Two office spaces are available. A large room at the end of the building can be used as conference room, lounge, and kitchen; two restrooms are adjacent. The overall space available is estimated to be approximately 500 m². A preliminary floor plan is shown in Figure 9.

5. Recoating Facility

The arrangement of the recoating facility is illustrated also in Figure 9. The corresponding view in section is shown in Figure 11. The recoating facility consists of two major parts:

(1) *A 50-ton capacity platform elevator to transport the mirror from the enclosure's floor level (7 m above ground) to the recoating facility (6 m below ground)*. This requires a vertical range of approximately 14 m. A roof cover is normally to be positioned over the elevator while it is not being used. This can slide away, allong the elevator to reach to the telescope floor level. The elevator also can be used as a service elevator between the telescope floor level and the ground-level storage area, allowing heavy or bulky equipment to be transported to and from the telescope safely. Furthermore, this platform elevator will also serve as an area for stripping and cleaning the mirror before it is recoated. This operation can be done with the elevator locked at the parking area level, for good access and ventilation.

(2) *A cubical building measuring 13 m on a side, to house the vacuum tank*. Part of this building is to be built underground, with its roof approximately 7 m above the ground level. The exposed part of this building will be insulated and covered with wall cladding

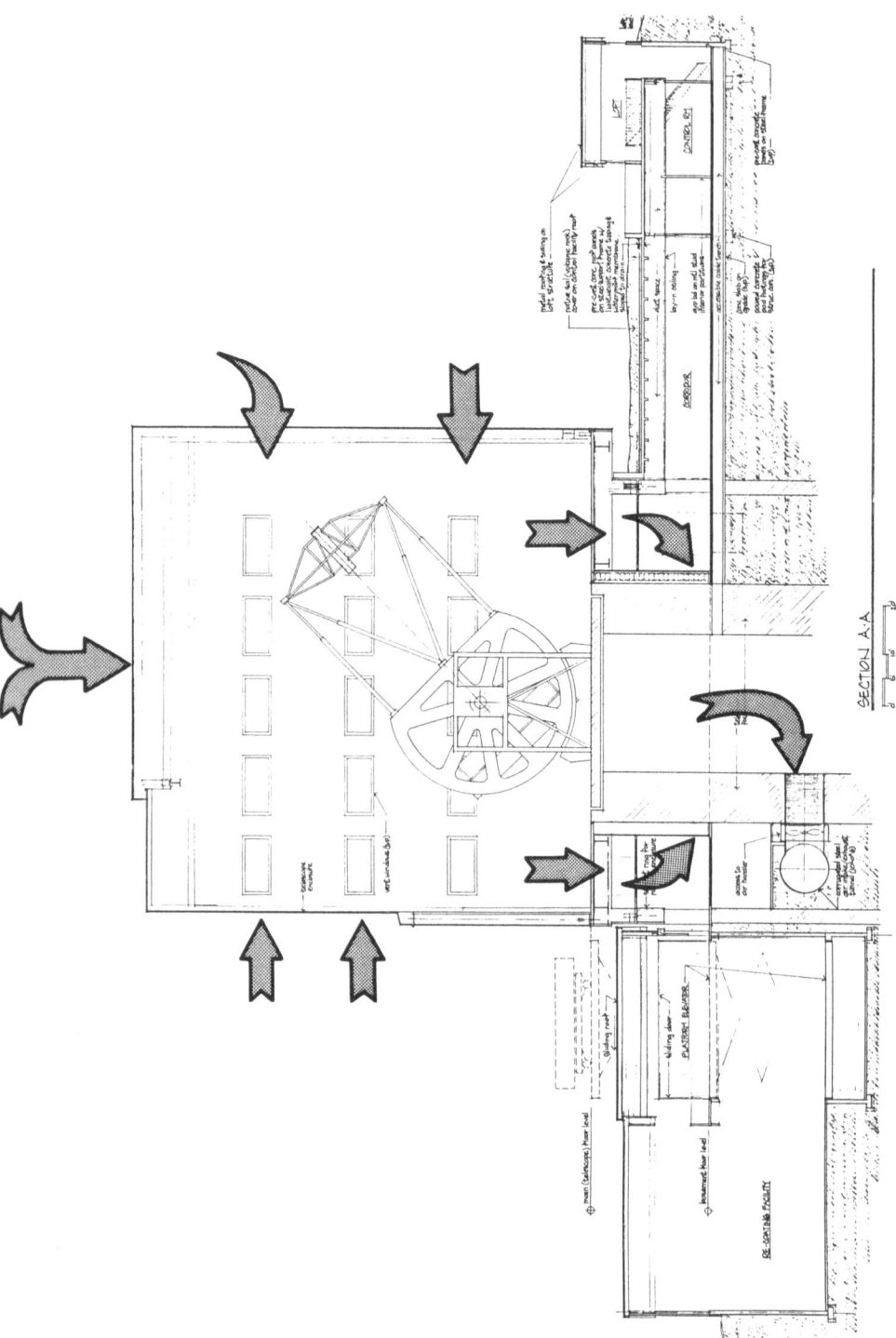

Fig. 11. The forced air ventilation system will replace up to 10 enclosure volumes of air per hour. A 3-m diameter conduit is oriented east–west direction to take advantage of the prevailing wind at the Manu Kea site. This alignment is approximately 90 deg from the CFHT and UH facilities, minimizing the disturbance to them. The conduit is approximately 76 m long, 3 m below the graded level.

material similar to that on the telescope enclosure, to insure that the surface temperature of the building follows the ambient air temperature variation.

6. Forced Air Ventilation System for the Enclosure

The thermal study done for the Keck telescope indicates that an air replacement rate of 10 enclosure volumes per hour is adequate to maintain thermal equilibrium between the inside and the outside (Nelson *et al.*, 1985). Preliminary calculations indicate that the conduit for an 8 m telescope building is required to be about 3 m in diameter.

Wind measurement at the Mauna Kea site shows that the prevailing wind is from either the east or the west 73% of the time. To take advantage of this natural condition, the conduit is to be placed in an east–west direction. A T-connection allows the exhausted air (downward ventilation) or the fresh air (upward ventilation) to be directed through the vertical path between the basement and the shutter of the enclosure. A filter system is called for in case the upward ventilation mode is used.

Both the east and the west outlets of this conduit are pointed away from the existing telescope facilities close by. These two outlets are about at the midpoint between the CFHT and the UH 2.2 m telescope.

References

Abdel-Gawad, M. K.: 1983, *Proc. SPIE* **444**, 161.
Barr, L. D.: 1989, *Astrophys. Space Sci.* **160**, 67 (this issue).
Beckers, J. M.: 1982, 'Interium Report on MMT Seeing Tests', *MMTO Technical Memorandum* 82-9, May 21.
Gillingham, P. R.: 1983, *Proc. SPIE* **444**, 165.
Merrill, K. M.: 1986, 'NNTT Site Evaluation Project', *Draft Summary Report*.
Nelson, J. E., Mast, T. S., and Faber, S. M.: 1985, *Keck Observatory Report* No. 90.
NOAO: 1988, *Draft Report* No. 2, p. B-10.
Siegmund, W. A. and Comfort, C.: 1986, *Proc. SPIE* **628**, 369.

PROGRESS OF THE MAGELLAN PROJECT*

W. A. HILTNER and S. A. SHECTMAN

Mount Wilson and Las Campanas Observatories, Pasadena, U.S.A.

(Received 11 January, 1989)

Abstract. The 8 m-diameter Magellan Telescope will be located at Las Campanas Observatory, Chile. The project is a joint effort of the Carnegie Institution of Washington, the Johns Hopkins University, and the University of Arizona. Conceptual designs for the telescope, enclosure and aluminizing facility are described. A detailed site survey is underway.

The Magellan Project is a joint effort of the Carnegie Institution of Washington, the University of Arizona and the Johns Hopkins University, to build and operate an 8 m telescope at the Las Campanas Observatory in Chile. The primary mirror will be an $f/1.2$ borosilicate honeycomb, identical to the ones being developed for the Columbus project.

Most of the work on the project has been the result of three activities. A Science Working Group, including representatives from each of the institutions, has developed the specifications for the telescope and instrumentation. Conceptual designs for the telescope mount, enclosure, and aluminizing facility have been developed by the project office, in collaboration with L&F Industries of Los Angeles, California. Finally, a detailed comparison of wind and seeing conditions for specific locations at the Las Campanas site is in progress.

The principal working focus of the telescope will be the Cassegrain. The baseline figure for the primary is a paraboloid. Two secondaries will be available, for the optical and for the infrared. The optical Cassegrain focus will include an optional wide-field corrector. Two designs for this corrector have been made by Epps, both of which include three fused-silica lenses and an atmospheric dispersion compensator. One corrector includes aspheric surfaces but produces a flat focal plane, the other contains only spherical surfaces but produces slightly better images on a curved focal plane. Our current preference is for the latter. The field diameter will be 40 arc min. The corrected final focal ratio for these designs is $f/6.5$. Faster values are undesireable for two reasons; the required secondary is larger, and the design of satisfactory collimators for wide-field spectrographs is more difficult.

The focal ratio for the infrared Cassegrain focus will be $f/15$. This will permit using a smaller, chopping secondary mirror with lower emissivity. The optical design also provides for a Nasmyth focus by deflecting the beam behind the primary mirror to a re-imaged horizontal focus. This focus is not coincident with the altitude axis, but mounting instruments on a de-rotator achieves the critical requirement for gravity-invariance.

---

* Paper presented at the Symposium on the JNLT and Related Engineering Developments, Tokyo, November 29–December 2, 1988.

Design studies have been carried out for two kinds of alt-azimuth mountings. The first concept, based on a conventional yoke structure, has been superseded by a design which we call the alt-az disk mounting, originally suggested to us by Warren Davison of the University of Arizona. An artist's rendering of this mounting is presented in Figure 1. The altitude motion is provided by suspending the primary mirror cell between two large vertical disks which are supported by hydrostatic bearings. These bearings are in turn mounted on a large horizontal turntable, which provides the azimuth motion. The horizontal turntable is supported on separate horizontal and vertical hydrostatic bearings. Careful placement of the bearings results in the application of all forces to the azimuth disk exactly in the horizontal plane. The lowest resonant frequency for this mounting is 7.5 Hz, with a moving weight of 190 tons. The dynamical performance of this mounting is appreciably better than can be achieved with a yoke structure of comparable weight and cost. The construction of the mounting makes use of welded plate elements as well as space-frame style truss elements. For either the alt-az disk or the yoke type of mounting, this type of construction was found to be more efficient than a pure space-frame.

The large-diameter bearings for the azimuth disk will have to be finished onsite, but the kinematic mounting of the altitude disks, together with a relatively flexible out-of-plane structure for the azimuth disk, means that the precision required for these bearings is relatively low. The altitude disks can be fabricated in the factory and shipped to Las Campanas complete, but the azimuth disk must be shipped in two sections and assembled at the site.

The azimuth turntable will be mounted flush with the observing floor, and an instrument clearance of 3 m will be provided between the Cassegrain focus and the floor. The trusswork connecting the two altitude disks is not symmetrical, and provides unencumbered access to the Cassegrain area from one direction (the rear in Figure 1). We expect to use friction drives and encoders for both the azimuth and altitude motions. The top structure of the telescope incorporates a square secondary support ring, which avoids the use of any bending members above the altitude disks. The secondaries will be supported on a compound spider system, with the IR secondary (or possibly an optical prime focus corrector) always in place. Using a handle fixture on the observing floor, the optical secondary will be removable when the telescope is in the horizontal position. The optical secondary cell completes its own spider when it is installed in the telescope. Additional spider vanes, to resist torsion at the secondary, can be added if required.

The alt-az disk configuration results in very convenient access to the Cassegrain focus, but practical limits on the size of the altitude disks result in the primary mirror being mounted slightly forward of the altitude axis. This increases the mechanical clearance required in front of the telescope. The compound spider arrangement also increases the clearance requirement for the telescope, to a radius of 11.9 m. The enclosure is being designed to provide a clearance radius of 12.8 m.

Our initial design study for the enclosure was based on a co-rotating square building similar to the one constructed for the Apache Point 3.5 m telescope. The moving weight for this design was greater than 500 tons. We are presently investigating the design of a lightweight hemispherical dome. The moving weight for such a dome would be in the range of 100 tons. With a hemispherical dome, there is sufficient clearance between the observing floor and the dome rail to permit a handling cart for the primary mirror and cell to be rolled from he observing chamber to an aluminizing facility in an attached support building. Since the telescope pier raises the level of the observing floor 4 m above ground level, there is adequate space for control rooms, laboratories, and the necessary utilities in the lower level of the support building.

Our present concept for the aluminizing facility involves an aluminizing chamber consisting of three sections. The stationary top section would contain the pumps and filaments, and would be fixed to overhead supports. The center section would be the mirror cell, and the bottom section the mirror handling cart. The mirror cell would be required to withstand vacuum pressure only around its perimeter. The dirty vacuum of the cart and cell sections would be separated from the high vacuum of the top section by a low-pressure seal. In order to store the tank in vacuum, the handling cart section would be attached directly to the stationary upper section.

A detailed site survey is presently underway at Las Campanas, under the supervision of Dr Eric Persson. Wind velocities at heights of 9 and 18 m above ground level are being recorded at 4 sites. At the two prime sites, image motion telescopes monitor the seeing at a height of 8 m.

At the present time, the conceptual design of the facility is essentially complete. By mid-1990, we expect to have enough data from the site survey to permit the exact site for the telescope to be chosen. Rapid progress in the development of the mirror technology is being made at the University of Arizona, and we expect that the project will shortly enter the phase of detailed design and construction.

MACK – Is the real reason for using the cell as part of the aluminizing plant that you do not want to move the mirror out of the cell?

SHECTMAN – Yes.

KODAIRA – How large is the tolerance for the flexure bend of the top end for this fast optics?

SHECTMAN – The tilt of the secondary mirror must be held to 3 arc sec. This will require a modest amount of active correction.

THE UK LARGE TELESCOPE PROJECT*

RICHARD S. ELLIS

University of Durham, England

(Received 10 January, 1989)

Abstract. The current status and the organization of the UK Large Telescope programme are summarized. Some scientific and design issues are discussed on the basis of work done for the SERC Large Telescope Panel report. A national Phase A programme is now underway addressing more detailed aspects of these questions. This study will culminate in a final proposal for funding in 1990.

1. Introduction

Here I wish to briefly summarize the current status and organization of the UK Large Telescope (UKLT) programme, make some scientific and design comments appropriate to large aperture telescopes, and finally add some remarks relevant to the need for wide field spectroscopic survey telescopes.

Although design work for a UK 8 m telescope has been ongoing in the Royal Observatories for many years, the impetus for a national programme began with a future planning review of UK astronomy conducted by the *Royal Astronomical Society* in 1985–1986 (*The Scientific Priorities for UK Astronomical Research for the Period 1990–2000*, RAS 1986). In ground-based optical and infrared astronomy, this report argued for both a wide field spectroscopic telescope (see Section 4 below) and a share in a large aperture telescope. The report led to the establishment of the SERC *Large Telescope Panel* which conducted a more specific review of these facilities in 1986–1987 culminating in their recommendation for a 50% UK share of an 8 m class telescope (*Large Telescope Panel Report*, SERC 1987). The latter report is part of a larger financial planning document for all of ground-based astronomy (*The Ground-Based Plan*, SERC 1988). This plan includes the 8 m class telescope as one of four *key projects* for the future.

The LTP report recommended establishing a national Phase A study period to be led by a newly-appointed Project Scientist and funds for this stage were approved in late-1987. The aims of the Phase A study are to increase national awareness for the proposal (particularly on scientific aspects), to examine design issues and to cost the proposal more accurately. The Project Scientist, Dr Roger Davies, took up his appointment in October 1988. His team will coordinate technical studies in the Royal Observatories, the Rutherford–Appleton Laboratory and various University research groups. The Team will also provide the LTP with information necessary to aid in its continuing discussions on possible international partnerships on the two SERC sites – La Palma and Mauna Kea.

---

* Paper presented at the Symposium on the JNLT and Related Engineering Developments, Tokyo, November 29–December 2, 1988.

Astrophysics and Space Science **160**: 107–110, 1989.

2. Scientific Aspects

The LTP Report cited above was largely an outline scientific case. The technical aspects were largely a summary of work being done already on telescope and instrumentation aspects at the UK Royal Observatories and SERC Establishments.

The report showed that the movation for larger apertures is three-fold:

(1) In photon-limited regimes such as high dispersion spectroscopy faintness gains proportional to D^2 are realised for a fixed signal/noise and exposure and this has significant implications for many areas of stellar and QSO spectroscopy. The UK has a growing community of echelle users on both the 3.9 m AAT and, shortly, the 4.2 m WHT. This is a growth area of UK astronomy and thus one where a gain in aperture will be particularly effective in coming years.

(2) In the mid- and far-infrared ($\lambda > 5$ μm), the different dependence of diffraction and seeing with wavelength implies diffraction-limited gains in performance $\sim D^4$ may be realised for unresolved sources. Spectacular gains are likely in imaging, polarimetric and spectroscopic studies of star-forming regions and active galactic nuclei. The UK has already made a commitment to facilities at infrared/sub-mm wavelengths (UKIRT, JCMT) and thus again this is a relevant area for growth.

(3) In sky-limited regimes including studies of faint objects traditionally associated with large telescopes, large gains have recently been made with modern instrumentation and improved detectors. Future gains can now only be made with larger apertures. A significant fraction of AAT and WHT time is currently devoted to sky-limited programmes and these would no longer be competitive when 8–10 m telescopes are available elsewhere.

In all areas it is clear that significant gains in performance can only be realised with the combination of larger apertures *and* improved image quality together with efficient infrared performance. Target image qualities of < 0.5 arc sec are essential. The recent evidence (Laing, this issue) that superlative conditions may fairly frequently, e.g., on La Palma, augurs well for large telescopes optimised for such conditions.

3. Design Considerations

The major thrust of the design work relevant to the UKLT programme has come from the collaboration between the RGO and the Instituto d'Astrofisica de Canarias (IAC) (see Mack, this issue). This provided the bulk of the technical input into the LTP report. The Phase A design study will substantially expand the scope of this work.

For the purposes of discussion, the LTP report favoured a $f/1.8$ meniscus primary with active support. Fast secondary focii ($f/6$) are essential for wide-field fibre work. Nasmyth platforms are ideal for versatile use of instrumentation. A dedicated infrared focus of about $f/13$ was also proposed. The case for a prime focus and a slower chopping-secondary infrared focus remained unclear. The LTP costed the telescope largely on the basis of the recently-completed WHT (scaling up the fork-style mount and enclosing it within a carousel enclosure similar to the JCMT).

The Phase A study will now address these issues more rigorously. Of particular interest are the following items:

(1) A design and structural comparison of the fork mount adopted in the RGO–IAC study and the rocking chair mount proposed for the Magellan and NOAO proposals.

(2) Continued research on the active support system for the 8 m meniscus, following early studies at the IAC.

(3) Fabrication and support of aspheric convex scenarios. This is crucial to determine the fastest feasible secondary f-ratio.

(4) A comparison of the performance and cost of the carousel and dome enclosures.

(5) The infrared performance of the telescope and a determination of the heat budget.

(6) Hardware and software aspects of the design of the telescope control.

(7) The design of the instrumentation package, with particular reference to the optimum configurations for fibres and pixel matching for high performance spectrographs.

(8) Further site characterisation, particularly with regard to the frequency of superb image quality and stable infrared conditions.

4. Wide-Field Spectroscopic Telescopes

In addition to proposing a large aperture optical/infrared telescope, the LTP report also highlighted the need for a wide-field spectroscopy survey telescope as a successor to the photographic Schmidt telescopes. Such a facility would be important for those statistical programmes requiring large numbers of spectra, e.g., galaxy redshift surveys for large-scale structure, follow-up studies of satellite missions (IRAS, ROSAT, etc.) and dynamical/chemical studies of large star catalogues in studies of galactic structure. Again, such a telescope would exploit in a natural way the measuring machine catalogues that have been constructed within the UK in recent years.

The LTP specified the following parameters for such a telescope:

– an unvignetted field of view of at least 2 deg both to enlarge significantly over current 4 m fields (~ 40 arc min) and to match the surface density of rare objects;

– an aperture of ~ 4 m so that *detailed* spectroscopic studies (velocity dispersions, compositional studies, etc.) will not be prejudiced;

– an on-line capability for ~ 500 fibres.

Three solutions are being investigated to this problem. Firstly, a dedicated 4 m Ritchey–Chrétien telescope with a fast wide-field Cassegrain. This has the disadvantage that a large and costly achromatic corrector is required, but some design progress has been made to overcoming the problem. Secondly, a number of *special purpose* telescopes, such as the Willstrop and Columbia Survey Telescopes are being examined. The Willstrop telescope is a three-mirror telescope with a wide field, excellent image quality and a squat f/ratio. The focus is rather inaccessible and significant asphericities are involved. However, a 0.5 m prototype is nearing completion. The Columbia design is a single spherical primary with individual aberration correctors for each fibre.

The most exciting and immediate solution to this problem, however, follows a design by Wynne for a 4-element prime focus corrector for the AAT which, when optimised for 1 arc sec images adequate for fibre work, yields a field of 2 deg at about $f/3$. A detailed design study on this remarkable opportunity to *upgrade* the AAT is being carried out at the AAO at present (see Cannon, this issue).

Acknowledgements

The UKLT Programme would not have reached its current momentum without strong support from staff at both Royal Observatories and the Rutherford–Appleton Laboratory. I thank the panel members Mike Edmunds and Jim Hough for their dedicated effort and many colleagues worldwide, including those friends here in the JNLT Working Group, who have willingly offered advice.

KODAIRA – Do you have special preference between the northern and southern hemisphere when UK commits to a 8 m-class telescope project?

ELLIS – The UK has almost four times as much equivalent 4 m power in the northern hemisphere than in the southern hemisphere. For this reason we are only considering a northern site in the first instance.

CANNON – You mentioned the enclosure being either a dome or a carousel. Have you considered the ESO VLT type of enclosure, and if not, why not?

ELLIS – The ESO-style enclosure has not been considered so far but if it is thought to be cost-effective and appropriate for the two sites I mentioned, it could be examined. A critical aspect of ESO's discussion is the apparent constant wind direction and the expected savings overall on four telescopes. In the case of the UKLT, neither of these advantages are straightforward to assess at present.

BARBIERI – Regarding the option updating at AAT prime focus for wide-field spectroscopy, how efficient is the match between the $\sim f/3$ focus and the fibers? Where is the spectrograph located?

ELLIS – Fibers would probably feed spectrographs remotely located.
Losses in the long fibers would only be serious in the ultraviolet (< 400 nm), otherwise the matching at $f/3$ is the more efficient than at Cassegrain $f/8$.

A PROGRESS REPORT ON THE DESIGN OF THE 8 M APERTURE OPTICAL/INFRARED TELESCOPE*

B. MACK, D. HARMAN, S. ATKINSON, and C. GARCIA

Royal Greenwich Observatory, East Sussex, England

(Received 10 January, 1989)

Abstract. Progress of the design study carried out y the Instituto de Astrophysica de Canarias and Royal Greenwich Observatory joint project office for an 8 m aperture Optical/Infrared Telescope. This includes the development of an active pneumatic support system for both the radial and axial supports of an 8 m diameter thin meniscus mirror and the investigations of new technology encoding systems.

1. Introduction

The William Herschel Telescope (WHT) of 4.2 m aperture is the second largest single primary mirror alt-azimuth mounted telescope in the world and its image quality, tracking and pointing performance is exceptional. The technology developed for the telescope may, therefore, be used to provide the foundation for the design of an 8 m aperture optical infrared telescope.

The WHT primary mirror weights 16.4 tonnes, it is 0.5 m thick and has a focal ratio of $f/2.5$. The weight of an 8 m, 0.2 m diameter thickness meniscus mirror of focal ratio $f/1.8$ is 22.0 tonnes. The structure of the 8 m aperture telescope will, therefore, be less than 300 tonnes compared to the WHT of 212 tonnes.

The rotational inertias of both the altitude and azimuth axes do not scale up as the total weight. Therefore, the hydrostatic bearings, drive systems, computer software, aquisition and guidance, and servo systems developed for the WHT may, therefore, be used for the 8 m.

In the early design phase of the WHT various areas were recognised as areas which could be further developed. Not only for the enhancement of the performance of the WHT but to provide raw data for the future design of a larger more advanced instrument.

Two important areas were:

(1) Mirror Supports control.

(2) Incremental and absolute encoding systems using gears, friction driven and tape systems.

This paper, therefore, contains a brief description of the design of the primary mirror support system and the development work on encoding systems.

* Paper presented at the Symposium on the JNLT and Related Engineering Developments, Tokyo, November 29–December 2, 1988.

Astrophysics and Space Science **160**: 111–118, 1989.
© 1989 *Kluwer Academic Publishers.*

2. Primary Mirror Support System

This can be divided into:
  (a) Initial optimisation of the axial support system.
  (b) Design of the radial/transverse supports.
  In the final solution the axial support system must also provide the forces required to neutralise the effect of the out-of-balance bending moments inherent in a thin meniscus mirror when the telescope tube points to the horizon. However, the optimum distribution of axial supports is independent of the radial configuration and can, therefore, be analysed separately.

3. Axial Supports

The mirror support systems of the AAT, WHT, and the UKIRT all use pneumatic belophram types of axial support units and these are also to be used for the 8 m primary. These units are simple, light and can be independent of the mirror cell deflections. These qualities allow for many possible solutions to their distribution within the mirror cell.
  The analysis for the optimum design of the axial supports begins by calculating the optimum radii to reduce image blur. The magnitude of the reactions on the inner support annuli are then adjusted until the deflective curve gives the best surface and minimum focus shift, taking into account the practical limitations. There must also be sufficient numbers of support points to avoid secondary deflections between supports. This result has been achieved by an analysis assuming knife-edge supports. An 8-ring system of 384 pneumatic cylinders spaced as shown on Figure 1 will provide an acceptable solution. Each cylinder swill support approximately 65 Kg f of the primary mirror. Although the axial support system has been optimised for radial and circumferential distribution there is also a vertical and horizontal symmetry as shown on Figure 2. This feature allows for the optimisation of the structure of the cell which supports the mirror (Figure 3). This design allows for the axial actuators to be removable from the rear of the cell without disturbing the primary mirror.

4. Axial Support Control

As in all telescopes, the axial support forces required for the primary mirror must change as a function of altitude angle. The load per axial support has to be, therefore, controlled over a range of 5 to 65 Kg f. Each axial support of 75 mm diameter would have a volume of trapped air of 26.5 cc and it would be difficult to design a valve that would have the fine degree of control required and also have the capacity for the wide range of flow control necessary to accommodate the load variation. The proposed solution is to provide a flow control valve which is normally in a neutral position giving correct balanced flow required to maintain the pressure in the axial unit. The control of the air supply manifold pressure can easily be achieved by a conventional proportional control valve. Thus the individual control valve task for each unit is relatively simple. The valve

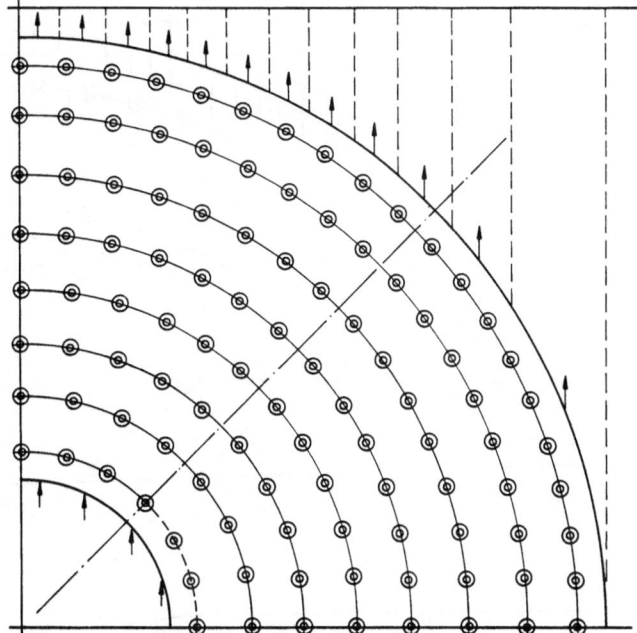

Fig. 1.   Pneumatic cylinder spacing for the 8 m primary.

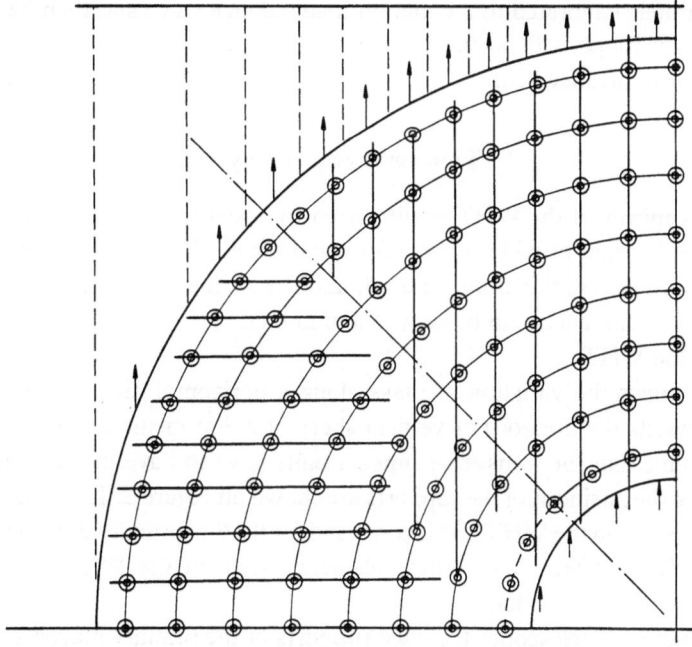

Fig. 2.   Vertical and horizontal symmetry in the support system.

B. MACK ET AL.

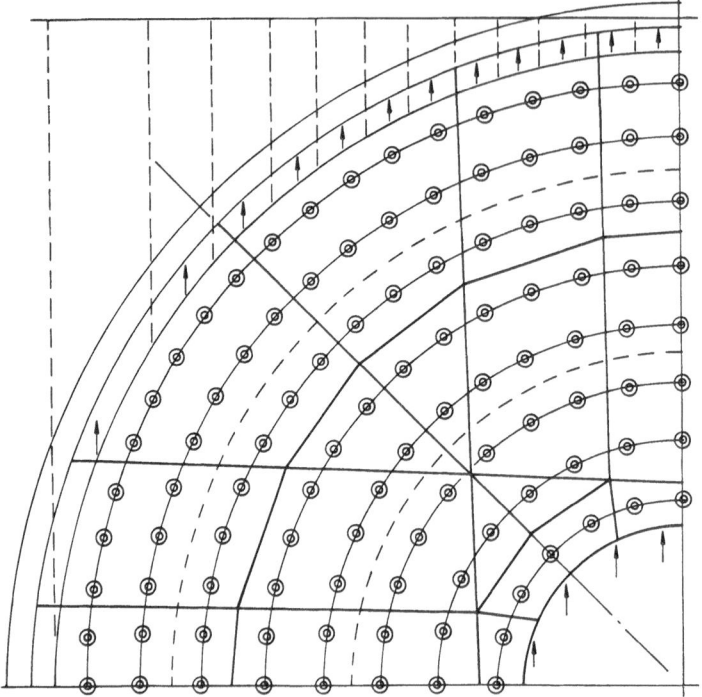

Fig. 3.   Optimization of the structure of the mirror-supporting cell.

can be accurately calibrated to a common load balance valve and then be used over a limited control range with a line performance. Figure 4 shows a schematic of the pressure control arrangement.

5. Transverse Support System

The primary mirror of the WHT is supported by equal transverse vertical forces and each lever and weight provide support for equal slices of mirror. A common design of a level and weight can, therefore, be used, and the local surface deformations induced by the Poisson ratio effect can be kept to a minimum. This scheme has been used with success on the WHT.

Figure 5 shows the variation of mass along a horizontal diameter (telescope tube pointing towards the horizon) of vertical slices of a 8 m meniscus mirror.

If we assume that the transverse support units have to have the same load-bearing capacity then the positions of the supports are shown on Figure 2. This concept has been used with success on the WHT and is proposed for the 8 m primary transverse supports. But the mechanical lever and weight elements used on the WHT are replaced by push-pull pneumatic cylinders.

In the majority of telescopes the edge supports of the primary mirror are positioned so that the line of action of all the support forces pass through the center of gravity of

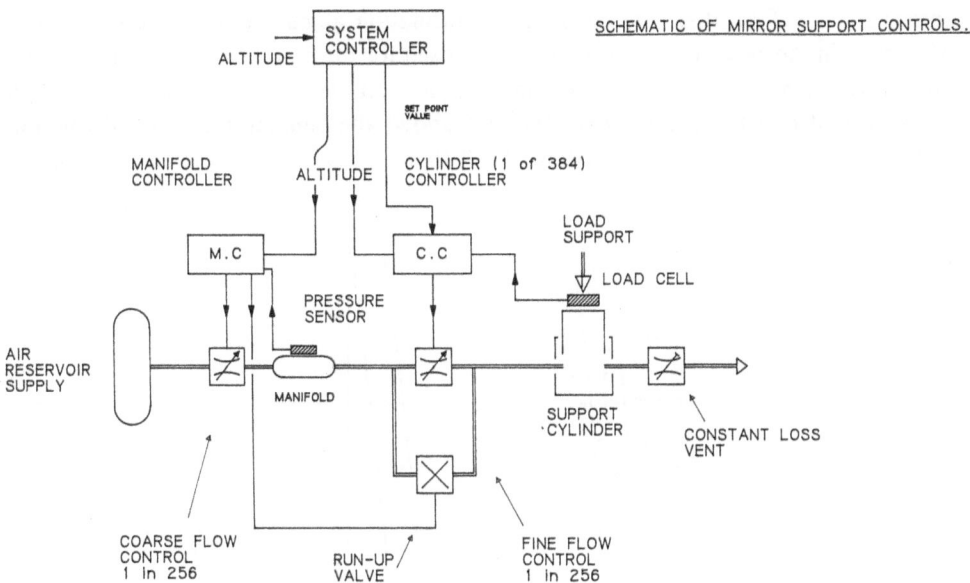

Fig. 4.   Schematic of mirror support controls.

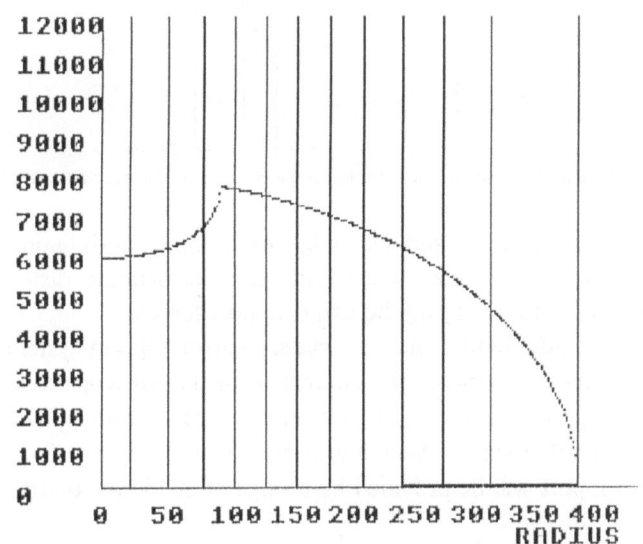

Fig. 5.   The variation of mass along a horizontal diameter.

the mirror. This condition is relaxed in an alt-azimuth telescope, where the sum of the moments of the forces about horizontal line passing through the center of gravity of the mirror must be equal to zero to achieve balance.

In relatively thick mirrors the vertical component applied by one support has an equal and opposite force provided by the corresponding support at the top portion of the mirror, it, therefore, does not cause any deformation along the horizontal diameter. However, in the case of a thin meniscus proposed for the 8 m mirror this will not hold true. It is not possible to have the mounting points on the same plane passing through the centre of gravity of the mirror. Figure 6 shows the bending moment distribution induced by a transverse support system. It will, therefore, be necessary to have transverse

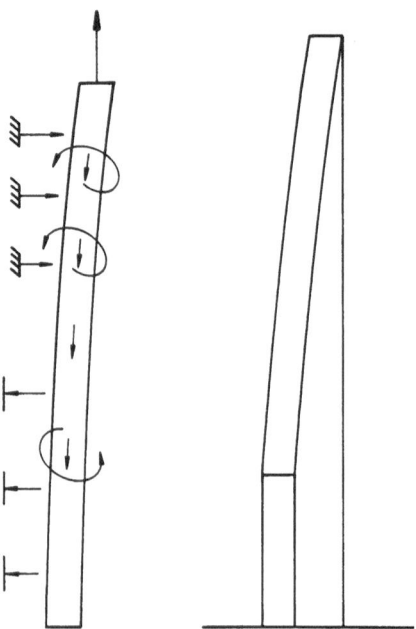

Fig. 6.   The bending moment distribution induced by a transverse support system.

supports acting on both the inner and outer edges of the mirror and eliminate the residual bending moments by applying an equal and opposite bending distribution. These moments can be applied by varying the axial support forces.

The design criteria adopted for the 8 m primary mirror support system is, therefore:

(a) The axial support system will be optimised for the telescope tube at zenith.

(b) Each axial support unit will be of the pneumatic belophram type and have a load cell incorporated into that part of the unit in contact with the rear surface of the mirror.

(c) The axial support will be provided by 8 rings of 384 load-controlled pneumatic cylinders.

(d) The shear forces caused by the temperature variation of the cell and the mirror and the small transverse forces in the belophram material will be eliminated by the inclusion of a low friction bearing between the mirror and the load cell of the axial supports.

(e) It is proposed to support the mirror transversely by a system of pneumatic push-pull cylinders each acting near the gravity plane of its designated slice of the mirror. Out-of-balance bending moments being compensated for by the controlled variation of the axial support forces.

The active control of the 8 m primary mirror support forces and, therefore, pressures will be based on the following criteria:

(a) The support load of the cylinders will be servo-controlled by the mirror supports computer.

(b) A constant pressure controller will give coarse control of the air pressure in the main air supply reservoir which compensates for telescope tube position.

(c) Fine control of the pressure in the axial support units will be by means of high response, flow valve, servo-controlled by the signals from the load cell. This system will enable the forces on each axial support to be varied as required.

(d) The transverse supports will use similar load-controlled pneumatic cylinders, however, they will be modified to give a push-pull loading capability.

6. Encoding Development

Gear-driven absolute and incremental encoders are used on the Anglo-Australian telescope at Siding Springs and as early as 1975 studies proposed that a gear-driven absolute encoder be retained for the WHT for both azimuth and altitude drives. At that time a friction-driven roller incremental encoder was considered but the information available on the performance of this type of encoder was based on other users unconfirmed results.

Limited investigations were carried out by Marconi Radar and S.E.R.C. which included tests undertaken in an attempt to establish the effect of creep in roller drives. The question of whether to use a single encoder for rate and position or to use a separate encoder for each function which affects the control philosophy was investigated and resolved. Studies indicated that the use of separate encoders would comply with the mechanical suitability, realiability, and cost for the telescope.

The use of separate encoders appears to be extravagant but there are cases where two separate encoders are essential. A roller-driven incremental encoder must always be accompanied by a separately driven absolute measuring device, typically a gear-driven encoder. Conversely, a gear-driven encoder can be used for both rate and position but this does not give as smooth an incremental drive as the roller.

The relative costs of the different encoder systems prior to the final design of the WHT were investigated against their technical realiability, stability, and accuracy and the decision was taken to use both gear-driven absolute and incremental encoders for the WHT. At that time there was not sufficient technical information available to prove that a friction-driven incremental encoder could provide the accuracy required for the servo system. However, it was possible at minimal cost to incorporate various features which could be used for the future development of telescope drives and provide valuable information for future telescope concepts.

7. Encoder Tests

The WHT uses both incremental and absolute gear-driven encoders on the azimuth and altitude axes. The bull gears by which these are driven are 3.6 m in diameter and Maag S1 quality. This gear is also used for the drive and the torque motors and gearboxes are positioned in the column and the base of the telescope. The specification for the drive gear had to be accurate to allow it to be used for encoding purposes.

In 1981 the design of the drive system was fixed and although it was realised that the use of a very large accurate gear would be expensive, to ensure that the telescope would reach its drive specification a large encoding gear was specified for the telescope.

However, it was decided that development work should continue to determine whether a less accurate gear could have been used.

The WHT main bull gears, therefore, have two features which allow them to be used with either a friction-driven or a wrap-around-grating incremental encoding system.

Near to the gear-driven encoders is a friction-driven incremental encoder.

The WHT is now in operation and development work on the friction-driven encoders and wraparound tape systems is now be carried out. This program will not affect the observing runs on the telescope and, therefore, using the sky as a reference and the excellent pointing and tracking now being achieved by the WHT accurate information on the design and working of the friction-driven encoders and tape encoding compared to the gear-driven systems will become available.

The computer will monitor the output from both the gear-driven and friction-driven encoders and comparisons will be made on the smoothness, hysteresis, backlash, and response of the systems.

Pre-load induced hysteresis tests, random creep due to slow speed errors, consistancy of ratio under dynamic conditions and the errors induced by misalignment will be investigated.

Although the WHT has shown that the use of gear-driven encoders can provide a solution within the accuracy specifications. It is hoped that this continuing work will provide factual hard evidence of the encoders performance so that in the detail design for the drive system of the telescope a smoother encoding system can be developed. Although the development work should eventually lead to an increased tracking accuracy for the 8 m telescope, experience now gained on the quality of the drives of the WHT does indicate that a very accurate gear is essential not only for encoding but for the smoothness of the drive. A Maag S1 quality gear will, therefore, be used for the drive of the 8 m telescoe and will also feature a smooth track for the future development of wrap-around tapes, friction encoders, and friction drives.

THE GALILEO PROJECT*

A 3.5 m Italian Telescope Facility

CESARE BARBIERI

Astronomical Observatory, Padova, Italy

(Received 2 December, 1988)

Abstract. The Galileo project comprises the design, building, and operation of a 3.5 m Italian telescope the main elements (diameter, mechanical structure, active optics, etc.) of which consist of a duplication of the ESO New Technology Telescope (NTT). Modifications have been introduced in order to allow, beyond the $f/11$ Nasmyth foci, a prime focus $f/2.2$ station, a trapped $f/6$ focus, and a small Cassegrain $f/20$ facility. Other changes with respect to the NTT have been made to the control and data acquisition system, and to the service building. The telescope could be operational at the end of 1992.

The Galileo project is a duplication of the ESO New Technology Telescope (NTT) intended to serve the needs of the Italian astronomical community in the 4 m class instruments, in the northern hemisphere. Italy is one of the few European countries which does not have a large national facility. Therefore, a comprehensive development plan has been put together by the Consiglio delle Ricerche Astronomiche (CRA) in 1988. The main elements are the participation in the Columbus collaboration (a 2×8 m binocular telescope) and the 3.5 m Galileo telescope. In order to capitalize on the experience gianed by the Italian industries in the design and construction of the NTT, the CRA decided to direct the Galileo study toward a duplication of that instrument. However, in order to satisfy a large scientific community, changes were allowed in order to add more capabilities than permitted by the $f/11$ Nasmyth foci of the NTT. The present paper will describe the results determined during Phase A study between April and November 1988.

The characteristics of the NTT have been described by Tarenghi (1986); the telescope is the first example of the active optics concept; other new features can be found in the support and movement of the telescope, and in the rotation of its building. The Galileo project has adopted as baseline all these elements, introducing the following changes and additions:

– The top end has been made interchangeable, so that a prime focus unit, or a different secondary mirror M2, can be mounted. No deterioration to the excellent performances of the NTT in resonant frequencies has been caused by this modification.

– The Nasmyth flat M3 unit can also be removed in order to direct the light to a small Cassegrain facility or to a trapped $f/6$ focus (dedicated M2's are needed in both applications).

* Paper presented at the Symposium on the JNLT and Related Engineering Developments, Tokyo, November 29–December 2, 1988.

Astrophysics and Space Science **160**: 119–122, 1989.
© 1989 *Kluwer Academic Publishers.*

– The angle of the spider has been closed from 90 to 60 deg, in order to simplify operations of dismounting and storage of the several top ends. The structural rigidity of this solution is very high (first frequency at 30 Hz).

– The movements of the telescope are ensured by brushless motors in both axes.

– The controls of the telescope and of the active optics have been clearly separated from the data acquisition controls, through two independent and standard Ethernet lines.

– The IHAP environment has been abandoned.

– The height of the central part of the rotating building has been augmented by 2 m in order to add a crane to remove the M2 and M3 units. A working space has been derived inside the central pier. A platform has been added in order to service the prime focus.

– The annex building has been enlarged to store in it up to four M2 and M3 units. It has been detached from the rotating part, granting access to the latter with an external platform. However, because the site of the Galileo will be decided only in Spring 1989 (three options are considered at moment, namely Mt. Graham in Arizona, La Palma on the Canary Islands, and Mauna Kea at Hawaii), the design of the buildings is still preliminary.

The main elements of the Galileo telescope are shown in Figure 1.

There are several original studies made by the Galileo project. One of the major scientific programs for the telescope is deep photometry of faint objects in crowded fields, through large bandpass filters (say *UBVRI*) and with CCD detectors, in a moderate field. Examples of these programs are studies of H–R diagrams in globular clusters to very faint magnitudes. This interest stems also from the availability of one

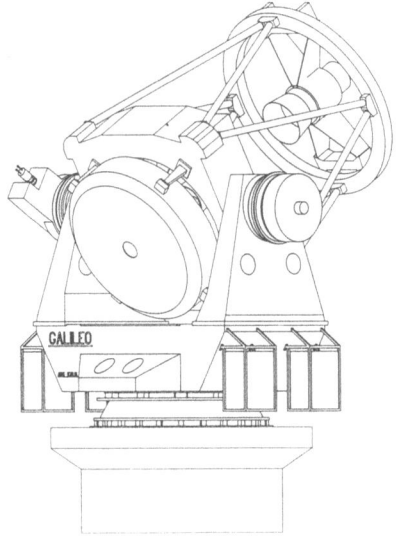

Fig. 1.   The Galileo telescope. The top frame can be removed and several units exchanged, from prime focus to a small Cassegrain $f/20$ unit.

of the most powerful softwares for this application: namely, Romafot (see, for instance, Buonanno *et al.*, 1983). The scale of the $f/2.2$ prime focus, coupled with the pixel size of available 1024 × 1024 CCD detectors such as Thomson's, is ideal for this application, provided the image quality can be kept better than 0″.5 over large bandpasses from U to I. Several correctors have, therefore, been studied to achieve these goals, from simple Gascoigne plates to aspheric triplets. A comparison of the results is shown in Figure 2. We stress that these performances, not yet optimized, have assumed the same optical figure for the primary mirror of the NTT; future studies will explore the capabilities of the active optics with respect to the deformation of the primary mirror or the advantages from a small modification of its figure.

Another program of great interest for our community is multi-object spectroscopy with fibres. The quality of the $f/6$ focus, affected mostly by field curvature if no additional optical element is introduced beyond M2, is such to render this focus extremely interesting to this application. The complexity of the spectrograph has, however, suggested a delay for its study.

Given a prompt availability of the funds, and a decision about the site not later than March 1989, the Galileo telescope could be operational before the end of 1992, as demonstrated by the construction phases of the NTT.

Our study has greatly benefited from documents and information from the ESO/NTT staff. Many Italian astronomers have substantially contributed to the success of the Phase A study, performing work in all areas from scientific priorities to

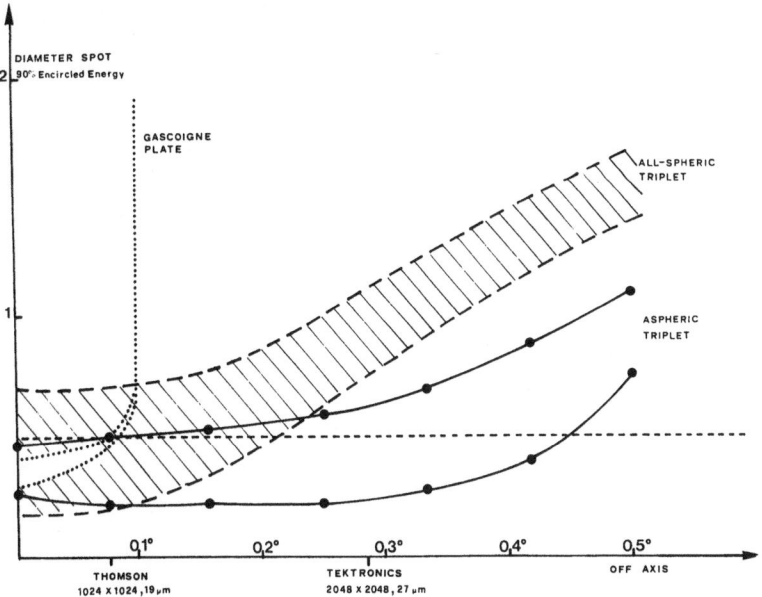

Fig. 2. A comparison of the performances of the three correctors studied for the prime focus, Gascoigne aspheric plates, all-spheric triplet, 2 aspheric lenses triplet, assuming the same primary conic constant of the NTT. The vertical line gives the field of the Thomson 1024 × 1024 19μ chip; an additional field of a few arc min is required by the active optics sensor and by the guiding unit.

optics, controls, site selection, etc.; their names are given in the Final Report, Vol. 1, that can be obtained from the underwriter. Engineering contracts have been assigned to the following Italian firms: ADS (Lecco) for the structure, Officine Galileo (Firenze) for the correctors, Zollet (Belluno) for the building. The following industries have also greatly contributed, providing information about elements they have built for the NTT: Ansaldo, INNSE, Macdit, Mecnafer, Schott, Zeiss.

References

Buonanno, R., Buscema, G., Corsi, C. E., Ferraro, I., and Iannicola, G.: 1983, *Astron. Astrophys.* **126**, 278.
Tarenghi, M.: 1986, in *The Second VLT Conference*, Venice 1986.

A DECADE OF COST-REDUCTION IN VERY LARGE TELESCOPES
(THE SST AS PROTOTYPE OF SPECIAL-PURPOSE TELESCOPES)*

HARLAN J. SMITH

McDonald Observatory, Austin, U.S.A.

(Received 10 January, 1989)

Abstract. Many design and technical innovations over the past ten or fifteen years have reduced the costs of very large telescopes by nearly an order of magnitude over those of 'classical' designs. Still a further order of magnitude reduction is possible if the telescope is specialized for on-axis spectroscopy, giving up especially the luxuries of wide field, multiple focal positions, and access to all the sky at will. The SST (Spectroscopic Survey Telescope) will use eighty-five 1 m circular mirrors mounted in a steel frame composed of hundreds of interlocking tetrahedrons, keeping a fixed elevation angle of 60° with rotation only in azimuth. Using an optical fiber it will feed as much light to spectrographs as can be done by a conventional 8 m telescope, yet has a target basic completion cost of only $6 million.

1. Introduction

Why has one of astronomy's most powerful instruments, the 5 m Palomar telescope, remained effectively unsurpassed for 40 years? The principal reason, of course, has been *cost*. Light-gathering power increases only with the square of the aperture, whereas for any given design the mass – and to first order the expense – of a telescope grow more nearly with the cube of the aperture. Thus, for example, a simple scaled-up 8 m copy of the 5 m design might cost as much as $500 million.

Over the past decade four quite different solutions to this problem have been under active development. Examples of each are discussed at this meeting. All of them depend on two primary efforts. First is a great reduction in mass of the primary mirror over simple scaling up of classical mirrors. This permits comparable reduction in mass and, hence, of cost for the telescope. Reduction in mirror mass for a given light-gathering power can be achieved by using:

– an assembled array of smaller and/or thinner or honeycomb mirror telescopes (e.g., the Arizona MMT, or the Columbus project),

– an array of independent telescopes whose beams can be combined (e.g., the ESO 16 m project),

– a primary mirror composed of a mosaic of thin small mirrors each controlled to produce its segment of the total figure (e.g., the Keck telescope),

– a very thin monolith (e.g., the erstwhile Texas 300-inch project, or the JNLT).

* Paper presented at the Symposium on the JNLT and Related Engineering Developments, Tokyo, November 29–December 2, 1988.

The second major way to reduce cost is to keep the dome as small as possible, since its expense is a substantial fraction of that of the telescope. This is achieved by requiring a very short focal ratio in the case of single-mirror primaries, or more easily by the intrinsically shorter focal lengths of the smaller mirrors in multiple-mirror or multiple telescope designs. As a result of these and other developments, costs are now expected to average only about $50 million for 8 m-class telescopes. This order-of-magnitude reduction has opened the way for an explosive proliferation of plans for great new telescopes.

It does not seem likely that major general-purpose telescopes will become very much less expensive in the foreseeable future. However, it *is* possible to achieve still another order of magnitude reduction in cost of a large telescope, by specifying a unique function and by accepting some appropriate compromises in performance. In particular, the SST as discussed in this paper has expected basic completion costs of only about $6 million. A number of design novelties permit this truly drastic reduction in cost. They are primarily based on the decision to restrict use of the telescope solely to spectroscopy – the most important function of modern astronomy.

2. The SST

2.1. HISTORY OF THE PROJECT

The SST (Spectroscopic Survey Telescope) was formally proposed in 1985 by Daniel Weedman and Larry Ramsey of the Pennsylvania State University. Their preliminary concept was supported by an NSF grant to pursue the design. Penn State and shortly thereafter the University of Texas at Austin each separately matched the modest NSF development grant. The two universities are now supporting the project jointly. They have agreed to raise funds for and to share in the cost of building and operating the telescope at McDonald Observatory of the University of Texas. Design and development responsibilities have been divided, with Penn State (Weedman and Ramsey) undertaking the optical work including the initial spectrograph, while Texas is responsible for mechanical design and construction. Chris Sneden is Project Scientist for the Texas group, with Frank Ray as Chief Engineer and Designer.

2.2. PHILOSOPHY OF THE SST

Why are large telescopes so expensive? Basically because astronomers understandably would like them to be able to do all things for all people, and to do them exceedingly well. Mirrors are specified to produce images as good or even better than the atmosphere will transmit from space, and to do so under all foreseeable extremes of temperature and orientation. Fields of view are desired to be very large, with images good to the edge, voer a wide range of colour. Thermal emission into the focussed beam must be minimal. A number of focal positions should be provided, if possible with a selection of focal ratios available. Tubes must maintain perfect alignment of secondary optics at all elevation angles and have minimal response to wind buffeting. Immensely massive mountings must point to near-arc sec accuracy and track to fractions of a tenth of an arc sec, allowing the astronomer to set on and follow as long as desired any object ore than a few degrees above the horizon.

Such features are undeniably desirable, if one can afford them. However, the question may reasonably be asked whether *all* great new telescopes need to be general purpose. Might it be more cost-effective, perhaps more astronomically effective as well, to concentrate one or more large telescopes toward specific important functions? In fact, this has already begun to happen, with several recent 3 m-class telescopes having been optimized for work in the thermal infrared.

Spectroscopy appears to offer the next opportunity for such specialization. In particular it offers relief from the need for extremely high quality images and for large field. For most spectroscopic applications maximum light-gathering power is the only real driver. Once this point is accepted, radically new approaches become feasible.

2.3. SSI DESIGN

2.3.1. *Primary Mirror*

This is usually the single most difficult and expensive item in a veyr large telescope. Very high quality, well-supported 8 m mirrors are currently in the $10 million class, weigh typically 10 or more tonnes, and in addition require multi-million-dollar handling fixtures and coating facilities. Yet the same number of photons can be gathered by a large number of small, thin mirrors feeding their light to a common focus.

If the many small mirrors are to form part of a large parabolc primary, the problems of generating and aligning off-axis figures in all but the central mirror are serious and expensive to solve. On the other hand, if the small mirrors are part of a large sphere then all can be identical, each with a line-of-sight curvature requiring only a minimum amount of grinding and polishing with a single large tool – the easiest possible task for an optician.

If the primary is to be effectively a continuous reflecting surface, the small mirrors must be tilted with virtually no interstices. This requires cutting each finished circular mirror to a hexagon, which in turn normally leads to release of strain in the glass and loss of the high quality of figure previously polished into the circular blank – a difficult situation to correct. But if the goal is simply to collect light, it is best to keep the individual mirrors circular, and to add about 10% more mirrors to compensate for the loss of light at their intersections.

Large mirrors need to be of quite appreciable depth in order to be sufficiently rigid. If even the thinnest reasonable solid (non-honeycomb) structure is chosen, the glass must be very low expansion and consequently expensive. In addition, the time constant to follow air temperature changes becomes many hours, and this condition of thermal imbalance with the ambient air leads to local convection at the surface of the mirror and disturbance of seeing. On the other hand, if a deep honeycomb structure is chosen, fabrication and thermal-control problems become quite severe and expensive to solve. These problems disappear with the choice of very small individual mirrors, since they can be so thin as to be made of relatively cheap glass, and by virtue of their thinness they follow closely the temperature changes of the surrounding air.

Depositing a highly reflecting coating on a very large mirror requires an even larger vacuum tank with heroic handling fixtures – a major investment. It is also difficult to deposit, with sufficient uniformity, materials other than aluminum on so large a surface from sources which are not very far from the surface being coated. Conversely, small mirrors are easy to handle and to coat with only very modest evaporation chambers. In addition, commercial facilities can now put relatively exotic super-reflecting coatings on small mirrors at quite modest cost.

To achieve the effective light-gathering power of an 8 m conventional mirror, the SST will use 85 identical spherical figures 1 m-diameter circular mirrors, mounted together to form a single spherical primary mirror 10 m in diameter with 26 m spherical radius of curvature (13 m paraxial focal length) (Figure 1).

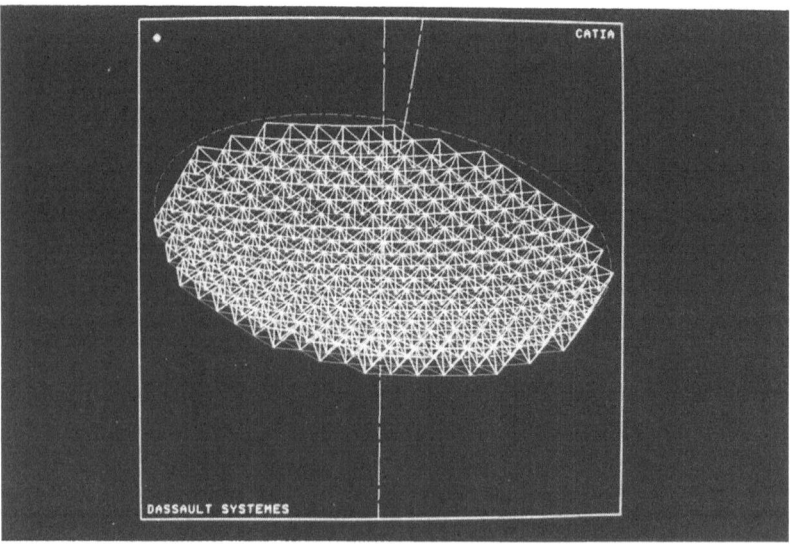

Fig. 1.   Computer model of SST primary mirror; its eighty-five 1 m circular elements will be mounted on this truss which constitutes the structural part of the mirror.

Each small mirror is made of ordinary low-cost production-run Corning Pyrex glass 5 cm thick, with rough-polished flat back and 26 m radius of curvature front surface giving a sagittal depth of 4.8 mm, almost invisible to the eye. Furthermore, since the SST is designed for spectroscopy rather than extreme quality imaging, the required optical tolerance is relatively relaxed. We have set 0.5 arc sec as the target performance for each mirror, and experience shows that this is achievable with the simple mirror polishing techniques in use. Focal length must be the same for each mirror to about 0.5 mm; polishing proves to change the focal length by about this amount in an hour's time. It should also be noted that overcoated silver surfaces are easy to apply to these small mirrors; by using overcoated silver the primary will have reflectivity around 98% at wavelengths longer than 4500 Å, while progress with multilayer coatings offers the prospect of good reflectivity deep into the ultraviolet.

After circularizing, rough grinding and fine optical annealing at Glass Fab Inc. (Rochester, New York), the mrrors are optically finished in a small optical shop made available in the Penn State Physics Department. A test mirror has been completed; the first ten of the final mirrors are currently being polished. Although mirrors are worked in groups, it appears that when full-scale production is undertaken mirrors can be completed with an average of two weeks grinding and polishing for each.

2.3.2. *Primary Mirror Support*

In several respects the SST will be the world's first large metal-mirror telescope. By themselves the many small thin mirrors are only a reflective facing, and have no mutual orientation or cohesion whatever. Accordingly, the primary is truly formed by the metal frame on which the reflecting elements are mounted. The design is based on interlocking identical tetrahedrons oriented spherically. Each tetrahedron is composed of 1.6 cm diameter rods of Invar (steel with ultra-low coefficient of expansion). The junction points, and their protruding stubs to which the Invar rods attach, are of ordinary stainless steel with its much higher coefficient of thermal expansion. The length of each element in the structure is calculated so as to produce the required overall spherical shape of the front surface of the frame, and simultaneously to average out the small and large thermal coefficients of Invar and stainless steel so as to give the entire frame an average coefficient of expansion which is the same as that of the Pyrex mirror elements.

The open-truss design of the frame brings several additional advantages. It is extraordinarily easy and cheap to build, consisting mainly of 2247 short Invar elements cut from ordinary rod stock. These are attached to the stainless steel junction stubs by simple collet fixtures, easy to assemble or to disassemble if needed. The frame is extremely light, weighing only 10 000 kg, yet its myriad interlocking triangles guarantee the great rigidity necessary to allow this frame to hold optical tolerances over a 10 m span. The high strength of steel, coupled with the rigidity of the frame, permits this very large 'metal mirror' to be axially supported by only 27 pressure points, three of which are kinematically defined, thus decoupling it from the unavoidable large thermal expansion and contraction of the underlying steel base of the telescope. Its completely open structure allows air to circulate freely around the back of the glass-reflecting elements, still further improving the ability of these mirrors to follow air temperature changes. A seven-mirror full-scale prototype – 8% of the final primary mirror – has been produced in the U.T. Austin Engineering and Astronomy shops, to test the design and verify costs (Figure 2).

Each mirror is supported about half a centimeter above a thin perforated aluminum plate by twelve springs which can be individually adjusted to take their proper share of the mirror weight. In turn each plate is supported above the metal mirror frame by three kinematic points which are springloaded to take all but about a kilogram of the load. The residual load is borne by micrometrically adjusting motors which can tilt each mirror and adjust its distance to the chosen common focus of all the mirrors. Radial

Fig. 2.   A full-scale prototype (8%) of the open-truss frame which, by supporting the 85 thin circular glass
elements, will really define the SST primary mirror.

support of the plates is also provided through their supporting pins, fitting into shallow
recesses int e bottom surfaces of the aluminum plates. Radial support of each mirror
segment is by a single central pin bonded to the segment which fits into a diaphragm
in the center of the cell plate.

2.3.3. *Telescope Tube*

As with any large telescope it is necessary to hold the secondary structure in an
adequately defined position and orientation, and to resist windshake. Since no concern
over image diffraction patterns is necessary for the SST, unusual opportunities are open
for the tube concept. Ray's design for this tube features a relatively conventional braced
outer structure, hexagonal in cross section, but this is both lightened and stiffened with
the aif of thin internal cross-members passing between adjacent columns of the tube.
The tube is 13 m in extreme diameter and 17.2 m long, but weighs only about 60 tonnes.
Analysis for transient vibrational response is in progress, using NASTRAN finite
element methods. Very low amplitudes for wind shake are anticipated (Figure 3).

The weight of the big secondary structure is largely taken by air springs exerting a
constant force on what are in effect conventional fins, but positioning of the secondary
is controlled by six carbon-fiber metering rods which emanate from the base corners of
the hexagonal tube and cross over the mirror to converge on the three defining points
for the secondary structure.

2.3.4. *Secondary Structure*

As with the Arecibo radio telescope, the use of a spherical primary opens up novel
possibilities. In particular, it creates a gigantic focal surface, in which any desired object

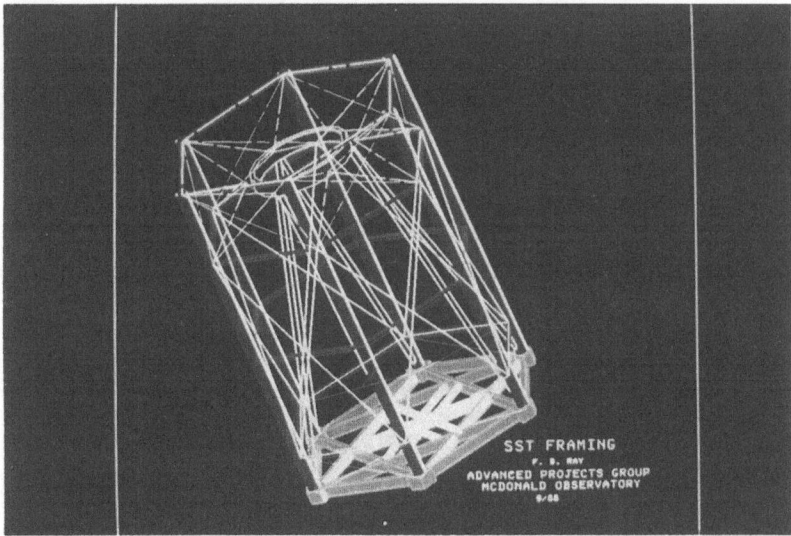

Fig. 3. Tension-braced structure of the 13 m diameter SST telescope tube. Note the extremely large secondary and, crossing over the primary mirror, the defining rods which control the position of the secondary with respect to the primary.

can be accessed by putting a detector at the appropriate place. Furthermore, by letting the detector travel along the focal surface, it obviates the necessity of moving many hundreds of tonnes of precision machinery to track the sidereal or other motions of the object being observed.

The design of the SST incorporates a 12° focal surface, inscribed within a 5 m diameter secondary ring structure. Cross-slide rails carry the detector assembly in x- and y-coordinate motion to allow selection of the object to be observed. Because the focal surface is convex spherical, the detector package must also track in the z-direction and have two axes of tilt with respect to the secondary structure. Linear encoders are required to report the location of the detector package, but – because of the long focal length of the primary mirror – only a relatively relaxes 30 μ precision is required in order to achieve, for example, accuracy of half an arc sec in position (Figure 4).

2.3.5. (Non)-Alt-Azimuth Mounting

The extreme simplicity and economy of the mirror support structures described above would be impossible if the inclination of the telescope tube to the vertical were allowed to change. Accordingly the SST is designed to operate at a fixed zenith distance of 30°. This means constant gravity aloading at all times, and should allow the mirror adjustments, once adequately made, to remain constant except for slow metal creep and any uncompensated thermal effects. Calculated performance of the mirror frame indicates that even extreme diurnal thermal changes should not disturb the total mirror figure by as much as a second of arc, but it will of course be very interesting to get real experience on this point. Current plans also include a small tower due north of the telescope,

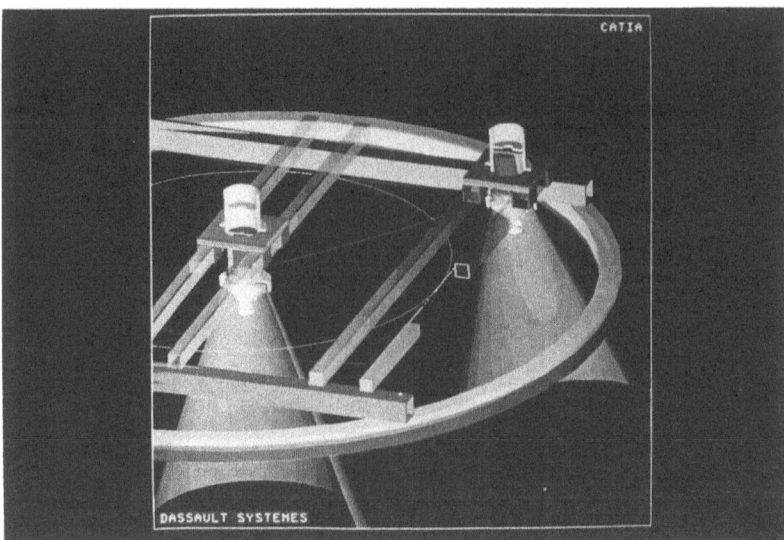

Fig. 4.   General design of the secondary structure with its x, y, z-coordinate stage carrying spherical aberration correctors.

containing a laser at the center of curvature of the primary mirror. As often as necessary it will be possible, by interrogating each mirror and noting with a CCD the location of the returned spot, to order the mirror adjustment motors to co-align all the mirrors to a high accuracy.

An azimuth air pad bearing permits a full 360° rotation. The McDonald Observatory is at latitude 30°. Recalling the constant 30° zenith distance, this means that when the telescope is aimed due north its 12° focal surface includes the declination range + 54° to 66°. Due south pointing comprises the declination range − 6° to + 6°. Intermediate declinations can be reached by intermediate azimuth settings.

Since the telescope remains fixed during any observation, and since there is freedom of orientation of the secondary $x-y$ stage, it is unnecessary to have either a conventional azimuth drive or all azimuth angles available. Instead, present plans envision simply lifting the azimuth base ring with air bearings, and using a small motor to rotate the telescope to the near-vicinity of any of 180 fixed azimuth positions, where it can settle into fixed kinematic defining points. When the true azimuths of these positions have been adequately determined and entered into the control computer, one of the necessary object-finding parameters will be under control.

2.3.6. *Spherical Aberration Corrector*

The SST's spherical primary, operating faster than $f/1.5$, produces extreme spherical aberration. Alternative two-mirror corrector designs were examined initially by S. Shectman, and independently by H. Epps. Our initial corrector will be a Gregorian probably featuring 9 m 'illumination' of the primary mirror and a 2 arc min field-of-view, produced by a two-element system with 35 cm and 22 cm elements. Its output will be at about $f/3.5$ (Figure 5).

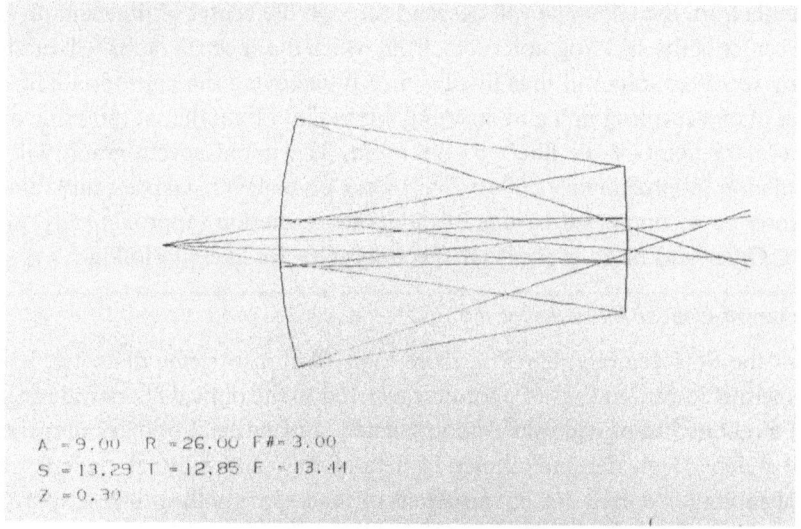

Fig. 5.   Approximate configuration of the initial spherical aberration corrector.

To function properly the corrector must not only be movable in the x, y, z coordinates noted above, but must also have two additional degrees of freedom of tilt, in order to be always accurately aligned with the 'central' (auto-collimated) ray from the object being observed. All five degrees of freedom are of course to be under computer control.

2.3.7. *Fiber Optics Feed and Spectrographs*

There is no present intention to mount instruments on the SST. Instead, because of the restriction to spectroscopic observation, it is efficient and cost-effective to place the input of an optical fiber at the focus of the spherical aberration corrector. This 30 m fiber brings the light to one of several permanently mounted horizontal optical-bench spectrographs in a large temperature-controlled room under the SST. Fibers are currently available having good transmission from about 3800 Å to beyond a micron; developments are likely fairly soon to open up the UV down to the ozone cutoff and out to at least 2.5 μ. Thermal emission from the regions between the small mirrors in the primary makes it inefficient to try for longer wavelengths with the SST.

Routine operation will probably be with a 300 μ (2.5 arc sec) fiber. This should represent a good compromise between the desire to get as much starlight as possible into the fiber even under indifferent seeing conditions, while keeping sky contamination to a reasonable minimum. However, provision is being made to have a variety of sizes and transmission characteristics of fibers available in the fiber cable leading to the spectrographs. The observer will be able to dial in any desired fiber by commanding rotating of a tiny turret or motion of a slide carrying the input ends of the various fibers.

Finding and guiding are to be done by allowing the light fro the remainder of the field to pass the fiber and be collected by a lens onto a CCD.

The spectrograph room will be arranged like a railroad-locomotive roundhouse. The fiber bundle from the telescope will descend through the center of the azimuth bearing into the center of the spectrograph room, from which the spectrographs will radiate out. Changing spectrographs will thus involve merely removing the appropriate fiber from the input of one spectrograph and inserting it into that of another at the same distance from the entry point of the fibers to the room. The initial spectrograph will be for low-resolution (approximately 500 to 2500) work on quasars, galaxies, and faint stars. It will soon be accompanied by one of moderate resolution (approximately 50000 to 100000). Other and more specialized spectrographs are likely to follow.

2.3.8. *Overall Optical Efficiency of the SST*

Although the SST primary mirror is more than 10 m in extreme diameter, its actual light-gathering power in terms of photons delivered to the optical fibers will range from about 7 to about 9 m of equivalent unobstructed conventional primary mirrors. Light losses will depend on the final choice of field-of-view and internal vignetting of the spherical aberration correctors, on the details of shadowing of the primary by structures above it, and on the degree of 'illumination' of the primary associated with any specific observation which is not almost central on the primary. However, these losses are partially compensated by the high reflectivity of the primary compared with the conventional aluminum coatings which will need to be put on most if not all large single primary mirrors. The SST will have an average light-gathering power delivered to the focus about the same as that of a conventional 8 m telescope.

2.3.9. *The 'Second' SST*

The enormous focal surface produced by the spherical primary makes it possible to have a second spherical aberration corrector which moves completely independently of the first, feeds its own spectrograph, and observes an object which may be up to 12° away from the object being studied with the other corrector. In effect, while the first SST will cost around $6 million, the second one will cost only a few hundred thousand dollars.

2.3.10. *The 'Dome'*

The fixed elevation angle of the SST allows it to dispense with the usual dome having a more than 90°-long slit opening. In line with the goal of maximum economy for each function, our initial study visualizes a simple rectangular building with roll-off roof, and with all external dimensions being a bit over 30 m.

An excellent site near the other McDonald Observatory telescopes on Mt. Locke has been chosen for the SST (Figure 6).

2.3.11. *Observing Routine*

Efficient observing with the SST will require scheduling not very different in character from that of the Hubble Space Telescope. The telescope will seldom be assigned to a

Fig. 6. Artist's rendition of location and probable general appearance of the SST at McDonald Observatory.

single observer for a run. Rather, desired spectroscopic observations from a large number of observers and programs will be prioritized and entered into a computer memory, from which an artificial intelligence program will recommend, for any set of observing conditions at any hour of any night, the optimum azimuth of the SST and order of objects to be observed. Resident observers will make the appropriate spectrograph settings and record the data, sending them to the astronomers at the completion of each run.

3. Summary of Disadvantages and Advantages of the SST Design

Con: A number of prices must be paid to achieve so remarkably economical a design:
   – The SST will have no capability for wide-field imagery or for work in the thermal IR.
   – The small field of at least the initial corrector will restrict any possible multi-fiber ('Medusa-type') work to a few objects in a very small area of the sky, or to a 'densepack' arrangement in which a fiber bundle brings the entire small field down to an instrument for analysis.
   – A small area (24° radius) around the North Pole is forever invisible to the telescope, as is all the sky south of $-6°$.
   – It will be necessary to wait for desired objects to enter an accessible setting of the telescope.
   – Exposure times on any object will be limited to about 40 min near the equator and up to about twice that for work at high declinations, making it impossible to get continuous data on variable sources.
   – Extinction data for even crude spectrophotometry must come from their telescopes.

Pro: We believe these penalties are far outweighed by the positive factors:

– The SST is *affordable*, as a special-purpose telescope with great power for general spectroscopic work – the most important source of astrophysical information.

– In addition, the SST is really *two quasi-independent telescopes*, each with light-gathering power in the class of nearly all of the largest telescopes now being built or designed.

– By virtue of its dedication to spectroscopy the SST should be *uniquely productive* in this vital area of observational astronomy.

– The SST should achieve *operational efficiency* of high order because of its dedicated staff of highly trained and skilled observers (similar to IUE), and because its instruments will not require the time-costly setup and frequent need for repair normally encountered with instruments which are frequently changed, nor much of the time usually needed for calibrations at the start of a run.

– The SST offers these advantages at a cost per telescope more than an order of magnitude less than that of a general-purpose telescope of comparable size.

Acknowledgements

It is diffcult to assign adequate credit for innovative ideas. However, the SST project has benefitted immensely from constructive suggestions by many people, among whom we wish particularly to acknowledge and thank Roger Angel, Norman Cole, Jerry Nelson, and Steve Shectman.

KODAIRA – How large is the image size after you correct the system for the spherical aberration?

SMITH – Final corrector parameters have not yet been chosen, but 1 arc sec will probably be about 150 μ.

GILLINGHAM – On the AAT, we use an invar wire to monitor the length of the telescope tube, for the purpose of focus control. Have you considered using wires rather than the carbon fibre reinforced struts you described for the controlling your top end position?

SMITH – Not yet.

ELLIS – Does $6 million include provision of your instruments? My impression is that such a telescope will, by virtue of its single object mode of operation, be most effective for high dispersion work.

SMITH – The $6 million estimate includes only a relatively low-resolution spectrograph. I agree that the SST is likely to be most effective for high dispersion; we plan to seek NSF funding for such an instrument.

NARIAI – Yamashita and I worked on three-mirror telescopes several years ago (*Ann. Tokyo Astron. Obs.*, series II, Vol. 19, p. 375, 1983).

SHECTMAN – The corrector consists only of two reflecting surfaces. Wider-field versions rely upon restricting the final focal ratio to an optimum value and adding additional refracting elements near the final focal plane.

HALL – Are you able to match the throughput of a 3 arc sec fiber from an 8–10 m telescope to the proposed spectrograph?

SMITH – With some difficulty, and large optics; also probably using image-slicing.

BECKERS – How good is your pointing and how do you acquire your object?

SMITH – We hope, by using well-calibrated encoders, to achieve 'pointing' accuracy of ± 1 arc sec, and to verify acquisition with a TV camera focused on the field surrounding the fiber input.

EDMUNDS – You implied that use of a second aberration corrector required a second spectrograph – won't you first of all just feed both fibres into the same spectrograph?

SMITH – Yes – a good way to begin. But we will need, and want as soon as possible, to have a high-resolution spectrograph as well.

SHECTMAN – The residual image size is due entirely to fabrication and alignment errors of the optical parts. The optical design produces on-axis images which are in principle perfect.

THE JNLT PROJECT

OUTLINE OF THE JNLT PROJECT*

KEIICHI KODAIRA

National Astronomical Observatory, Tokyo, Japan

(Received 18 January, 1989)

Abstract. The Japanese National Large Telescope (JNLT) is a 7.5 m reflector with a monolithic thin meniscus main mirror, having the candidate construction site on the northwest cone of the Mauna Kea, Hawaii.

The present concept of JNLT has the characteristics of a 'third generation' infrared telescope, which should be capable of various observations of high spatial resolution in the optical-infrared region.

Although the project is still under examination from the financial and administrative point of view, a wide range of technical studies were carried out by the JNLT Working Group with the collaboration of specialists from the academic and the industrial sectors.

1. Introduction

The Japanese National Large Telescope (JNLT) is a new-technology telescope which has a monolithic thin meniscus main mirror with a diameter of 7.5 m and a thickness of 20 cm, made of material of extremely low thermal expansion coefficient. The mirror shall be actively supported, and the telescope tube shall be set on an altitude-azimuth mount. The conceptual model of JNLT at the present stage of development is shown in Figure 1. The telescope shall be installed in a thermally controlled solid enclosure. The candidate construction site is located at the northwest cone of Mauna Kea, Hawaii.

This paper describes the historical background of the JNLT project, organizations promoting it, scientific motivations, and the special technical features of JNLT. Finally the possibility of international collaboration around the JNLT project will be discussed in some detail.

2. Historical Background

The largest telescope in Japan is the 188 cm reflector at the Okayama Astrophysical Observatory which was brought into operation in 1960. After ten years of operation of this telescope, the astronomical community in Japan started to consider a plan to construct a 3–4 m class reflector to meet the quickly increasing observational demand. This plan, however, could not be realized early enough, and instead they built first a 105/150 cm Schmidt telescope at the Kiso Observatory in 1975. The first priority during this period was given to the project to construct large radio astronomy facilities. In 1978 when the financial endorsement to the radio project became sure, the Japanese astronomy community started a revision of the future strategy, led by the National

---

* Paper presented at the symposium on the JNLT and Related Engineering Developments, Tokyo, November 29–December 2, 1988.

Astrophysics and Space Science **160**: 137–144, 1989.
© 1989 *Kluwer Academic Publishers.*

Fig. 1.   Conceptual model of JNLT. For colour reproduction of this figure see colour section.

Committee for Astronomy (NCA) of the Science Council of Japan (see Section 3). The large radio astronomy facilities at the Nobeyama Radio Observatory were completed in 1982. Meanwhile the Group of Optical and Infrared Astronomers (GOPIRA; see Section 3) embarked on the feasibility study of the JNLT project, and this went over to the phase of the conceptual study around 1984. During the feasibility study phase, many discussions and research activities were devoted to a few of the key items such as the size and the type of the main mirror and the observatory site which could be either domestic or abroad.

The NCA produced a revised report on the future strategy in astronomy in 1984, and recommended the JNLT project in a form very close to the present one. Nearly at the same time the community became aware that a kind of a new national inter-university organization was necessary in order to promote the JNLT project, along with other requirements to activate astronomy research in Japan.

The results of the conceptual study of JNLT were critically reviewed in early 1986, and the engineering studies were started for the technical items identified in the critical review. Some of the results of the engineering studies will be presented in the succeeding papers in this symposium.

In the summer of 1986, the Memorandum of Understanding concerning the construction of JNLT was exchanged between the Tokyo Astronomical Observatory and the University of Hawaii. According to this memorandum, the candidate observatory site was fixed to the west side of the northwest cone of Mauna Kea (Figure 2), where a half year long site testing was carried out in 1987.

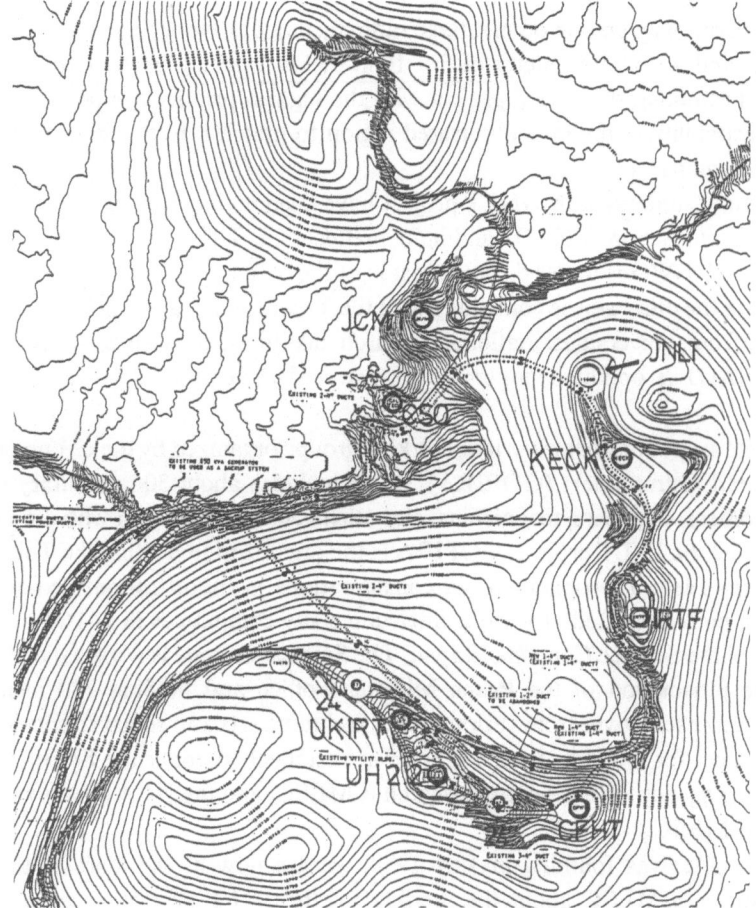

Fig. 2.   Candidate construction site for JNLT.

The National Astronomical Observatory of Japan (NAOJ), a new national inter-university research institute, was founded in July 1988, as the result of reorganization of the Tokyo University's Observatory, Mizusawa Latitude Observatory, and a part of the Atmospheric Research Institute, now to promote the JNLT project.

3. Promoting Institutes and Groups

There are thirteen national inter-university research institutes which are serving as the center of various research fields in Japan. In the case of natural sciences they are responsible to construct and run major research facilities. The first of them was the Institute for High-Energy Physics (KEK) founded in 1971. There are now two institutes directly related to astronomy. One is the Institute for Space and Astronautical Science (ISAS) and other is the NAOJ. ISAS launched the X-ray satellite GINGA and is going to launch an X-ray solar mission, Solar-A in 1992. ISAS further intends to aunch an X-ray, non-solar mission, SXO around 1993, and a space VLBI experiment, VSOP.

NAOJ is running, among others, large radio facilities at Nobeyama consisting of one 45 m dish and an interferometer of 5×10 m dishes working in the mm wave range. NAOJ is promoting the JNLT project with the highest priority to upgrade the observational capability in the optical-infrared region to match it to those in the radio and the X-ray regions.

Preliminary studies for the JNLT project have been mainly carried out by the JNLT Working Group set up in NAOJ and in its predecessor, Tokyo Astronomical Observatory. The JNLT WG is under the supervision of the JNLT Committee in NAOJ on one side, and on the other side is consulting with the Technical Study Group which includes engineers from universities, national research institutes, as well as from industrial sectors. NAOJ is operated under the Ministry of Education, Science, and Culture (MESC).

Other study groups are the various working groups organized by the Group of Optical and Infrared Astronomers (GOPIRA), which includes about 300 voluntary members of astronomers, engineers, technicians, and others across the nation who are interested in promoting the JNLT project. GOPIRA is the organization hosting the Tokyo JNLT Symposium 1988. The opinions and considerations in the astronomy community in Japan can be represented at various inter-university committees in NAOJ and also at the NCA of the Science Council of Japan. The Science Council is a special organization under the jurisdiction of the Prime Minister's office and holding 210 members of academic representatives from various fields, including arts and social sciences. National Committees for individual science fields are established under the Science Council. The funding agency for astronomy research is MESC, and the NCA makes recommendations to MESC.

Since NOAJ was founded in July 1988, we are now at the start of the official negotiation with MESC about the funding for the JNLT project. According to the current model schedule, we need at least seven years to complete JNLT after the fund becomes available. Although MESC is not yet officially committed to the JNLT project at this stage, I optimistically hope that JNLT will come into operation around the mid-1990's.

4. JNLT Concept and Scientific Objectives

The scientific objectives projected to JNLT by the astronomy community in Japan cover a very wide range, and I will present here only the points specifically relevant to the JNLT concept at this stage.

First of all, JNLT shall be an optical-infrared telescope. The Earth's atmosphere has two major windows for the cosmic electromagnetic wave, one in the radio wave range and the other in the optical-infrared range. When the first-rank telescopes of today were in the phase of the conceptual study, the main detector in the optical region was still photographic emulsion, and that in the infrared region was the single pixel photometer. The situation has largely changed since CCD and then IR array detectors have become available, technically to merge both regions. Along with the advance in the observational cosmology, the increasing redshift also eliminated difference between the two regions. JNLT shall be optimized as the optical-infrared telescope in this sense. In contrast to the current infrared telescope, JNLT might be characterized as a 'third generation, ground-based, infrared telescope', which has a large collecting area, designed for infrared imagery with a high spatial resolution up to $\sim 0''.2$. JNLT shall posess high capability in polarimetry, high-speed photometry, and spectroscopy with a moderate field-of-view.

Located on Mauna Kea, JNLT shall make the best use of the high-quality seeing and the low residual of vapour in the atmosphere. In spite of the high altitude of 4200 m at Mauna Kea, we see much merit in operating JNLT as a ground-based telescope on which complicated instruments can be more flexibly operated compared with space facilities. JNLT shall be optimized for functions complementary to the space facilities such as HST, ISO, and other missions of the near future.

We anticipate that JNLT as such will be highly capable to study the structure, kinematics, chemistry, and physics of star-forming region, planet-forming region, and galaxy-forming region (among others). The last means the domain of the space and time where the galaxies were born.

Since the detection of a dust disk around β Pic by Smith and Terrile (1984), planet formation became one of the exciting objectives for a large optical-infrared telescope. β Pic is the 4th mag star of type A5V at a distance of 16 pc, and we may expect many other stars to be detected at various stages of proto-planetary disk formation within the reach of JNLT.

Recent N-body simulations, such as by Carlberg (1988) and Frenk (1987), demonstrate the formation process of large structures in the Universe and of galaxies. These simulations and the detections of possible very young galaxies such as 4C 4117 ($z = 3.8$) suggest that JNLT can be a powerful tool for exploration of the galaxy-forming region. No less interesting than the primeval galaxies themselves are the intergalactic clouds in the early epoch, which can be investigated in the same way as is used for the Lα forest if we can take spectra with high enoughS/N ratio and dispersion. Spectroscopic studies of multiple images generated by gravitational lenses will provide the information of the tangential extent of intergalactic clouds.

In order to make JNLT highly capable of such a research, the JNLT foci are designed as shown in Table I at this stage. JNLT shall have the primary focus of $f/2.3$ and $30'$ field-of-view with a red and a blue wide-field corrector, the Cassegrain and Nasmyth foci of $f/12.5$ with a $6'$ field-of-view, the secondary and the tertiary mirror being

TABLE I

JNLT foci

Primary + WFC	$f/2.0$		$D = 7.5$ m
	$f/2.3$	FOV = $30'$	blue, red
Cassegrain and Nasmyth	$f/12.5$	FOV = $6'$	IR: $D = 100$ cm
			opt.: $D = 126$ cm
	$f/35$	chopping	IR: $D = 50$ cm
Coudé		(for interferometry)	

separately made for the optical and the infrared use. The design consideration of the wide-field corrector will be given by K. Nariai in the following talk. For the thermal infrared, a $f/35$ chopping secondary shall be provided. JNLT shall have a Coudé optics train leading to a laboratory in the basement which will be used for interferometric experiments.

In spite of the fact that the scientific objectives mentioned above support the present JNLT concept, the most exciting results may come from unexpected discoveries which become feasible through the newly opened observational possibility by JNLT.

5. Technical Features

We have decided the primary mirror to be a monolithic thin meniscus type made of ultra-low expansion material in a relatively early phase of the preliminary study, because of its simplicity in the mechanical and thermal responses. The main mirror with diameter of 7.5 m and thickness 20 cm shall be actively supported at about 400 points on the basis of the floating principle. The servo actuators shall control the force, not the position. The oscillation characteristics of such a mirror system will be reported by Y. Yamashita in what follows. Each active support mechanism locally provides both axial and lateral force for a specified portion of the mirror, without edge supports. In order to realize this kind of support, the mirror blank shall have holes on the backside to accommodate the socket for the active supporting system (Figure 3). Although grinding and boring glass material are now common technology, one must be extremely cautious in this work; and this may be a draw-back of the present JNLT concept. The lateral force is provided by the lever-and-counter weight mechanism while the axial force is provided by the newly developed actuators which are composed of a motor, a ball screw, and a spring with a load cell monitoring the acting force. The smaller the number of the supporting ponts is, the more sensitive must be the actuator. The JNLT Working Group succeeded in developing an actuator with a sensitivity of 10^4. Details of the supporting system will be given in a succeeding paper by M. Iye.

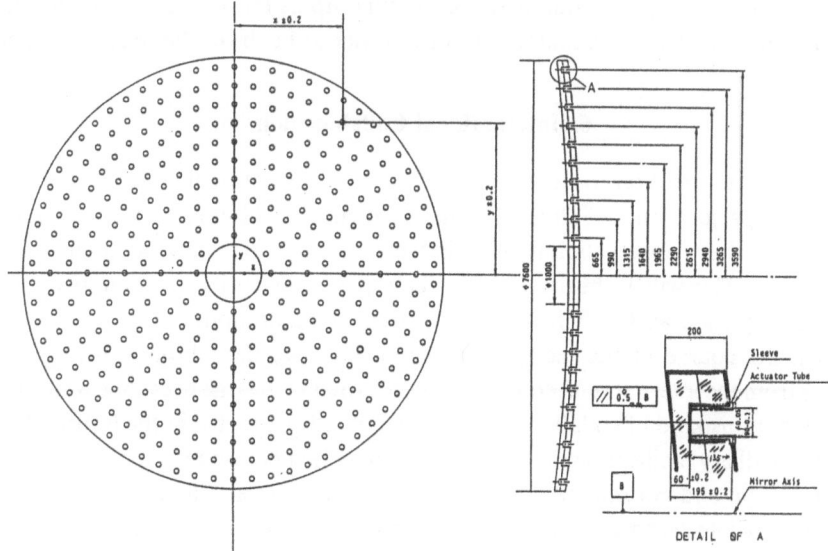

Fig. 3.   Thin meniscus design of a mirror blank for JNLT.

Since we adopt the principle of the floating support, any external forces disturbing the thin mirror might be fatal in controlling the mirror form. This point must be taken into account when the JNLT dome is designed. The candidate construction site of JNLT is at the west side of the W. M. Keck Observatory on the northwest cone of Mauna Kea. The results of site testing will be given in one of the subsequent reports by H. Ando. At the altitude of 4200 m, the average wind speed at Mauna Kea is relatively high (7 m s^{-1}), and we prefer a solid dome enclosure of relatively traditional type. The results of site testing seems to indicate a necessary dome height to avoid the serious effects of the boundary turbulence layer as law as 15 m, except for the special situation that the W. M. Keck Observatory stands about 130 m apart in the ENE direction. This circumstance, together with the topographical circumstances on the northwest cone, may force the JNLT dome to be as high as 50 m to the top of the roof. This concept of a large-size dome causes thermal problems which may interfere with the requirement of high-quality images. I. Mikami will discuss about the thermal aspects of the JNLT dome concept in one of the following papers.

As described in this paper, the present concept of JNLT aims at a highly capable new-technology telescope, and not necessarily at a low-cost telescope which often inevitably suffers from trade-off's. The present cost estimate of the JNLT just hits the line of the prediction for the case of *Conventional* by Meinel and Meinel (1980). The JNLT Working Group is still making efforts to make the concept and the design of JNLT more cost-effective by keeping the full functions as far as possible and by suppressing the accompanied risks as small as possible. I would like to point out here the differences in the infrastructures of NAOJ compared to those of NOAO or ESO;

NAOJ, which started just in this year, has to build up its infrastructures along with the construction of JNLT. This factor certainly works to push up the telescope cost.

6. International Collaboration

Finally I wish to mention to the importance of the international collaboration related to the JNLT project. First of all JNLT shall be located on Mauna Kea, where is known as one of the best observational sites and the home of the internationally-operated observatories. This will be the first time for the Japanese scientific community to operate a large facility such as JNLT in an international collaboration.

During the course of the technical developments we have been enjoying various supports from the international community concerned with large telescope projects, and I believe that this kind of collaboration will be mutually extended in the future. We may share the auxiliary facilities such as an aluminizing plant common among observatories at Mauna Kea, and coordinate various experiments between different telescopes. The possibilities of interferometric experiments using JNLT will be discussed by M. Ishiguro and S. Isobe in this symposium. In the construction of JNLT we may call for the top-level technologies all over the world. S. Okamura and T. Maihara are going to discuss about the JNLT observational instruments on the third day of this symposium, and some of those instruments shall be developed and constructed under the inter-national collaboration. We regard the JNLT project as the corner-stone to strengthen the international ties of the Japanese astronomical community, which were already initiated by the Nobeyama facilities in the radio astronomy and by the ISAS activities in the space astronomy.

In conclusion I should like to express our sincere thanks to the international community represented by the foreign guests to this symposium for their constant support for the promotion of the JNLT project.

References

Carlberg, R. G.: 1988, *Astrophys. J.* **324**, 664.
Frenk, C. S.: 1987, in S. M. Faber (ed.), *Nearly Normal Galaxies from the Planck Time to the Present*, Springer-Verlag, New York, p. 421.
Meinel, A. and Meinel, M.: 1980, in A. Hewitt (ed.), *Optical and Infrared Telescopes of 1990's,* KPNO, Tucson.
Smith, B. A. and Terrile, R. J.: 1984, *Science* **226**, 1421.

FINITE ELEMENT ANALYSIS OF A MENISCUS MIRROR*

Y. YAMASHITA

National Astronomical Observatory, Tokyo, Japan

(Received 10 January, 1989)

Abstract. Finite element analyses were carried out for a 7.5 m meniscus mirror of 20 cm thickness. Calculations were made for deformations of the mirror surface due to the gravity and the effect of a hole through which a lateral supporting mechanism would be installed. Vibrational eigenmodes were also calculated when the mirror is fixed by three axial and three lateral hard points.

In this report, I summarize finite element calculations of mirror deflections carried out by several members of our group.

Deformations of the mirror surface due to self-weight are calculated by Watanabe (1987) for a 7.5 m meniscus mirror of 20 cm thickness. The mirror is supported axially by a number of support points at the back surface, and laterally by the same number of support points located at the middle surface of the mirror thickness. Because of symmetry, actual calculations are done for a half of the model.

It is shown that about 350 support points are necessary in order to keep the peak-to-valley surface deformation within $0.1\ \mu m$ ($\sim 17\ nm$ r.m.s.). To maintain the 80% encircled energy within the diameter of $0.''1$, the relation between the surface error of mirror and its wave number is established by Ando (1988). According to his relation as shown in Figure 1, the peak-to-valley deformation of $0.1\ \mu m$ corresponds to a linear scale of about 1 m. In the actual model, 390 support points distributed in ten rings are adopted. Correspondingly, by taking into account of error distribution of 390 forces, the accuracy of axial supporting force is obtained as $150/\sqrt{n} = 7.5$ gf in the dynamic range of 0–50 kgf. As the number of support points n is increased, the relative force accuracy is relaxed in proportion to $\sqrt{n}$. In these considerations, the model of 20 cm thickness with 390 force actuators and the relative force accuracy of 1×10^{-4} seems close to optimum for our purpose.

For the purpose of comparison, deflection calculations are also made by Watanabe (1985) for several models of honeycomb mirror. Here, 80 support points are adopted. It turned out that gravity deformations are rather large, ranging 0.5 to 0.8 μm according to the models. These large deformations seem to come from a lack of circular symmetry of support points inherent from the honeycomb structure when the mirror is horizontal, and from complicated bending of rib structure when the mirror is vertical. The deformation cannot be reduced much by thickening the rib or the front and back plates. When the mirror is horizontal, the peak-to-valley deformation may be reduced to 0.1 μm by

---

* Paper presented at the symposium on the JNLT and Related Engineering Developments, Tokyo, November 29–December 2, 1988.

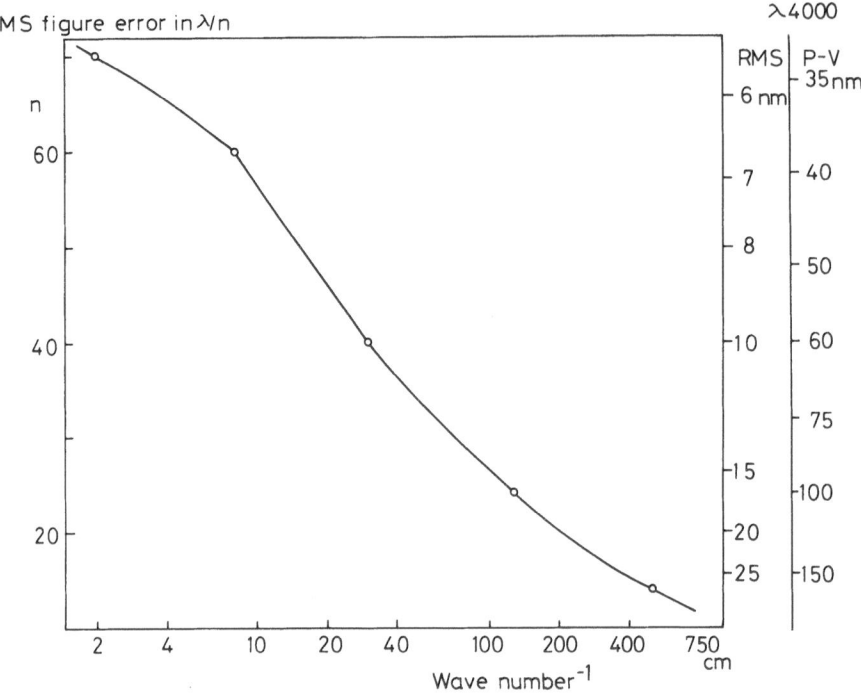

Fig. 1.   The relation between the surface error of mirror and its wave number.

properly adjusting supporting forces; however, when the mirror is vertical, the deformation of about 0.5 μm will remain, because we are considering an actuator of push
only. This is one of the reasons that we have adopted a thin meniscus mirror rather than
a honeycomb mirror.

To support the meniscus mirror laterally at the middle surface of mirror thickness,
we are considering to bore holes from the back. The effect of holes on the mirror
deformation is calculated based on a local model by Tsukui and Itoh (1986). Here,
support points are assumed to locate equidistant with the separation of 330 mm, and
periodic boundary conditions are applied. The hole has the diameter of 60 mm and the
depth of 130 mm. The model mirror is supported axially along the hole from the back,
and laterally at two points inside the hole. When the movement of centre of gravity of
0.6 mm due to the removal of mass inside the hole is properly taken into account, the
peak-to-valley surface deformation is less than 10 nm for all elevation angles.

As is well known, vibrations of thin mirror may be excited by wind loads or accelerations of the telescope. Natural vibrational modes have been calculated for the 8 m VLT
mirror of 20 cm thickness (Wilson *et al.*, 1988). In practice, six degrees of freedom of
the mirror must be fixed. Then, eigenmodes are calculated by Yamashita and Nishino
(1989) for the 7.5 m meniscus mirror with three axial and three lateral fixed points.
Patterns calculated up to mode 25 are presented in posters; they are quite different from
normal vibrations. Because of a symmetry of fixed points, all the modes are either
symmetric or antisymmetric with the reflection at the symmetric plane including the

mirror axis, or symmetric with the rotation of 120 degrees about the mirror axis. In the present model, the lowest mode is a tilting vibration of the mirror disk with the frequency of 15.5 Hz, while the second lowest is a bending vibration like an astigmatic deformation with 15.7 Hz. The frequencies of these two modes are very close; therefore, if they are excited, the beat appears with a period of a few seconds. This should be noted in practice.

The remaining problem is to obtain a response when a certain force is applied to each actuator. The response will be developed by Zernike polynomials. Then, if an error in the actual mirror figure is measured, for example, by a Shack–Hartmann method, and is also developed by Zernike polynomials, the error can be corrected by a superposition of responses for the respective actuators.

References

Ando, H.: 1988, *JNLT Technical Workshop*, OT-30-T3 (in Japanese).

Tsukui, K. and Itoh, N.: 1986, *JNLT Technical Workshop*, OT-20-M4 (in Japanese).

Watanabe, M.: 1985, *JNLT Technical Workshop*, OT-8-T4 (in Japanese).

Watanabe, M.: 1987, *Ann. Tokyo Astron. Obs.* **21**, 241.

Wilson, R., Franza, F., and Noethe, L.: 1988, *Proc. Very Large Telescopes and Their Instrumentation*, ESO, Garching.

Yamashita, Y. and Nishino, Y.: 1989, *Publ. Nat. Astron. Obs. Japan* **1**, 1.

ACTIVE OPTICS EXPERIMENTS FOR THIN MENISCUS MIRROR*

MASANORI IYE

National Astronomical Observatory, Tokyo, Japan

(Received 1 February, 1989)

Abstract. The concept of the active support system for JNLT is summarized. Performance of the force sensor, the optical wavefront analyzer, and the actuator under development for JNLT is reported. The results of a series of active optics experiments carried out by assembling these elements to support, measure, and actively correct a 62 cm thin mirror are described.

1. Active Support for JNLT

The 7.5 m Japanese National Large Telescope (JNLT) planned for construction on Mauna Kea, Hawaii, will have a thin $F/2$ primary mirror actively supported by 390 mechanical actuators to maintain the optical imaging performance of giving 80% encircled energy within $0\overset{''}{.}1$ diameter. A finite element analysis (Watanabe, 1987) has shown that in order to keep the peak-to-valley surface error less than 100 nm to achieve this imaging performance the allowable error of the axial support force is only ± 7 gf for each actuator carrying 0–50 kgf load. Figure 1 shows the proposed structure of the mirror cell housing 390 actuators to support the primary mirror.

Figure 2 illustrates the basic principle of the active support system for JNLT. The measurement of the primary mirror surface is accomplished in two independent ways.

(1) The Shack–Hartmann analyzer measures the mirror surface error by monitoring the deviation in the distribution of hundreds of spot images of a star produced by a microlens array. This is a direct optical method for measuring the wavefront error.

(2) The 390 force sensors attached to actuators measure the distribution of the supporting force of the primary mirror. This is a rapid mechanical measurement from which the exact shape of the mirror surface can be calculated.

The information obtained by these measurements is processed to generate the distribution of the correction force necessary to maintain the best imaging performance. The fast mechanical servo loop can be intermittently calibrated by the direct optical loop.

We identify three key ingredients that are necessary to achieve the active support of the primary mirror of JNLT. They are (1) high precision force sensor (relative error $\leq 10^{-4}$), (2) Shack–Hartmann wavefront analyzer (measuring error ≤ 10 nm), and (3) high precision actuator (relative error $\leq 10^{-4}$, dynamic range 0–60 kgf).

* Paper presented at the symposium on the JNLT and Related Engineering Developments, Tokyo, November 29–December 2, 1988.

M. IYE

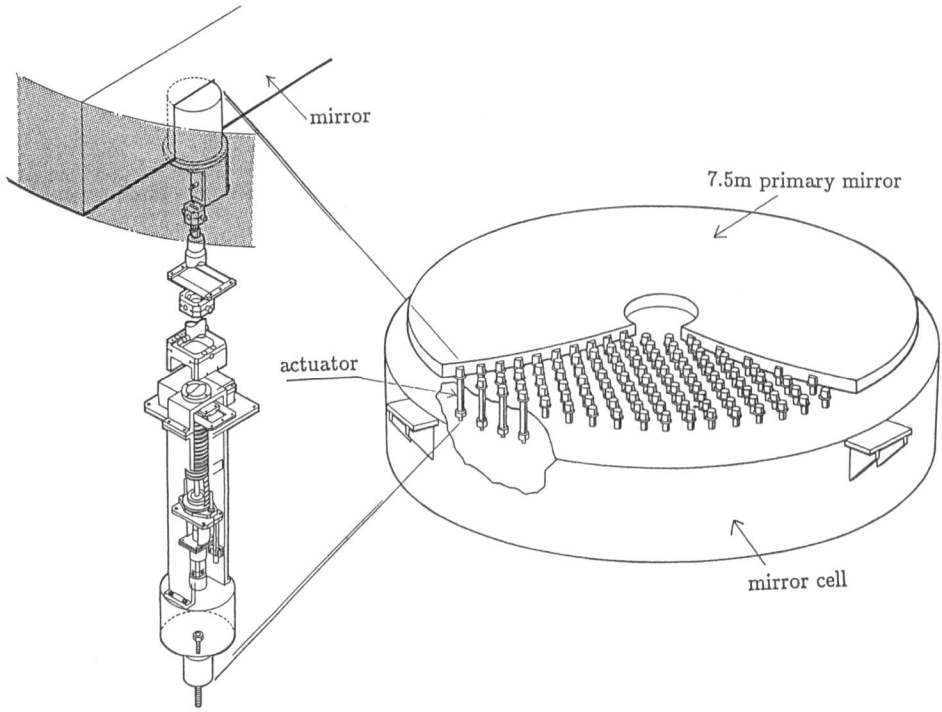

Fig. 1. Proposed structure of the support system for the 7.5 m primary mirror of JNLT.

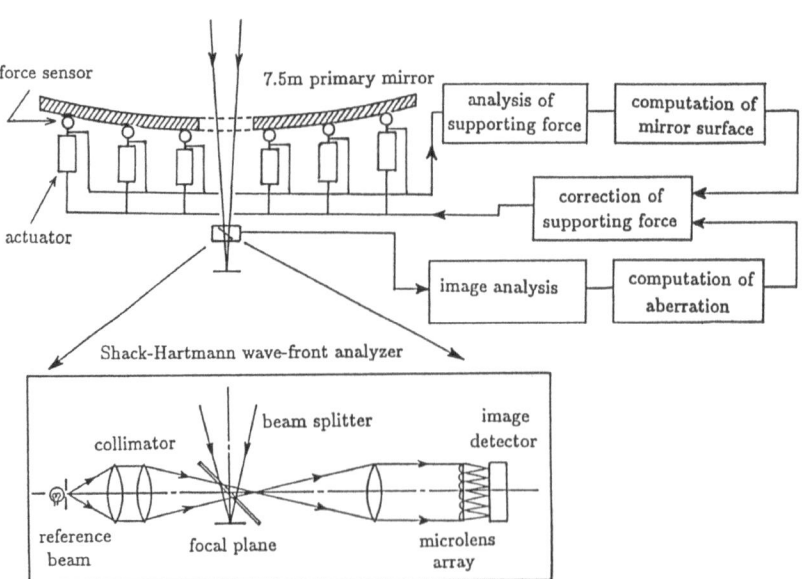

Fig. 2. Principle of the active support system for JNLT.

2. Key Ingredients for Active Support

2.1. FORCE SENSOR

A new type load cell has been completed by MELCO to attain very high precision of
± 2 gf in a large measuring range of 0–60 kgf. The working principle of this load cell
is that it measures the frequency modulation of a small tuning folk which is subject to
a strain force externally applied. It has a correctable temperature drift of about
1.5 gf/°C.

2.2. SHACK–HARTMANN WAVEFRONT ANALYZER

The details of the Shack–Hartmann wavefront analyzer developed at NAOJ is reported
by Noguchi *et al.* (1989). Figure 3 shows the optical layout of the constructed analyzer.

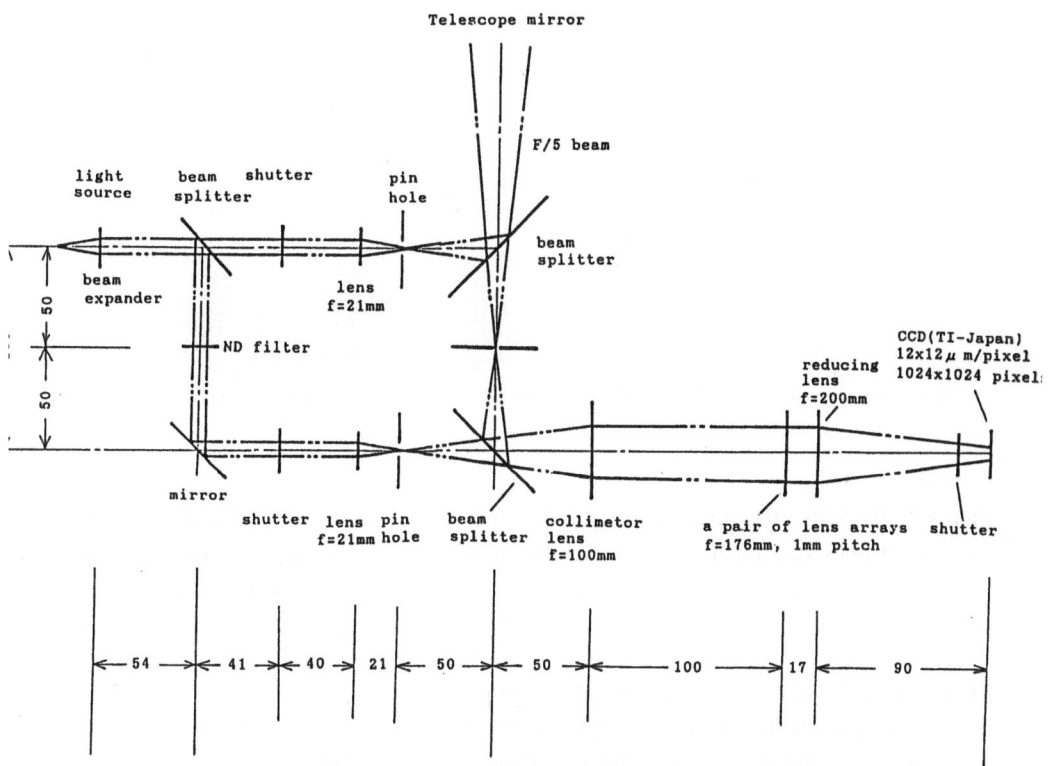

Fig. 3.   Optical layout of the constructed Shack–Hartmann wavefront analyzer.

One can select the optical path either from the reference light source or from a star by
a set of electric shutter without moving any optical element. This rigid configuration
assures a high stability (≤ 2 nm for each Zernike coefficients) independent of the
orientation of the analyzer. The microlens array placed at the image position of the
primary mirror produces hundreds of spots on the CCD. The positions of these spots

relative to the corresponding position of the reference spots carry the information of the surface slope at the corresponding part of the primary mirror. The measured distribution of displacement vectors in $X - Y$ plane on CCD (cf. Figure 4) is transformed to a Z-displacement of the mirror surface, which in turn is expanded in terms of Zernike polynomials.

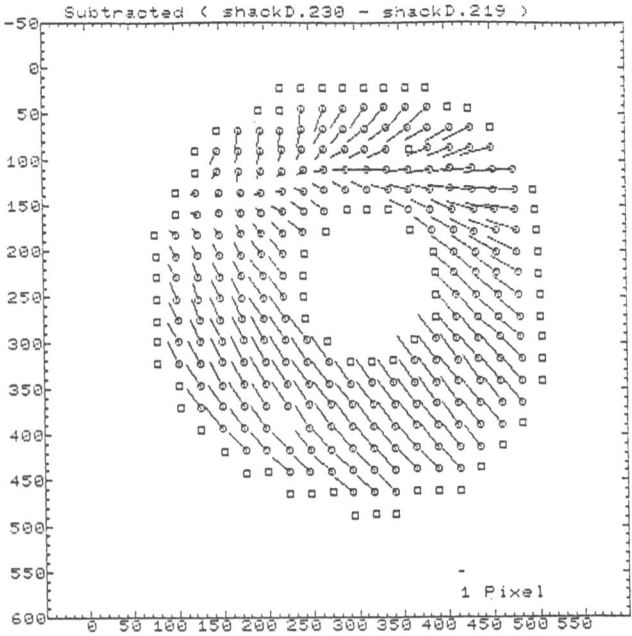

Fig. 4.   An example showing the displacement vectors derived by the Shack–Hartmann analyzer.

2.3. ACTUATOR

Figure 5 illustrates the principle of the full size actuator developed for JNLT active support system (cf. Nishimura *et al.*, 1988). The correction signal produced by a controller drives a motor to adjust the axial support force that is generated by a compressed spring. The attained accuracy is very high, $\pm$ 5 fg, over the entire operating range of 0–60 kgf. The radial support is achieved passively by a lever and a counter weight at the local center of gravity of the primary mirror.

As a summary of this section, we conclude that all the three necessary key ingredients for the active support system of the JNLT are now available.

3. Active Support Experiment of a 62 cm Mirror

In order to verify the feasibility of the active support concept for JNLT, a series of experiments was carried out in the spring of 1988 by assembling the three ingredients into a self-contained system. An $F/5$ pyrex mirror of 62 cm diameter and 2 cm thickness was produced and supported on an active support mechanism that consists of 9

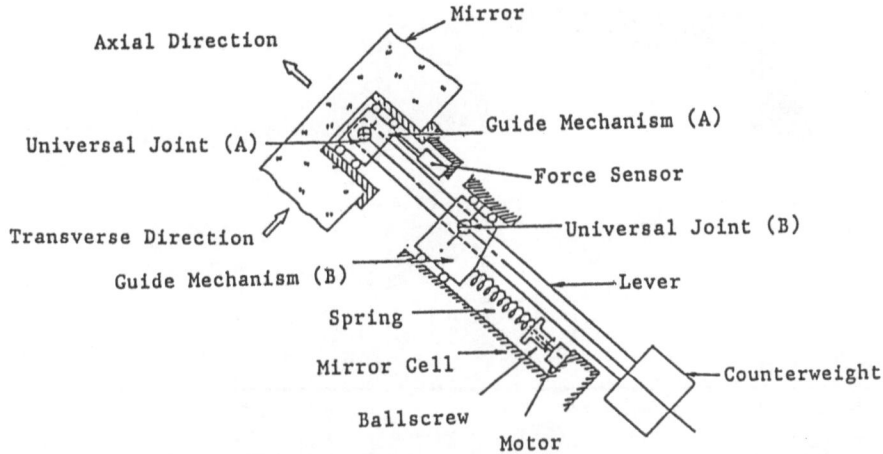

Fig. 5. Operating principle of the actuator that supports the primary mirror actively in axial direction and passively in radial direction.

actuators, 3 fixed points, and 12 load cells. It was combined with the Shack–Hartmann wavefront analyzer to measure and actively deform the mirror surface. The entire setup was mounted on a flat bench that can be tilted at $0° \leq EL \leq 60°$ (Figures 6 and 7). The details of the experiments are reported in Iye *et al.* (1989).

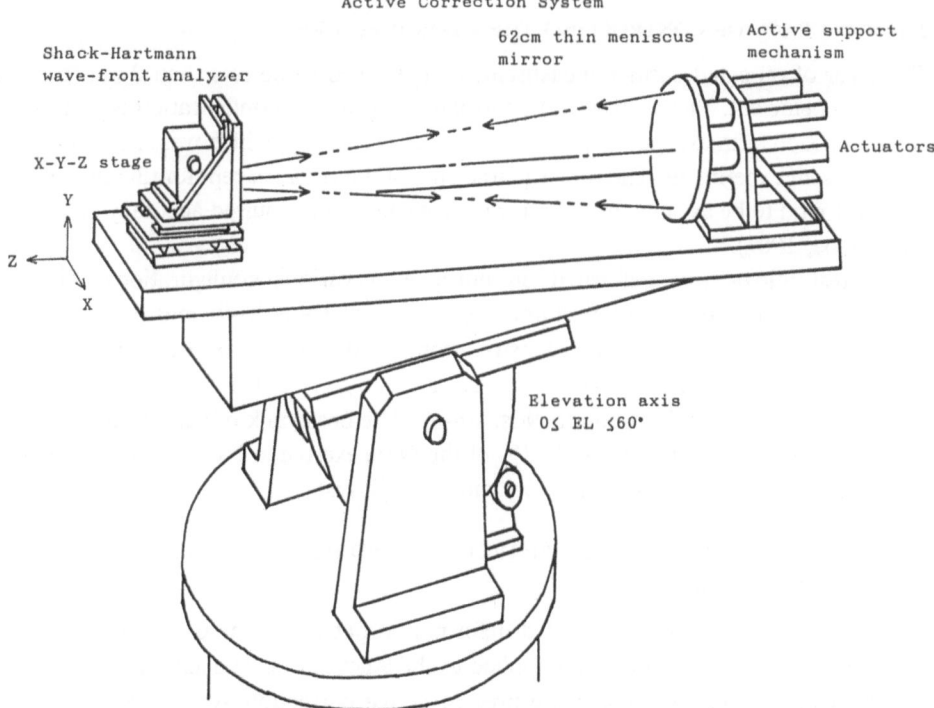

Fig. 6. Schematic drawing of the test setup for active support experiments of a 62 cm mirror.

Fig. 7.   Active support experiment at a tilted orientation. For colour reproduction of this figure see colour section.

3.1. ACCURACY OF CONTROLLING THE MIRROR FIGURE

The error of repeated surface measurements is proved to be less than 2 nm for each Zernike coefficient. The accuracy of controlling the surface configuration is better han 50 nm for each term except for the tilt and the defocus terms, which were not controlled in the present experiment. The root-squared sum of the errors except for tilt and defocus terms is found to be as small as 65 nm, which is within the required accuracy of 100 nm (Watanabe, 1987).

The test mirror was deformed on purpose in various configurations. Figure 8 illustrates the distribution of 9 supporting actuators (Nos. 1–9) and 3 fixed ponits (Nos. 10–12). The responsive force distribution is shown for a case where the actuator No. 1 was driven to generate an excess force of + 506 gf. It is clear that this excess force is compensated by the fixed points Nos. 10–12. The cross talk of supporting forces is largest at actuator No. 6 but is only 2% of the force exerted at No. 1. The interference among the actuators is, therefore, very mild in this system.

3.2. COMPARISON WITH MODEL PREDICTIONS AND HOLOGRAPHIC MEASUREMENTS

Figure 9 shows an example of forced deformation pattern of the 62 cm mirror where equal excess forces of – 300 gf are applied to the 9 actuators. The left panel shows the pattern actually measured using the present active correction system. A total of 20 contour curves are shown at 39 nm interval. The middle panel shows the corresponding

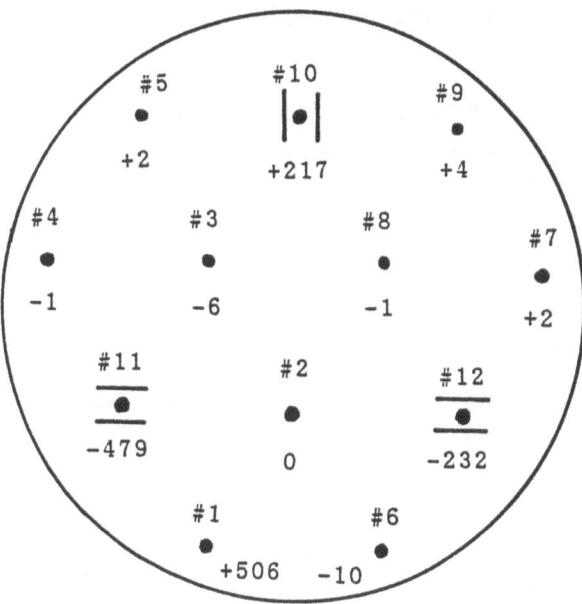

Fig. 8.   Distribution of 9 supporting actuators (Nos. 1–9) and 3 fixed points (Nos. 10–12).

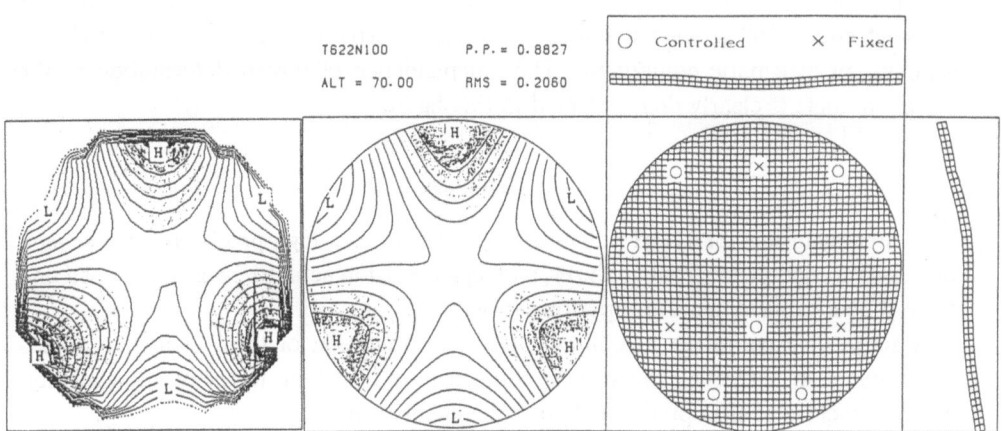

Fig. 9.   Comparison of forced deformation patterns actually measured (left) and theoretically predicted (middle and right).

pattern predicted with a finite element model. The cross-sections of the forcibly bent mirror are supplemented to the right panel in an exaggerated scale. This figure shows that the actual deformation is quantitatively in reasonable agreement with that expected by a model calculation.

A supplemental experiment using a holographic method to measure the forced deformation of the 62 cm mirror was carried out (cf. Itoh *et al.*, 1989). The holographic method was shown to give a slightly larger error than the Shack–Hartmann method in measuring the mirror surface. These two methods, however, gave very consistent results to each other.

3.3. FIGURE CONTROL AT VARYING ELEVATION ANGLE

Finally, it was checked whether the present system is capable of controlling the mirror figure independent of the elevation angle of the system. Figure 10 shows the deformation

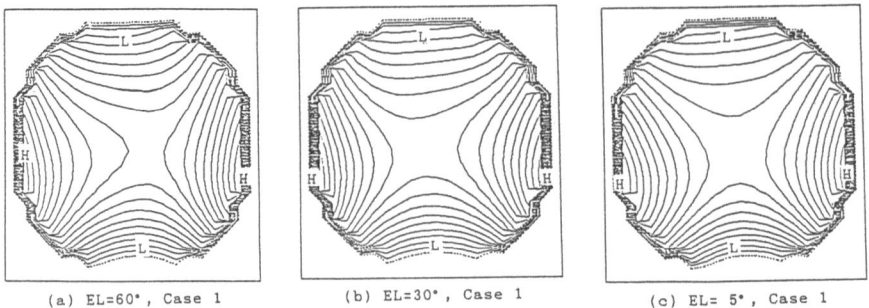

(a) EL=60°, Case 1   (b) EL=30°, Case 1   (c) EL= 5°, Case 1

Fig. 10.   Comparison of forced deformation patterns at various elevation angle of the system.

patterns measured at different elevation angles for the case where excess forces of − 500 gf and − 282 gf were assigned to actuators Nos. 6 and 5, 9, respectively, to generate an astigmatic component. The independence of forced deformation to the elevation angle is clearly demonstrated in this figure.

4. Summary

The three requisites for JNLT active support system, i.e., high precision force sensor, Shack–Hartmann wavefront analyzer, and high precision actuator, were developed and were shown to have necessary performances.

A series of experiments has shown that a combination of these elements gives a performance in correcting the surface configuration of a thin mirror at any elevation angle very close to that predicted by a finite element model.

Acknowledgements

The present paper is a summary of the joint effort for active optics experiments carried out under a collaboration of the JNLT working group of National Astronomical Observatory and the Mitsubishi Electric Corporation supported by a grant-in-aid for special project research by the Ministry of Science, Education, and Culture. The contributions of the following members were indispensable and are acknowledged. H. Ando, N. Itoh, H. Kawakami, I. Mikami, K. Miyawaki, T. Noguchi, Y. Norimoto, N. Oshima, H. Shibasaki, W. Tanaka, M. Tabata, Y. Torii, and Y. Yamashita.

References

Itoh, N., Sasaki, A., Mikami, I., Miyawaki, K., Tabata, M., Ando, H., Iye, M., Nishino, Y., Noguchi, T., and Yamlashita, Y.: 1989, *Publ. Nat. Astron. Obs.* **1**, 57.

Iye, M., Nishino, Y., Noguchi, T., Tanaka, W., Yamashita, Y., Itoh, N., Mikami, I., and Miyawaki, K.: 1989, *Publ. Nat. Astron. Obs.* **1**, 63.

Nishimura, S., Yamashita, Y., Iye, M., Itoh, N., and Mikami, I.: 1988, in M. H. Ulrich (ed.), *Very Large Telescope and their Instrumentation, Proc. of ESO VLT Conference*, p. 577.

Noguchi, T., Iye, M., Kawakami, H., Nakagiri, M., Norimoto, Y., Oshima, N., Shibasaki, H., Tanaka, W., Torii, Y., and Yamashita, Y.: 1989, *Publ. Nat. Astron. Obs.* **1**, 49.

Watanabe, M.: 1987, *Ann. Tokyo Astron. Obs.*, 2nd Ser., Vol. 1, No. 3, p. 241.

SMITH, H. – You have described the static situation, but how do you plan to treat rapidly varying and sometimes strong wind forces on the mirror surface?

IYE – It is necessary to reduce the wind speed perpendicular to the mirror surface to a level less than 1.5 m s^{-1} in the present design. We plan to ensure the reduction of wind speed by appropriately designing the dome opening and the structure around the mirror blank.

SIEGMUND – Does your support system respond to wind forces? Have you considered using position actuators rather than force actuators?

IYE – Current design incorporates three axially fixed points, where we can measure the wind load by force sensors. Although our primary intent is to reduce the wind load by proper protections, the secondary procedure we plan to install is to exert extra force distributed at 390 actuators, or less if possible, to cope with the wind load with a frequency response up to about 1 Hz. We have considered using position actuators but decided not to use them. We plan to use position sensor for monitoring the position of fixed points.

BARR – In order to cope with wind forces on the mirror, I would suggest considereation of hydraulic style support mechanisms. The analysis we have done at NOAO indicates that interconnected hydraulic pistons, arranged in three sectors, are capable of withstanding 10 m s^{-1} wind forces with less than 0.003 arc sec of image degradation. Active forces can still be applied to the mirror with appropriate lever systems.

IYE – Thank you for your comment. The hydraulic system and mechanical system with springs have different stiffness. It's an important point to reconsider.

FAST OPTICS FOR LARGE TELESCOPES*

KYOJI NARIAI

National Astronomical Observatory, Tokyo, Japan

(Received 10 January, 1989)

Abstract. Reports on two optical designs studied in connection with the JNLT project: namely, the primary corrector and the camera for a spectrograph, are presented.

1. Telescope

The diameter of the main mirror of JNLT is 7.5 m. In order that the dome structure and the telescope tube have reasonable sizes both from the technical and from the economical points of view, a small value of 2.0 is required for the F-number. The requirement that the diameter of the secondary mirror does not exceed 1.3 m determines the composite F-number at the Ritchey–Chrétien focus to be less than about 12.5.

The eccentricity of the main mirror is determined to be 1.010667 by the conditions for the Ritchey–Chrétien focus; the composite F-number of 12.5 and its position (B.F. = 3.75 m).

2. Primary Corrector

Wynne designed many primary focus correctors consisting of three or four lenses for paraboloidal and hyperboloidal main mirrors (Wynne, 1967, 1968, 1974, 1979). No aspheric surface was used in these correctors.

Three-lens corrector with one aspheric surface was studied by Faulde and Wilson (1974), Richardson *et al.* (1984), Epps *et al.* (1984), and Cao and Wilson (1984). Probably because of the additional freedom in the optimization introduced by the asphericity, the first lens in these designs has meniscus shape. Therefore, the flexure of the first lens may not be so serious as it was in Wynne's design. Nariai *et al.* (1985, 1987) studied three-lens corrector with two aspheric surfaces (see Figure 1).

In the first phase of these designs, the target for the image quality was between 0.5″ and 1″ which was the averaged seeing size at the Palomar Observatory or at the Kitt Peak. In the 1970's, telescopes of 4 m class were constructed at the sites with better seeing such as Cerro-Tololo, La Silla, and Mauna Kea. So the target changed to between 0.25″ and 0.5″ from the middle of the 1970's. At Mauna Kea, extensive efforts have been made on removing the cause of degradation of seeing, and it was found that the best image quality may be less than 0.2″. Therefore, we aimed at 0.1″ for the image quality given by the optical design.

* Paper presented at the symposium on the JNLT and Related Engineering Developments, Tokyo, November 2–December 2, 1988.

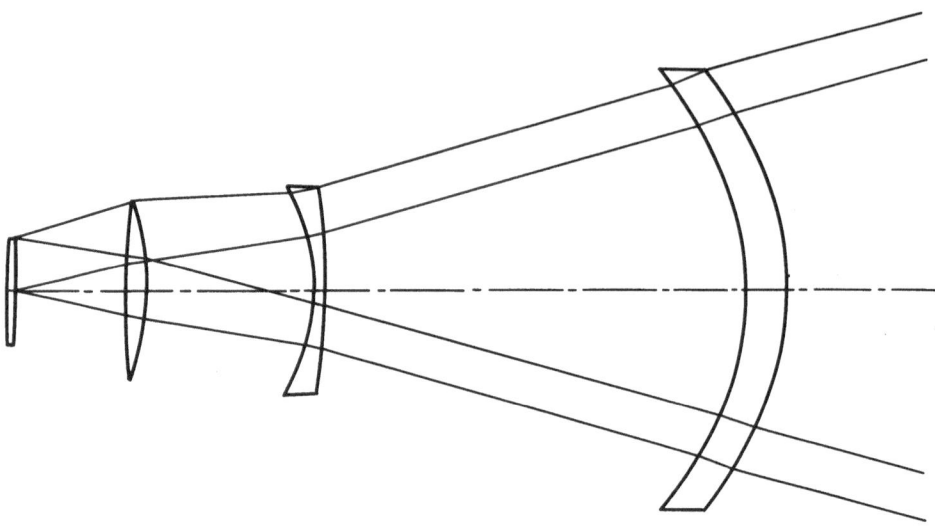

Fig. 1.   Cross-section of the primary corrector.

In our optimization, we use three wavelengths and three angles of incidence. The second-order term of the longitudinal chromatic aberration cannot be removed; so images of red and blue wavelengths are evaluated on a plane on which the image quality in these wavelengths is best. The spherical aberration has linear dependence on wavelength, which cannot be removed. Therefore, the spherical aberrations at the standard wavelength, of both the third- and fifth-order, are kept to be zero. Weights are put to images at 10' and 15' in the standard wavelengths (to correct coma) and to images at 0' for blue and red wavelengths (to correct sphero-chromatic aberration). Ratio of weights of coma to sphero-chromatism is not important as long as the second lens is fixed. Weights for image position are determined so that the results does not have large lateral chromatic aberration. Six radii of curvature in three lenses and the position of the third lenses are used as variables. The positions of the first and the second lenses are given as parameters, and are kept constant during optimization. Aspheric coefficients on the rear surface of the second lens and on the front surface of the third lens are also used.

Figure 2 shows position of the best image, difference of the center of gravity of an image to that in standard wavelength, and image size as functions of the angle of incidence. Results for 2 values of distances between the first and the second lenses, 673 mm and 450 mm, for the position of the first lens at 13.9 m and the radius of field-of-view of 15' are shown in Figures 2(a) and 2(b). It is seen in Figure 2(a) that the position of the best image has quite a steep dependence on the angle of incidence near 15'. This is interpreted as field of curvature of the fourth order. The curvature of field of the third-order aberration (square term) has a value that makes the average field as flat as possible. When the second lens is pushed forward, there is a solution in which

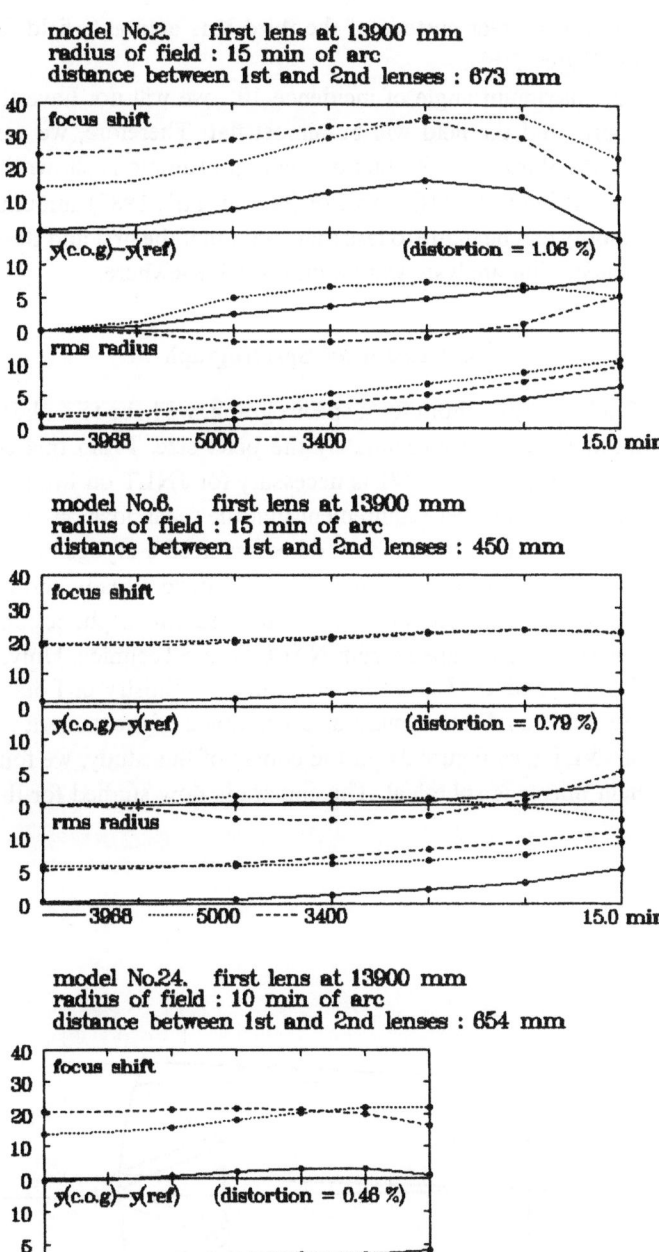

Fig. 2. Characteristics of images of primary corrector for various parameters in the optimization.

the aspheric term on the rear surface of the third lens acts as a field flattener of the fourth-order (see Figure 2(b)).

If we make the maximum angle of incidence 10′, we will not have the steep slope shown in the figure, and the field will be almost flat. Therefore, we can remove the aspheric term on the third lens. Results of such an example is shown in Figure 2(c).

The aspheric terms in our BLUE corrector (Nariai *et al.*, 1987) amounted to 300 μm. The aspheric quantity can be made to less than 100 μm if the first and the second lenses are properly located. Full analysis will be published elsewhere.

3. Camera for Spectrograph

In order to have a large throughput for a spectrograph, we have to make the seeing size projected on the detector is approximately the pixel size. From this condition, it is concluded that a camera of about $F/1$ is necessary for JNLT on Mauna Kea. Such a camera must also satisfy another requirement that the focal position should be located outside of the camera because the modern detectors need a cryogenic system which is considerably larger than the size of a detector. We have already shown that an $F/1$ camera can be designed with one corrector plate and two aspheric mirrors. A team consisting of scientists and engineers from NAO, Tokyo Technical Univ., and Cannon Co., obtained in 1987 and 1988 grant-in-aid from the Ministry of Education, Science and Culture, and made an $F/1.4$ camera as a prototype of the camera to be used in a spectrograph of JNLT (see Figure 3). In the course of this study, we found a solution in which the main mirror is spherical. The camera is now studied for the final check.

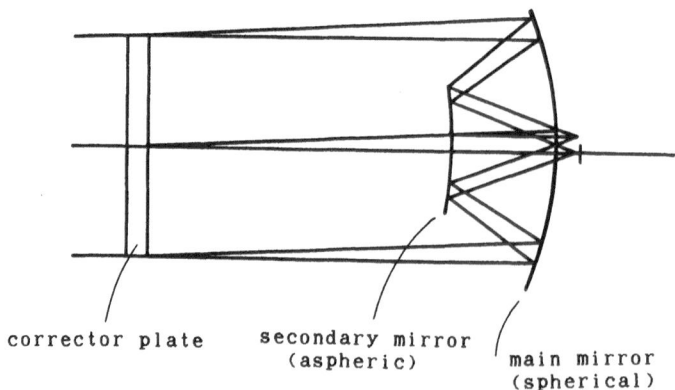

corrector plate   secondary mirror
                    (aspheric)   main mirror
                                      (spherical)

Fig. 3.   Cross-section of $F/1.4$ camera.

References

Cao, C. and Wilson, R. N.: 1984, *Astron. Astrophys.* **133**, 37.

Epps, H. W., Angel, J. R. P., and Anderson, E.: 1984, in M. H. Ulrich and K. Kjar (eds.), 'Very Large Telescopes, their Instrumentation and Programs', *IAU Colloq.* **79**.

Faulde, M. and Wilson, R. N.: 1974, *Astron. Astrophys.* **26**, 11.

Nariai, K. and Yamashita, Y.: 1987, *Publ. Astron. Soc. Japan* **39**, 505.

Nariai, K., Yamashita, Y., and Nakagiri, M.: 1985, *Ann. Tokyo Astron. Obs.* **20**, Ser. II, 431.

Nariai, K., Yamashita, Y., and Nakagiri, M.: 1987, *Ann. Tokyo Astron. Obs.* **21**, Ser. II, 358.

Richardson, E. H., Harmer, C. F. W., and Grundman, W. A.: 1984, *Monthly Notices Roy. Astron. Soc.* **206**, 47.

Wynne, C. G.: 1967, *Appl. Optics* **6**, 1227.

Wynne, C. G.: 1968, *Astrophys. J.* **152**, 675.

Wynne, C. G.: 1974, *Monthly Notices Roy. Astron. Soc.* **167**, 189.

Wynne, C. G.: 1979, *Monthly Notices Roy. Astron. Soc.* **189**, 279.

Wynne, C. G.: 1980, *Monthly Notices Roy. Astron. Soc.* **193**, 7.

GILLINGHAM – For the prime focus corrector have you considered: (a) the design of atmospheric dispersion correctors? (b) the possibility of a field wider than the half degree, as described?

NARIAI – (a) We have made some calculations of ADC of Epp's type and make also of Chinese type (Su Ding-Qiang and Yi Mei-Liang). We have to include ADC in our final design. (b) In order to have a wider field, you may need a larger diameter for the first lens; which may cause a larger longitudinal chromatic aberration of the second-order. For JNLT, the optimum field-of-view seems a little less than a half minute of arc.

HÜGENELL – At your second overhead-picture you showed the design of the primary corrector, with perhaps 70 or 80 cm diameter of the largest corrector lens. It is a great problem to produce those corrector lenses with increased diameter? I ask it because we need perhaps an even larger corrector lens for the ZAS project.

NARIAI – A lens whose diameter exceeds 1 m may be a great problem to a manufacturer. I think you have to limit the field-of-view in order to make the size of the first lens within the limit of fabrication.

MECHANICAL STRUCTURE OF JNLT*

N. ITOH and I. MIKAMI

Mitsubishi Electric Corporation, Amagasaki, Japan

and

Y. YAMASHITA and T. NOGUCHI

National Astronomical Observatory, Tokyo, Japan

(Received 27 February, 1989)

Abstract. The Japanese National Large Telescope (JNLT) requires mechanical performance of high tracking accuracy to achieve good image quality and a mechanical configuration to provide several kinds of focus modes. Under these requirements, a conceptual design for the JNLT mechanical structure has been performed. This paper presents the results of the conceptual design currently under consideration.

1. Introduction

The JNLT concept has features of high image quality and variety of focus modes.

The image quality performance goal is 0.2–0.3 arc sec enclosing 80% of light energy. To attain such a tight performance, 0.1 arc sec r.m.s. or less and 1 arc sec r.m.s. or less are budgeted to the tracking error and the pointing error, respectively. In addition, the minimization of thermal inertia of the tube and the structure will be indispensable to reduce the locally induced degradation of the seeing.

The focus modes of the JNLT are primary, Cassegrain, Nasmyth, and coudé. A mechanical configuration to provide these several kinds of focus mode is required.

Based on these requirements, a conceptual design has been performed for the tube and yoke structure and the drive mechanism.

2. Outline of Proposed Mechanical Structure

An outline of the proposed mechanical structure is shown in Figure 1.

3. Major Characteristics of Mechanical Structure

Major characteristics of the JNLT mechanical structure are as follows:

Focus modes: Primary (opt.), Cassegrain (opt. and IR), Nasmyth (opt. and IR), and coudé (opt. and IR).

Tracking accuracy: 0.1 arc sec r.m.s.

Pointing accuracy: 1 arc sec r.m.s.

---

* Paper presented at the symposium on the JNLT and Related Engineering Developments, Tokyo, November 29–December 2, 1988.

Fig. 1.   Mechanical structure for JNLT.

Locked rotor natural frequency: ≈ 4 Hz.
Bearing: Hydrostatic type for both Alt and AZ.
Total weight: ≈ 300 tons.

4. Tube Structure

The proposed tube structure is shown in Figure 2. It consists of upper truss of 2 stages, hexagonal center section and lower truss. The upper truss has the following features.

4.1. MINIMUM GRAVITATIONAL DISTORTION

The top ring inclination angle can be controlled appropriately, by adjusting the structural member stiffness distribution. The principle of this top ring inclination control is shown in Figure 3. The proposed structure has been optimized on the basis of this principle.

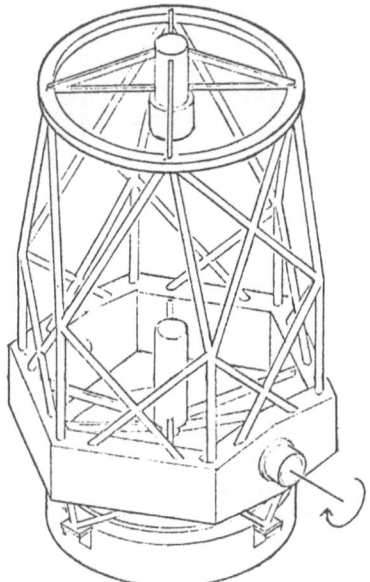

Fig. 2.   Tube structure.

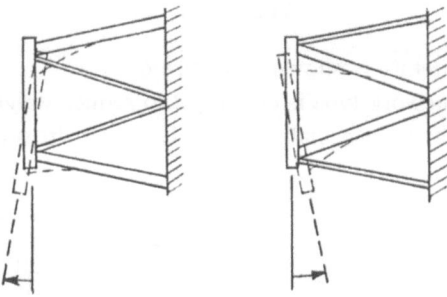

Fig. 3.   Principle of top ring inclination control.

Figure 4 shows a computer analysis result of the optimized structure.

The top ring inclination has been reduced to be 0.6 arc sec by adjusting structural member stiffness distribution of 2nd-stage truss.

4.2. STRUCTURAL CONFIGURATION SUITABLE FOR TOP RING EXCHANGE

The proposed structural configuration saves space required for top ring exchange. The top ring of the proposed structure can be exchanged by moving it in horizontal direction straightforwardly, while that for the conventional truss should be once lifted up as shown in Figure 5.

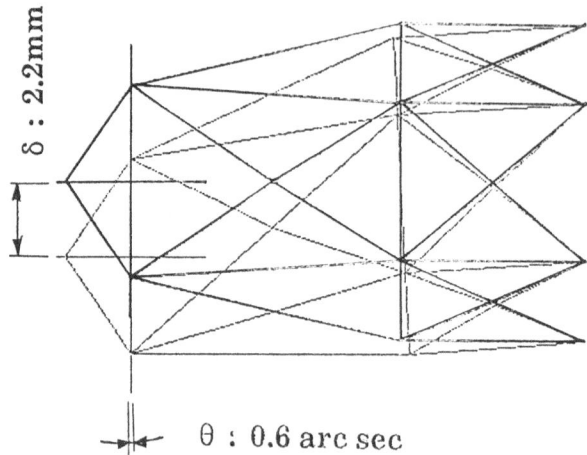

Fig. 4. Computer analysis result of upper truss.

The proposed concept saves also the weight of the structure. The proposed structure is stable by itself during top ring exchange, while the conventional one needs additional structural ring to make the structure stable by itself.

5. Yoke Structure

The proposed yoke structure is shown in Figure 6.

The structure of framework type is employed to reduce weight and thermal inertia. In order to increase stiffness especially along EL axis, hydrostatic pads are located at

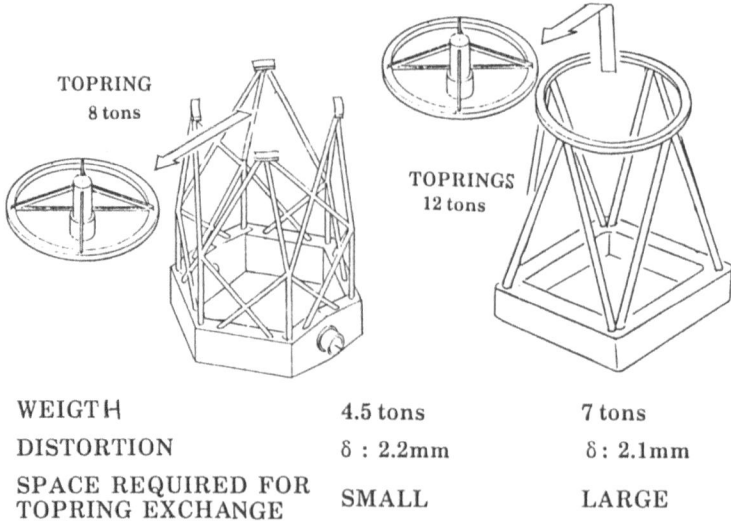

WEIGTH	4.5 tons	7 tons
DISTORTION	δ : 2.2mm	δ : 2.1mm
SPACE REQUIRED FOR TOPRING EXCHANGE	SMALL	LARGE

Fig. 5. Comparison between proposed and conventional structures.

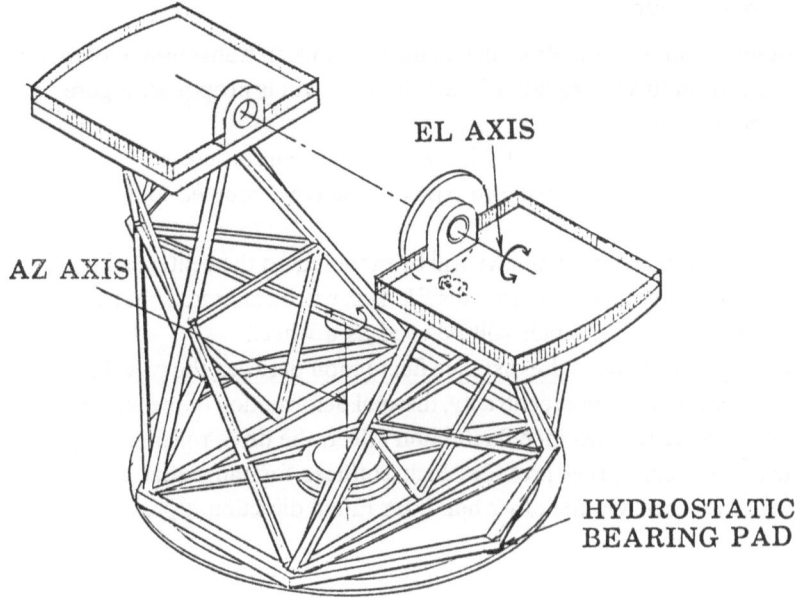

Fig. 6.   Yoke structure.

opposite sides. In total, six pads will be used. To make all the pads same size, the structure will be optimized so that the same loading is imposed to each pad.

6. Drive and Encoder Mechanism

Two types of drive mechanism are studied. One is friction drive and the other is gear drive.

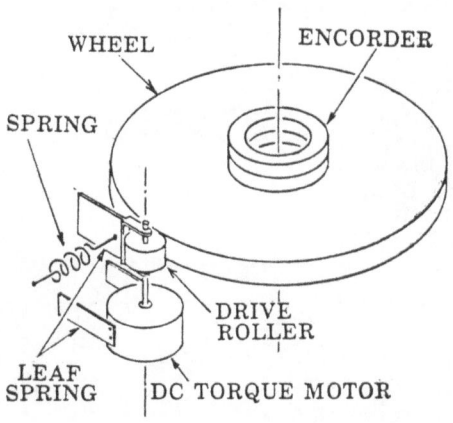

Fig. 7.   Friction drive and encoder mechanism.

6.1. FRICTION DRIVE

The most important point in drive mechanism is smooth transmission of driving torque to final axis. From this viewpoint, friction drive system is proposed. Figure 7 shows the proposed configuration.

It consists of an encoder, a wheel, a drive roller, and a motor.

The encoder of multi-pole synchro type or tape type, coupled to final axis directly is proposed.

The wheel of around 4 m diameter will be mounted on the center section for EL axis. For AZ axis, the wheel similar to that for EL axis, will be mounted on the yoke structure, or the circular track on the pier will be used as a wheel.

In such large mechanism, the distance fluctuation between the wheel and drive roller center will occur because of eccentricity, thermal deformation of the wheel, etc. To make it free from such center distance fluctuation affect, the drive roller is supported with a leaf spring. This leaf spring provides high stiffness for tangential direction to carry driving reaction force. On the other hand, for radial direction it provides low stiffness.

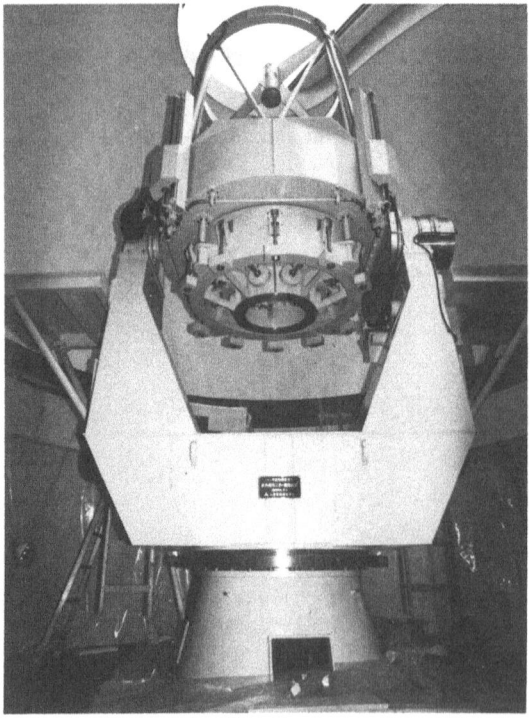

Fig. 8.   1.3 m telescope. For colour reproduction of this figure see colour section.

A spring to apply onstant pre-loading to avoid slippage is provided to the driving roller. If the variation of the driving roller bearing friction due to the pre-loading is proven to affect the servo error, hydrostatic type design will be adopted for it.

A large torque DC motor will be directly coupled to the driving roller shaft, to eliminate gear reduction which causes servo error.

The drive and encoder mechanism of the above configuration has employed for an actual telescope of 1.3 m aperture, installed at the Institute of Space and Astronautical Science. It is now under telescope analysis testing.

Figure 8 shows the telescope and Figure 9 shows details of its AZ drive mechanism. The encoder coupled to final axis directly is of multi-pole synchro type with resolution of 0.3″ (2^{22}). The wheel diameter is 1.4 m and the drive roller 50 mm. The torque motor is of brushless type with maximum torque of 50 Nm.

6.2. GEAR DRIVE

In design of a gear-drive mechanism, angle transmission error due to gear tooth profile error, pitch error and variation of tooth deflection due to anti-backlash opposing torque

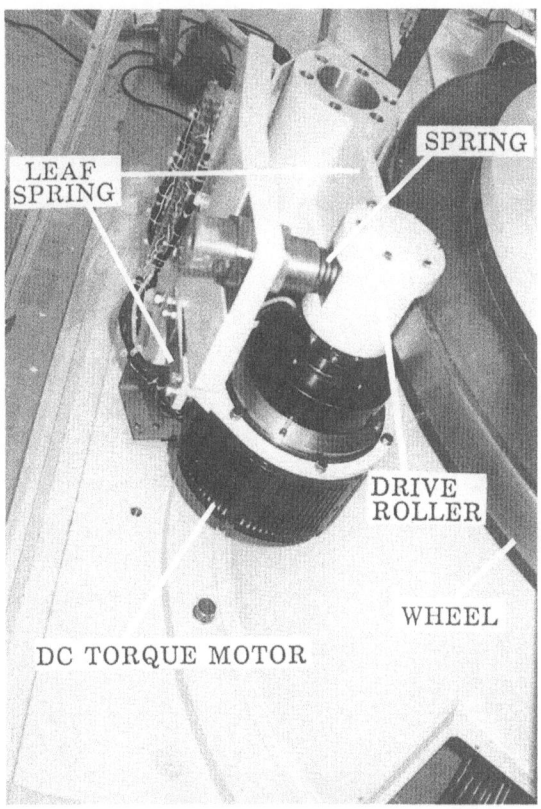

Fig. 9.   AZ drive mechanism. For colour reproduction of this figure see colour section.

should be studied. If these errors residual after compensated by servo system is proven to be so small as compared with the budgeted tracking accuracy, gear-drive system would be employed.

7. Conclusion

We have presented the concept of the JNLT mechanical structure currently under consideration.

DOME DESIGN OF JNLT*

I. MIKAMI, N. ITOH, K. MIYAWAKI

Mitsubishi Electric Corporation, Amagasaki, Japan

and

H. ANDO, T. NOGUCHI, M. NAKAGIRI, and E. WATANABE

National Astronomical Observatory, Tokyo, Japan

(Received 27 February, 1989)

Abstract. In recent years, it has become increasingly clear that the dome is often itself a major source of seeing degradation, due to its adverse thermal properties. In order to enable the Japanese National Large Telescope (JNLT) to attain the highest image quality as a ground-based telescope, 0.1″ r.m.s. on 80% light energy diameter basis is budgeted to the dome-induced seeing degradation as a design goal. Feasibility of this goal has been studied, using the latest design concept of the dome, in the following manner.

- Model the JNLT dome on the latest design plan for dynamic thermal analysis.
- Perform the dynamic thermal analysis to attain temperature distribution in the dome.
- Predict the seeing degradation induced by the dome using the thermal analysis result.

Although this study is just on an elementary level, the result indicates that our design goal is attainable on an extension of the current design concept and the dome cooling by thermal emission is a key factor we have to overcome to minimize the dome-induced seeing degradation.

1. Introduction

The Japanese National Large Telescope (JNLT) will use a large dome to isolate the telescope from hostile surroundings on Mauna Kea (see Figure 1). Initially, a semi-spherical dome mounted on a tall cylindrical building which contains all the operating facilities and the personnel areas was preferred by the JNLT working group. This concept, however, appeared to be mainly based on the microthermal effect of the boundary layer but to be little based on the locally-induced degradation of the seeing. In order to minimize the seeing degradation by both of the microthermal effect and the dome, a new concept comprising a dome with no major source of heat dissipation and a separate control building completely isolated in a sense of thermal effect is now under study for the final decision. The layout of the new concept, bases of the design, analytical modeling of the dome to estimate the seeing degradation and its result are mainly presented.

---

* Paper presented at the symposium on the JNLT and Related Engineering Developments, Tokyo, November 29–December 2, 1988.

Astrophysics and Space Science **160**: 173–181, 1989.
© 1989 *Kluwer Academic Publishers.*

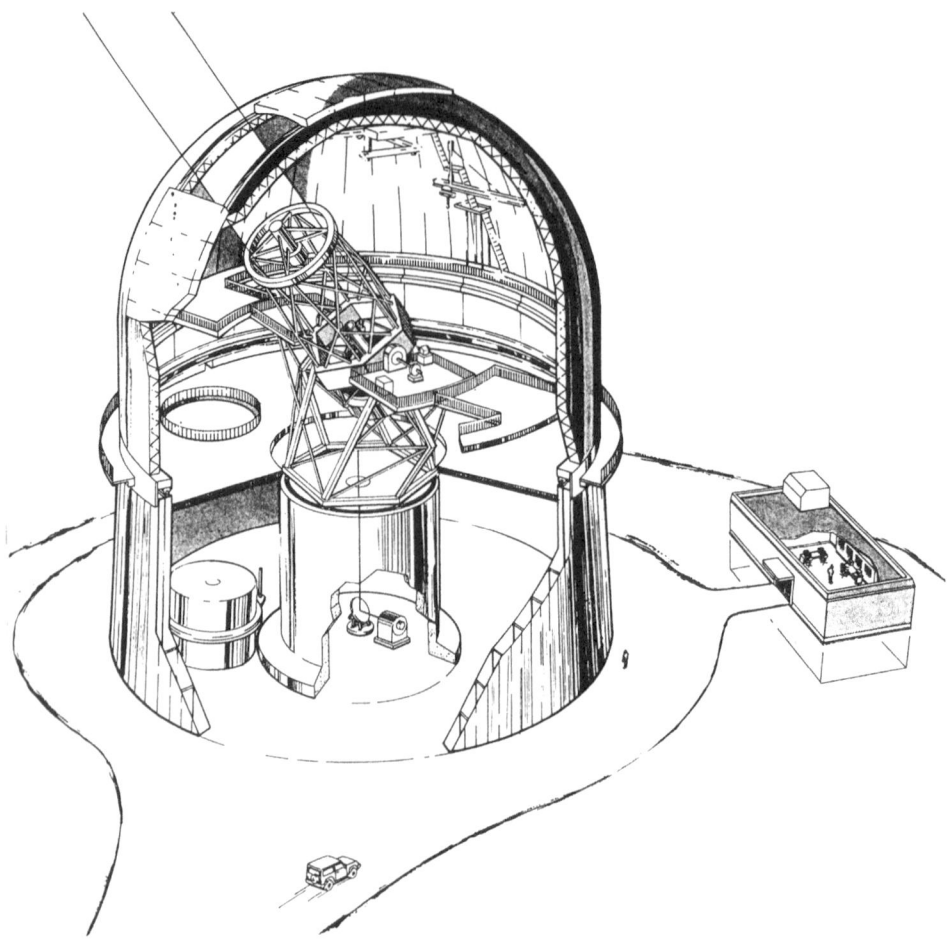

Fig. 1.   The latest design concept of JNLT.

2. Design Concept for Dome and Enclosure

2.1. INITIAL DESIGN CONCEPT

Figure 2 shows the initial design concept of the dome and enclosure [1]. This concept was established by the JNLT Working Group in 1983. It comprises a semi-spherical dome of approximately 40 m in diameter, a telescope pier mounting the telescope elevation axis at 27 m high from the ground level and a cylindrical building containing control facilities and personnel area. The telescope mounting height was determined through actual CT^2 (seeing indicator) measurement at the site so that the seeing degradation by microthermal turbulence can be lowered on the order of 0.6″ FWHM. This plan, however, seems to have little advantage in a sense of the dome seeing, because a large quantity of heat will be transferred from the building to the dome internal area, and not enough ventilation is provided in this concept.

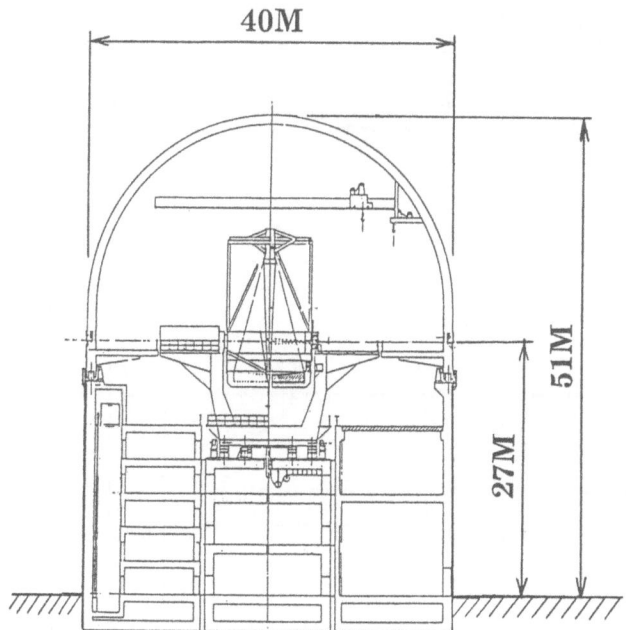

Fig. 2.   Initial design plan of JNLT dome and enclosure.

2.2. LATEST DESIGN CONCEPT

To improve the seeing performance of the dome and enclosure, other design plans have been studied. Figure 3 shows the latest (but not final) design concept of the dome and the enclosure. It comprises a semi-spherical dome mounted by a substructure of dome, a telescope pier and a building separated from the dome area.

In this figure, arrows show ventilation air flow to let the dome temperature track the ambient temperature. Design features of this concept are as follows.

– The separate building concentrating control facilities and personnel area and the dome substructure housing no major heat source will minimize the thermal effect to the dome and the telescope, and can simplify the thermal analysis greatly.

– The ventilation system will minimize temperature range distributed over the dome area. The direct ventilation of the observation floor will particularly give perfect separation in thermal sense between the dome and the substructure.

– The telescope mounting height unchanged, 27 m as the elevation axis height, will give only 0.06″ (FWHM) image degradation due to the microthermal turbulence, the value of which is estimated through the CT^2 measurement.

3. Analytical Estimation of Dome-Induced Seeing Degradation

The seeing degradation induced by the dome can be estimated analytically by
- dynamic thermal analysis to attain temperature distribution over the dome, and;
- conversion of the thermal analysis result to the seeing degradation.

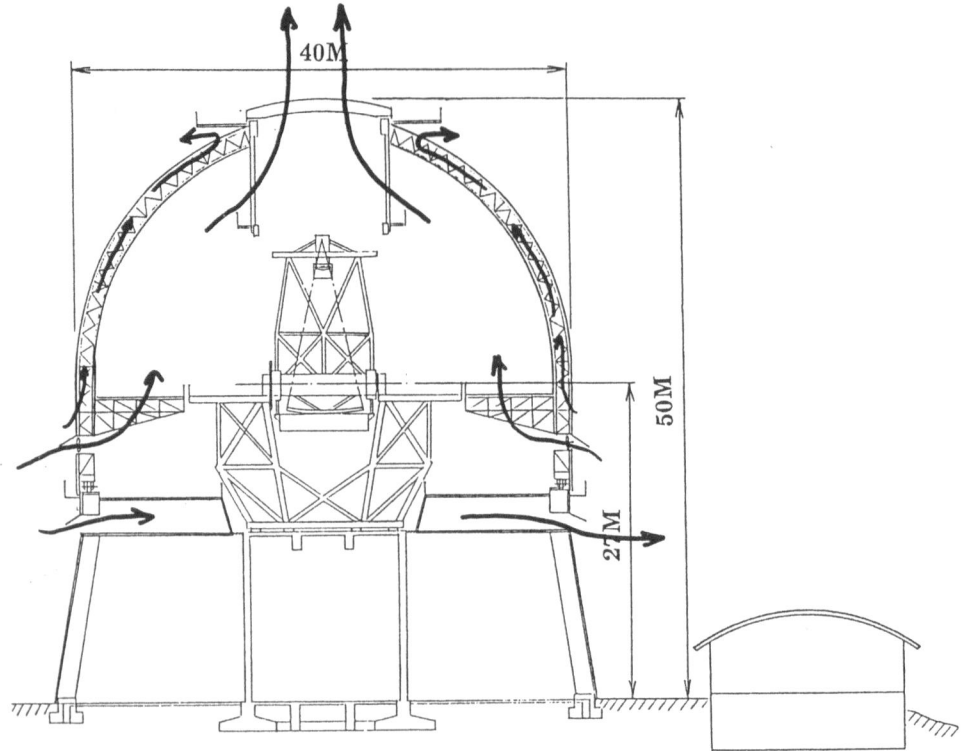

Fig. 3.   Latest dome design plan.

3.1. DYNAMIC THERMAL ANALYSIS

3.1.1. *Modeling – Daytime*

Figure 4 shows concept of daytime operation of the dome and labeling of every element of heat exchange. In the daytime, the observation slit and all the vents are shut and we will depend on the insulator, which is attached inside of the dome with the air gap, to

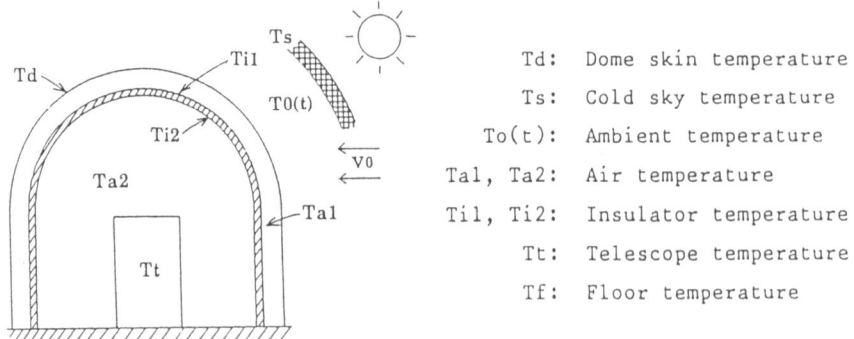

Fig. 4.   Dome modeling for daytime.

keep the internal temperature (T_{i2}, T_{a2}, T_t, and T_f) close to that of the end of observation.

Figure 5 shows simplified calculation model for daytime. Each of the elements is modeled as one mass having an averaged temperature for simplification purpose. As an

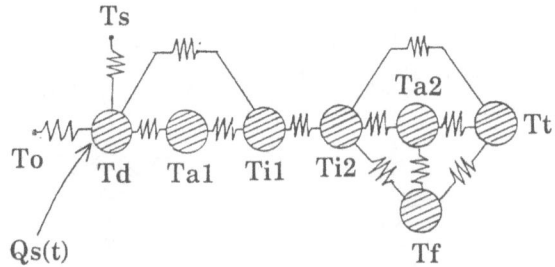

Fig. 5.   Simplified calculation model for daytime.

example, the dome skin is modeled as a mass (M_d) having an averaged temperature (T_d). And all masses are connected by thermal resistances (R_j), such as heat transfer, emission and/or conduction, each, respectively. The solar energy ($Q_s(t)$) is modeled to give heat calorie by emission to the dome outer skin and as a function of altitude of the Sun. The cold sky is modeled as an emission cooling object having a constant temperature of $-20\,°C$. The wind velocity (V_0) is modeled as 7.5 m s^{-1} constant. And as the ambient temperature, typical temperature data [2] measured in summer of 1966 are used as a function of time.

3.1.2. Modeling – Nighttime

Figure 6 shows concept of nighttime operation of the dome and labeling of every element of heat exchange. In the nighttime, all the elements except the Sun can be modeled with the same principle as that of daytime, but only the thermal resistances shall be replaced by those made by the ventilation. And as the telescope heat dissipation (Q), 1.5 kW is preliminary assumed to appear as a step function at sunset or the start of observation

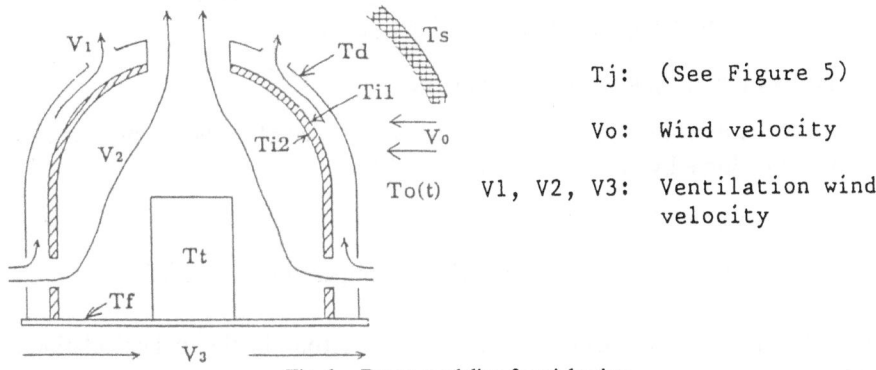

Tj: (See Figure 5)

Vo: Wind velocity

V1, V2, V3:   Ventilation wind velocity

Fig. 6.   Dome modeling for night-time.

and to disappear at sunrise or the end of observation. This calculation model is shown in Figure 7.

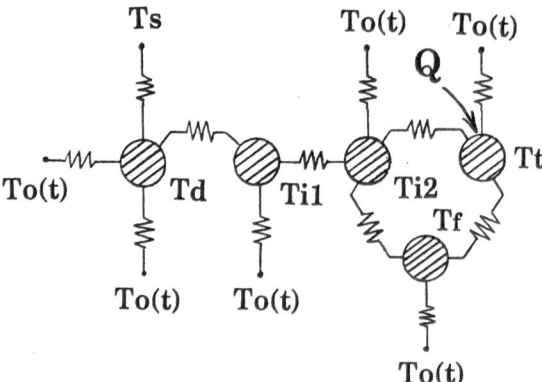

Fig. 7.   Simplified calculation model for night-time.

3.1.3. *Fundamental Equation*

Dynamic response of each element in the daytime and night-time model can be obtained by resolving the following simultaneous differential equations

$$Cp_j W_j \frac{dT_j}{dt} = Q_j,$$

where Cp_j, specific heat of j element; W_j, mass of j element; T_j, temperature of j element; Q_j, heat calorie supplied to j element.

3.1.4. *Calculation Result*

Figures 8 and 9 show result of the dynamic analysis for the dome of bare aluminum skin and for the dome of steel skin with TiO_2 paint each, respectively. The aluminum skin is chosen as a typical example of material having high absorptivity of the solar energy and low emissivity, and the steel skin with TiO_2 is chosen as a typical example of material having completely opposite character to aluminum.

3.2. CONVERSION TO SEEING DEGRADATION

The result of the dynamic thermal analysis can be converted to the seeing degradation induced by the dome by applying the equation [3]

$$S \sim 0.5 \times \Delta T_{io},$$

where S, star image diameter (arc sec); ΔT_{io}, air temperature difference between inside and outside dome (°C).

   Figure 10 shows the seeing degradation converted from the thermal calculation as a result of an aluminum skin dome.

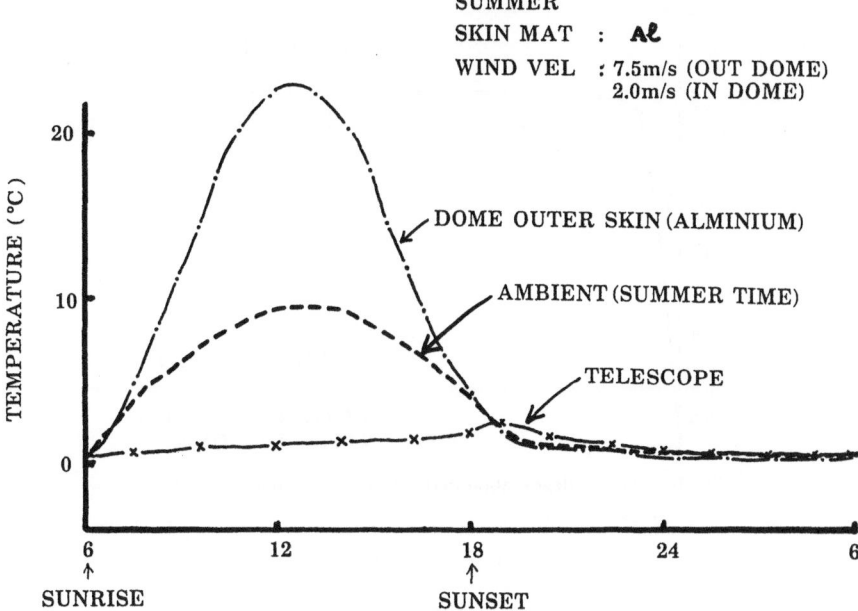

Fig. 8. Calculated result for Al skin dome.

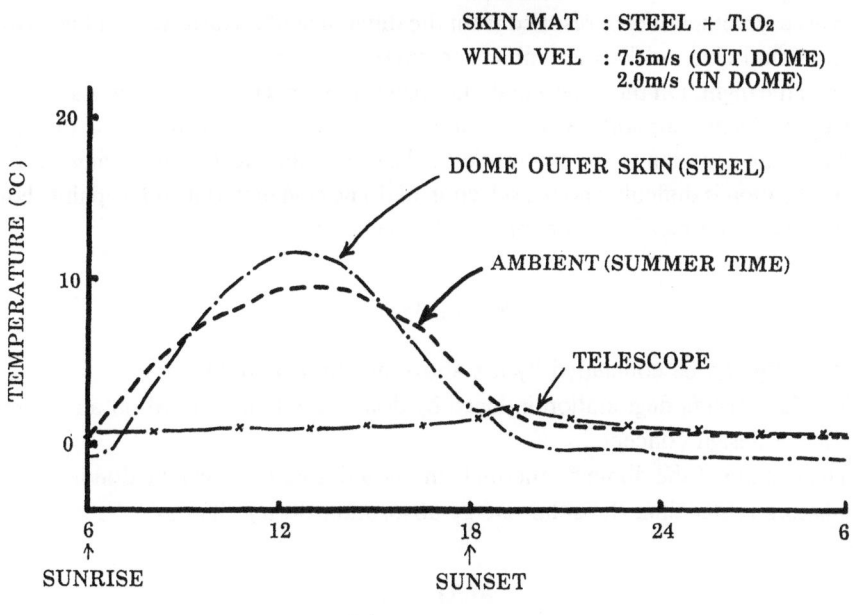

LOCAL TIME (HOUR)

Fig. 9. Calculated result for steel + TiO₂ skin dome.

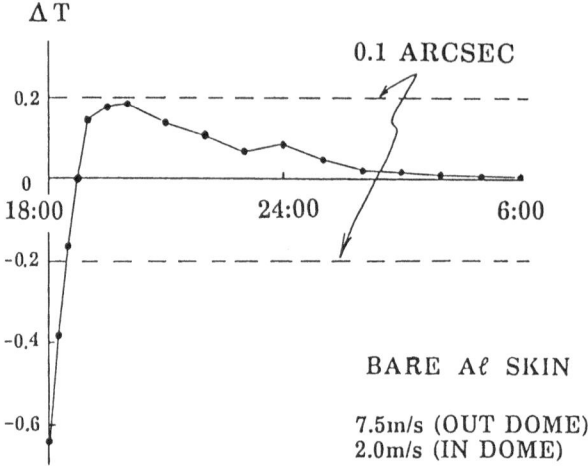

Fig. 10.   Seeing degradation derived from a thermal calculation.

4. Discussion

The telescope temperature response and the internal air temperature response of the aluminum skin dome shows good agreement with those of the steel skin with TiO_2 paint. This predicts that they will track the ambient temperature change fairly well with little regard to the skin material. In this respect, one can see no major difference in the seeing performance between the two cases during the observation.

The temperature response of the dome skin itself, however, shows great difference in two cases each other, which resulted from the difference of absorptivity and emissivity. The aluminum skin tracks the ambient temperature with difference within $\sim 0.2\,°C$ almost all the night. On the other hand, the steel skin with TiO_2 paint shows about $2\,°C$ below the ambient temperature almost all night. This can predict that the cooled dome skin due to the emission to the sky would be the dominant effect to the seeing, although its quantification is difficult, and the selection of dome skin material and its paint should, therefore, be made carefully to minimize the seeing degradation.

5. Conclusion

The following can be concluded by the above-mentioned study.
  – The 0.1″ seeing degradation induced by dome could be attained on an extension of the latest design concept.
  – The cooling of the dome by thermal emission during night will be dominant effect to the seeing rather than the solar energy absorption during the day.

References

[1] Tokyo Astronomical Observatory JNLT Working Group: 1984–1985, *Japanese National Large Telescope-Feasibility Study Interim Report*, pp. 7-1–7-36.

[2] Tokyo Astronomical Observatory JNLT Working Group: 1984–1985, *Japanese National Large Telescope-Feasibility Study Interim Report*, pp. 9–18.
[3] Gillingham, P. R.: 1983, *Proc. SPIE* **444**, 165.

WONG – Would you like to comment on the use of the oversized dome?

MIKAMI – This diameter of the dome is determined to accommodate the Al plant in the substructure (the mirror in the cell shall be lifted down to the plant through the observation floor). But I think this size has an advantage in terms of wind buffeting to the thin meniscus mirror, until we can find its effect to be negligible small quantitatively or until the other measure to eliminate the vibration by mirror support control is established through the study we are carrying out.

ENARD – It is not easdy to determine the effect of wind buffets in the mirror. Nevertheless it is easy to measure the average wind pressure of the mirror level on existing large domes. I could guess that the wind speed is much below 2 m s^{-1}.

MIKAMI – I hope so too. Regarding this aspect, JNLT working group is still studying to establish the method to control the flexible mirror with accuracy even against the wind attack, as an example, using active damping, passive damping, etc. Also, we've just finished the pressure measurement on CFHT mirror surface to utilize in our further study of JNLT.

BARR – What emissivity did you assume for the telescope in your analysis?

MIKAMI – I have no exact number at this moment. But we assumed the normal paint on the telescope, which I think would be 0.4–0.6 as emissivity to the sky.

HALL – (1) Why did you decide to exhaust air out of the slit of JNLT? (2) Have you considered using existing domes on Mauna Kea such as the aluminum skin UKIRT and the TiO painted steel domes to the test your thermal model prediction?

MIKAMI – (1) The Mauna Kea is a dusty place. So, if inhaling the air downwards, it may cover the primary mirror with dust. Therefore, I prefer this direction. (2) Thanks for the comment. The JNLT working group will ask such a test, subject to your approval, in the later stage.

McLAREN – At CFHT, the major heat sources in the dome are such things as hydrostatic bearings, dome drive, and electromechanical devices and not leakage from the working areas below in the building. Do you have a figure in mind for the permissable dissipated power in the dome?

MIKAMI – In the analysis, we tentatively assumed 1.5 kW as the dissipated power in the dome internal area. This value is a subject to confirm through further design of each drive mechanism of the telescope, and would be revised if necessary. At this moment, I have no value of the dissipated power as the permissible limit.

SMITH, H. – (a) If you exhaust air heat upwards out of the dome, could this cause any problem to the KECK telescope only 150 m distant when the wind blows in that direction? (b) How much heat is liberated by the many actuator motors under the mirror?

MIKAMI – (a) I have to say 'sorry', in this case. We are also trying to minimize the heat dissipation from the dome/telescope, which will help to reduce this sort of worry. (b) In summary, the total power required for the actuator motors, we tentatively assume the value tobe 1.3–1.5 kW. However, 10–20% of this power could be dissipated as heat during observation, if we do not use insulators to cover the mirror cell.

GILLINGHAM – A comment on the relative merits of upwards and downwards ventilation: in about 1980 we upgraded the AAT ventilation to give about 20 changes of air/hour. For the first year, we ran the ventilation upwards and dowards on alternate nights. As we expected, we found a small but statistically significant advantage in *upward* ventilation (out the observing aperture). We have since run the ventilation always in this direction, with the additional advantage that we can filter the air supply.

MIKAMI – Thanks for your comment.

SIEGMUND – It will require very large vent area to achieve an airspeed of 2 m s^{-1} inside the dome. Is that what you plan?

MIKAMI – I preliminarily plan to locate the normal-door-sized vents around the dome (probably at every 20 deg around the dome axis), having ventilation fans. The fans rotating speed, the number of vents to open, the area of vents will be controlled depending on the wind velocity outside to get 2 m s^{-1} inside the dome.

MACK – Are you refrigerating the oil for the hydrostatic bearings?

MIKAMI – Yes, we plan to refrigerate the oil for the bearings.

EVALUATION OF THE JNLT SITE*

HIROYASU ANDO, TAKESHI NOGUCHI, MASAO NAKAGIRI,
AKIHIKO MIYASHITA, YASUMASA YAMASHITA, KYOJI NARIAI, and
HIROYOSHI TANABE

National Astronomical Observatory, Tokyo, Japan

(Received 18 January, 1989)

Abstract. Two steps have been taken to decide at what place and altitude to set up the JNLT on Mauna Kea. First, the wind tunnel experiment has been made in collaboration with the Institute of Meteorology using the two models of summit area with the reduced scales of 1/1000 and 1/5000. This study tells us that the north–west cone is suitable for JNLT. Secondly, we have done the measurement of the microthermal activities in this area with a 30 m tower, which was continued for about 4 months in collaboration with the University of Hawaii. This experiment has given the mean vertical profile of C_T^2 over 4 months and its scale height in the boundayr layer on our site. By use of these measurements, the contribution of the boundary layer to seeing is estimated. The behaviour of C_T^2 under strong winds can be explained very well by topographic effects, which is in fairly good agreement with the results of our wind tunnel experiment.

1. Introduction

The astronomical seeing is composed of three factors; free atmosphere, boundary layer, and dome. The condition of free atmosphere is closely related to the global climate, and an important factor to select the observing site globally. For example, Mauna Kea in Hawaii, La Silla in Chile, La Palma in Canary, and Mount Graham in Arizona are acknowledged as the best observing sites in the world.

The boundary layer is greatly affected by local topography and a factor to select the observing site locally. This will be discussed here.

The thermal activities in the dome are now widely known to be a crucial factor for seeing. But this point will be discussed elsewhere in this proceedings (see Mikami *et al.*).

2. Wind Tunnel Experiment

The purpose of this experiment is to search for the optimum place for JNLT from the viewpoint of topographic turbulence in the boundary layer. This has been done using the two models of summit area on Mauna Kea with the reduced scales of 1/1000 and 1/5000. Wind speed in the tunnel is 3 m s^{-1} which corresponds to 30 m s^{-1} for the former model, and 50 m s^{-1} for the latter. The strength of turbulence is measured as standard deviation of wind velocity at each point.

Figure 1 shows the horizontal distribution of the topographic turbulence at 4193 m level in the easterly winds. The points with Nos. 1 to 3 indicate the candidate sites for JNLT. Site No. 1 is on the north–west cone just 150 m west of KECK dome. The 60 cm

---

* Paper presented at the Symposium on the JNLT and Related Engineering Developments, Tokyo, November 29–December 2, 1988.

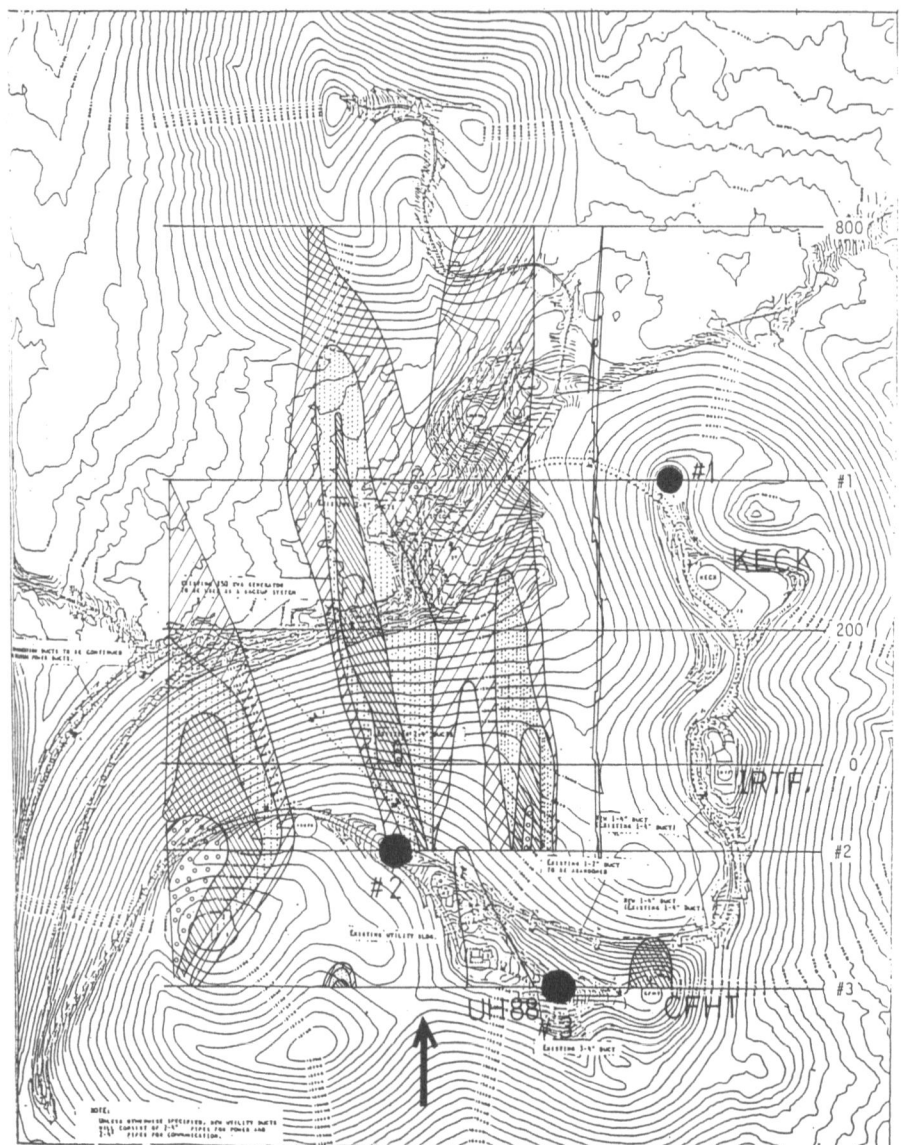

Fig. 1. Horizontal distribution of topographic turbulence at 4193 m level on Mauna Kea in the easterly winds, which was obtained by the wind tunnel experiment. The points with Nos. 1 to 3 indicate the candidate sites for JNLT.

telescope is at site No. 2. Site No. 3 is reserved for the 8 m telescope of the US. As easily seen from this figure, the turbulence is generated by the summit ridge and summit itself. Site No. 1 is not affected by this turbulence. But site No. 2 suffers from the turbulence from the summit.

In the south–east winds, turbulence from the summit ridge directly flows over site No. 1 (the worst case for site No. 1). Figure 2 shows the vertical structure of turbulence

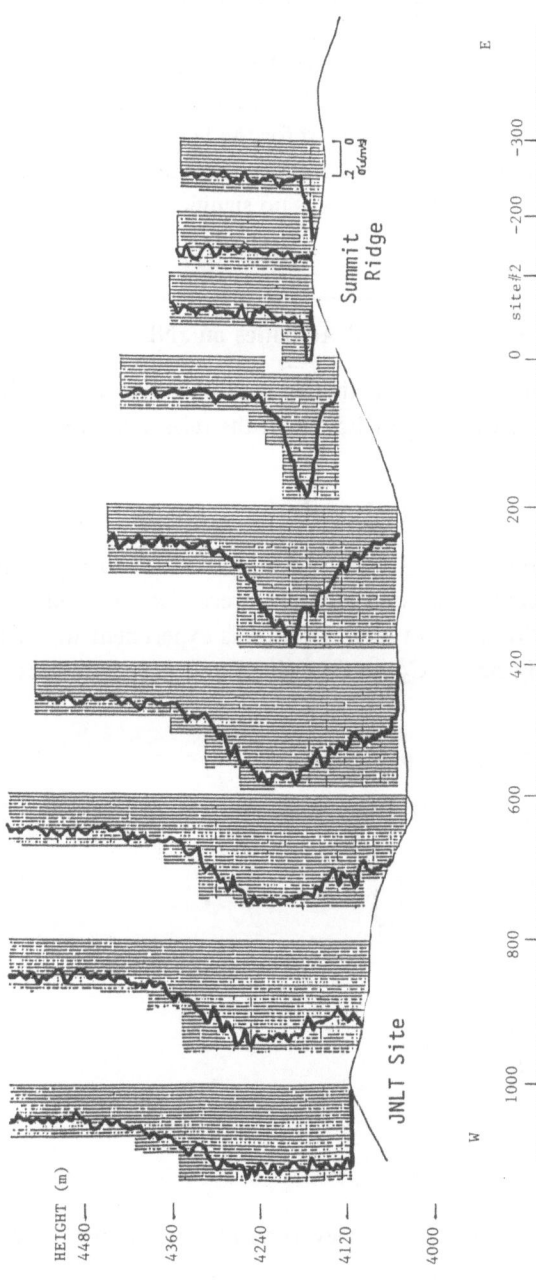

Fig. 2. Vertical evolution of topographic turbulence generated by the summit ridge (UH 88, CFHT, UKIRT). Turbulence extends up to 200 m high on JNLT site in the south–east winds.

along a line combining site No. 1 and summit ridge. This demonstrates the evolution of turbulence generated at the summit ridge. At the ridge, a narrow jet is created, and it expands vertically in the downflow. At site No. 1, turbulent region is extended about 200 m high.

In summary, topographic turbulence correlates well with the summit ridge and the nearby cinder cones under the strong winds. From the experiment, site No. 3 is found to be the best site, but it is already reserved for the US. We have chosen site No. 1 for JNLT. Site No. 1 is as good as site No. 3 except for the case in the south–east winds. We have also found that KECK dome gives no significant effect if the two domes have the same height.

3. Microthermal Activities on JNLT Site

The measurement of microthermal activities on JNLT site was conducted in August to November 1987 in order to know how high the telescope tube should be lifted up. Seeing indicator C_T^2 is defined by

$$C_T^2 = \langle [T(x + r) - T(x)]^2 \rangle / r^{2/3} \quad (r = 1 \text{ m here}). \tag{1}$$

This parameter was monitored at 3 levels (6.5 m, 13 m, 27 m) using 30 m meteorological tower. Atmospheric temperature and humidity were also sampled at 6.5 m level, and wind speed and direction at 30 m level. From this experiment we wish to give (1) the mean relation of seeing size θ, C_T^2, and height z, and (2) behaviours of C_T^2 under the strong winds.

3.1. SEEING SIZE θ, C_T^2, AND HEIGHT z

Seeing size θ(FWHM) and $C_T^2(z)$ are connected by

$$\theta = 5.3 \lambda^{-1/5} (\sec z)^{3/5} \left(\int_z^\infty C_T^2(z) \, dz \right)^{3/5}, \tag{2}$$

where

$$C_n^2 = \frac{7.9 \times 10^{-5} P}{T^2} C_T^2 \tag{3}$$

and

$$C_n^2 = \langle [n(x + r) - n(x)]^2 \rangle / r^{2/3} .$$

The observed mean value of C_T^2 at each level can give a relation of exponential type with regard to the height z, as shown in Figure 3; that is,

$$C_T^2(z) = C_T^2(z_0) \exp[-(z - z_0)/h], \tag{4}$$

where h is the scale height (16.4 m) of C_T^2.

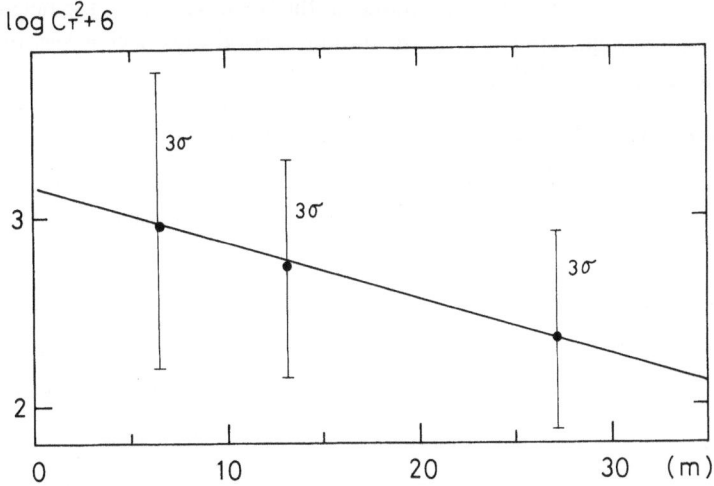

Fig. 3. $C_T^2(z)$-height relation derived from the measurement of C_T^2 at 3 levels (6.5 m, 13 m, 27 m). The vertical bar shows the 3σ level of measurement at each height.

Then, Equation (2) is reduced to

$$\theta(z_0) = 5.3\lambda^{-1/5}\left(\frac{7.9 \times 10^{-5}P}{T^2}\right)^{6/5}[C_T^2(z_0)]^{3/5}h^{3/5}\,. \tag{5}$$

This gives the relation of seeing size θ and C_T^2. With $\lambda = 550$ nm, $T = 270$ K, and $P = 600$ mb (typical values at Mauna Kea), this relation is illustrated in Figure 4. Combination of the relations $C_T^2 - z$ and $\theta - C_T^2$ leads to an interesting result. If we

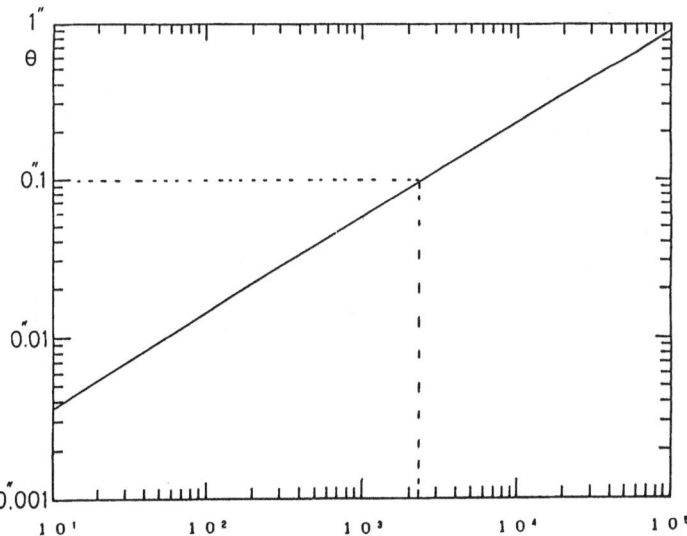

Fig. 4.   Relation of seeing size θ(FWHM) and C_T^2.

want to take seeing less than $0\rlap{.}''1$ on average in the boundary layer, the height of 6.5 m for telescope is enough. However, if we should keep always seeing less than $0\rlap{.}''1$, it should be lifted up to the height of 27 m.

3.2. BEHAVIOURS OF C_T^2 UNDER THE STRONG WINDS

Now we show an interesting correlation between seeing in the boundary layer and topography. As pointed out before, $\int_{6.5}^{27} C_T^2 \, dz$ is a good indicator of seeing in the

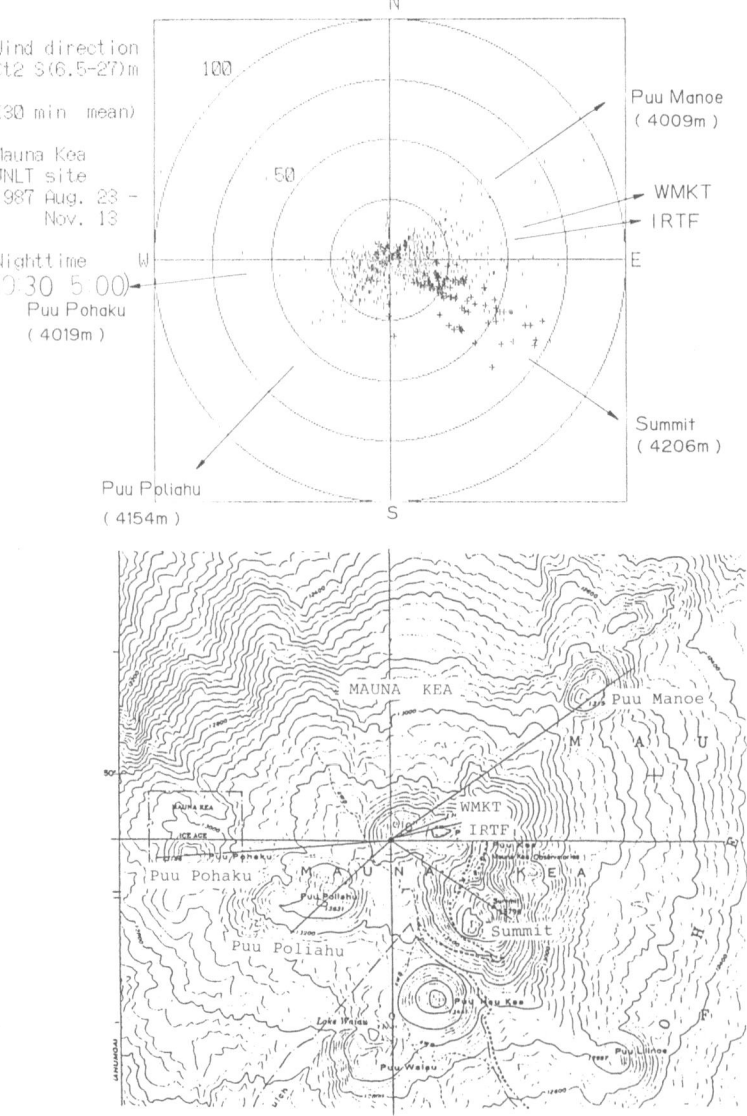

Fig. 5. The integrated values $\int_{6.5}^{27} C_T^2 \, dz$ as a seeing indicator and wind direction. The effects of topography (summit, its ridge, and larger cinder cones) on the indicator are clearly seen.

boundary layer. Figure 5 shows such a relation. The distance of each point indicates magnitude of this integrated value of C_T^2, and its direction corresponds to the wind direction.

The large integrated C_T^2 values are shown to be created in the particular directions, which closely relate to the summit ridge, summit itself, and the larger cinder cones. Particularly, this tendency is conspicuous under the strong winds. Thus, this fact reminds us of the results of our wind tunnel experiments.

Acknowledgements

We acknowledge Dr T. Fujitani for conducting the wind tunnel experiment of local meteorology at Mauna Kea, which has been done as a collaborative work with Institute of Meteorology. We are grateful to Drs D. Hall, A. Erasmus, Mr B. Barnes, and UH 88 supporting staff for their kind constant support and guidance, which have led to the successful fulfillment of our site test at Mauna Kea.

TELESCOPE ENGINEERING AND
FABRICATION OF LARGE MIRRORS

SPIN-CAST ZERODUR MIRROR SUBSTRATES OF THE 8 M CLASS AND LIGHTWEIGHT SUBSTRATES FOR SECONDARY MIRRORS*

HARTMUT W. HOENESS, RUDOLF MUELLER, ERICH W. RODECK and
FRIEDRICH B. SIEBERS

*SCHOTT Glaswerke, Mainz, F.R.G.***

(Received 10 January, 1989)

Abstract. For about 20 years SCHOTT has been supplying the glass ceramic ZERODUR, a material with very low thermal expansion. Besides many other applications, ZERODUR is excellently suited for the manufacture of mirror substrates for telescopes. About 80% of all telescopes in the western world with mirror diameters > 1.8 m have been equipped with ZERODUR during the last 10 years. The development of modern astronomical telescopes is aimed at larger primary mirrors and lighter secondary mirrors.

New techniques have been developed by SCHOTT for manufacture of thin monolithic mirror blanks of more than 8 m in diameter. The development of thin meniscus shaped shells using the spin-casting technique was successfully completed last year. During a test production several mirror substrates up to 4.1 m in diameter and down to 57 mm in thickness could be produced. The know-how has been acquired for the fabrication of mirror substrates of more than 8 m in diameter by the spin-casting technique.

SCHOTT has also performed considerable developmental work in the field of lightweighted ZERODUR mirror substrates which can be generated using different techniques: forming of the lightweighted structure during casting, fusion of individual components to a total structure and lightweighting of a massive block by various mechanical machining methods.

1. The Material ZERODUR

ZERODUR* is a glass ceramic with about 70% crystal phase (high quartz mixed crystal) and approximately 30% residual glassy phase. The thermal expansion of the crystal phase is negative, whereas it is positive for the glassy phase. The relations of the phase composition can be controlled by defined heat treatment such that the resulting coefficient of thermal expansion (CTE) is almost zero. The typical specification is CTE $= 0 \pm 0.05 \times 10^{-6}\,\mathrm{K}^{-1}$ (0 to 50 °C).

Table I demonstrates the low thermal expansion of ZERODUR compared to other substrate materials.

Therefore, ZERODUR is used in applications with extraordinary precision requirements in particular if temperature changes or temperature gradients within the workpiece have to be taken into consideration. ZERODUR has a good transmittance as a result of the small crystal size (approximately 50 nm). ZERODUR can be easily

---

* Paper presented at the Symposium on the JNLT and Related Engineering Developments, Tokyo, November 29–December 2, 1988.

** Registered trademark of SCHOTT Glaswerke, Mainz.

TABLE I

Comparison of different mirror materials

Ø 3.6 m mirror	ZERODUR*	Borosilicate glass	Aluminium
Change of diameter when changing the temperature from 0 to 50 °C	0.5 µm	500 µm	4000 µm

* Based on CTE = $-0.0026 \times 10^{-6}\,K^{-1}$ (mean value) of the 3.6 m mirror substrate delivered for the New Technology Telescope of the European Southern Observatory (ESO NTT).

machined with equipment and tools commonly used in the optical and precision instrument industry; it can be polished to the highest precision since the chemical properties and the hardness of both phases are very similar. Therefore, ZERODUR is an excellent material for mirror substrates of astronomical telescopes.

2. New Requirements Coming from Astronomy

The development of modern astronomical telescopes is aimed at larger primary mirrors (8 m class) and lower weight secondary mirrors.

2.1. PRIMARY MIRRORS

Mirror substrates of the mentioned size are very heavy if the conventional telescope design is adopted. Extremely high costs result for the manufacture of the mirror substrates including the necessary capital investments on one hand, and the on the other hand, the entire telescope will be very heavy and, therefore, expensive. Practically, such a telescope cannot be financed.

Therefore, based on an innovative concept, meniscus-shaped mirror substrates, being as thin as possible, are required. They need, however, an active support to compensate for their insufficient stiffness. An example for this requirements is the Very Large Telescope (VLT) of the ESO.

2.2. SECONDARY MIRRORS

The secondary mirrors should be very lightweighted. This can preferably be achieved by a thin shell which is either actively supported or receives its rigidity from a supporting structure on the backside. This structure can consist of ZERODUR or can combine the ZERODUR faceplate with other materials. The necessity for lightweighting also pertains to primary mirrors of smaller diameters for special applications, for instance, space telescopes.

3. New Manufacturing Techniques

3.1. SPIN-CASTING TECHNIQUE FOR THE MANUFACTURE OF THIN MENISCUS-SHAPED MIRROR SUBSTRATES FOR THE 8 m CLASS

The economic production of these mirror substrates requires the development and application of new manufacturing techniques. In 1987 SCHOTT developed the spin-casting technique for the production of mirror substrates for the 8 m class.

3.1.1. *Manufacturing Concept*

The basic idea of the spin-casting technique is to reduce the volume of the casting compared to the conventional method. This is achieved by generating the concave surface of the meniscus by spinning a concave casting mold which at the same time forms the convex backside of the mirror.

The success of the new technique is shown by the ratio between mirror volume and the casting volume calculated for the two examples shown in Table II.

TABLE II

Volume ratio between mirror and casting for different casting techniques and mirror diameters

Volume ratio for	NTT 3.6 m	VLT 8.2 m
Conventional technique	0.30	0.16 (performance not planned)
Spin-casting technique	not performed	0.53 (performance planned)

The result is a reduction in capital investment (smaller melting tank, smaller machines, smaller buildings) along with a reduction of the manufacturing cost (less material, less energy) and a shortened production time (shorter melting periods, shorter cooling and ceramization cycles, shorter machining cycles).

3.1.2. *Pilot Production to Develop the Spin-Casting Technique*

The goals of the pilot production were to generate the know-how for the spin-casting of ZERODUR, to ensure the high quality level of ZERODUR with the new technique too, and to guarantee the scale-up from experiments of <4.1 m in diameter to a production of diameters > 8 m.

Approximately 35 castings of 1 m in diameter (conventional) were performed plus additional 15 spin-cast blanks with the dimensions shown in Table III. One casting of each diameter was a geometrical model of the ESO VLT 8.2 m mirror.

Results of the pilot production: the spin cast mirror blanks meet the SCHOTT specification for the inner quality of ZERODUR following dimensional tolerances were achieved. The deviation from the theoretical parabolic curvature measures ± 0.8 mm for the 1.8 m casting, ± 1.2 mm for the 2.8 m castings and a deviation of -2.2 to $+3.0$ mm is observed for one 4.1 m casting. It can be inferred from this, that the accuracy for an 8 m blank can be expected to be better than 6 mm.

TABLE III

Spin-cast ZERODUR mirror blanks from the pilot production

Diameter (m)	Radius of curvature (m)	Thickness (mm)	Weight (t)	Number
4.1	23–14	130–110	4.3–3.6	3
2.8	33– 6	220– 90	3.4–1.3	4
1.8	14– 2.8	270– 57	1.7–0.4	8

3.1.3. *Production of 8 m Mirror Substrates*

The planning activities for the manufacturing equipment as well as the production logistics had progressed far enough that SCHOTT was and is in a position to submit qualified quotes. The 9/1988 SCHOTT received the order from ESO to produce 4 ZERODUR mirror substrates with a diameter of 8.2 m for the VLT.

The start of the project 'Manufacture of 8 m blanks' was in 10/1988. The goal of the project is to manufacture several blanks with a diameter of 8.2 m, a thickness of 177 mm, and a radius of concave surface of 28.8 m in the high quality of ZERODUR. For the realization of this project it is necessary to build up a new manufacturing plant. The new facilities will be ready for start-up at the end of 1990. The first blank will be delivered for polishing in 1993; further pieces will follow at intervals of approximately one year.

3.2. MANUFACTURING TECHNIQUES FOR THE PRODUCTION OF LIGHTWEIGHTED MIRROR SUBSTRATES

In order to meet the new requirements for the lightweighted mirror substrates mentioned in Section 2, several different techniques have been developed by SCHOTT.

3.2.1. *Forming of the Lightweight Structure by Casting*

Cores are inserted into the melt immediately after the feeding of the molten glass into the mold. Lightweighted mirror substrates can be generated by this technique with a rib structure which is open at the backside. Samples have been produced with various diameters and thicknesses.

3.2.2. *Manufacture of Lightweight Structures by Welding of Single Components*

This method consists of the assembly of single components which are subsequently welded together at the contact surface by thermal treatment. Complex structures can be produced applying this method (for instance, lightweight mirror substrate with a backplate). Special consideration had to be given to the fact that ZERODUR is a glass ceramic which is transformed from the glassy state by a special heat treatment. Single components can only be welded together as long as they are in the glassy state if the structure is to develop the specific properties of ZERODUR. Since the glass can crystallize at the temperatures reached during the welding, there are maximum thicknesses of weldable components: the time for homogeneous heating increases towards larger thicknesses resulting in the danger of premature crystallization.

Samples of lightweighted mirrors have been generated with and without a backplate, with plano and curved faceplates, with thicknesses up to 200 mm and diameters up to 500 mm; in certain cases weights per unit area of less than $60 \, \mathrm{kg \, m^{-2}}$ have been achieved.

3.2.3. *Manufacture of Very Thin Mirror Substrates by Slumping*

Plano plates are slumped to menisci with this method. During the process, a glassy plate of ZERODUR is positioned onto a slumping mold with the desired curvature. This setup is heated in an appropriate oven to temperatures allowing for plastic flow such that the plate can be pressed into the slumping mold. Extensive basic research has been performed to determine the process parameters and the process limits for the sumpling of glassy ZERODUR as for the welding techniques. Slumped plates of ZERODUR can be pressure welded to other ZERODUR components which are still in the glassy state, for instance, a lightweight structure. Thus the slumped plates form the faceplate and, if required, the backplate of a lightweighted mirror. However, slumped menisci may also be used as separate, for instance, actively supported thin mirror substrates or may be combined with a supporting structure from other materials, for instance, carbon fiber enforced plastics. Menisci have been generated of up to 1 m in diameter and thickness between 4 and 100 mm; their precision of curvature is approximately 0.1 mm.

3.2.4. *Machining of the Lightweight Structure from a Massive Piece*

By use of this method, the lightweight structure is manufactured from a massive piece by drilling, milling, or blind hole drilling with undercut. This technique allows the manufacture of monolithic lightweighted mirror blanks with and without a backplate. The samples generated have a weight per unit area of approximately $120 \, \mathrm{kg \, m^{-2}}$. In addition, it is under investigation since the beginning of 1988 to what extent the processes of water-jet and ultrasonic cutting are suitable for the manufacture of light-weight mirror substrates.

A thin high-pressure water jet with an added abrasive is used in the *water-jet cutting process* to cut pieces of virtually any shape from a glass plate which could be, for instance, the desired support structure for a lightweight mirror. Material removal during the *ultrasonic cutting* results from an abrasive suspended in water. In this case, however, the energy is supplied by a tool powered with ultrasonic waves. The tool may be, for instance, a thin-walled tube of circular, square, or hexagonal shape.

3.2.5. *Further Prospects*

Very complex structures, lightweighted mirrors among them, can be manufactured by combining the processes outlined in the preceding sections. Manufacturing is underway of lightweighted mirror samples of 500 to 700 mm in diameter with front and back plates. Weight reductions of more than 85% are achievable with this combination of processes. This translates into a weight per unit area of less than $35 \, \mathrm{kg \, m^{-2}}$ for a mirror thickness of 85 mm and a square or honeycomb structure with 50 mm cavity width.

LARGE MIRROR FABRICATION OF LOW EXPANSION GLASSES*

JULIE SPANGENBERG-JOLLEY

Corning Glass Works, Corning, U.S.A.

(Received 21 January, 1989)

Abstract. Low expansion glasses offer many advantages as mirror blank materials due to their thermal and mechanical properties as well as the flewibility they offer in design and fabrication. Fused silica, Corning Code 7940, and ULE (trademark) titanium silicate, Code 7971, produced by the flame hydrolysis process, are high purity and homogeneous glasses. Determination of the average and the variation pattern of the coefficient of thermal expansion (CTE) within ULE mirror blanks is readily accomplished by ultrasonic measurements.

The ability to fusion-seal of the glasses offers mirror manifacturing design freedom of shape and size. The hex-seal process has successfully produced solid monolithic mirror blank up to 4 m diameter and large thin plates up to an 80 : 1 diameter : thickness ratio. The fabrication of a large mirror blank such as the JNLT 7.5 m blank would utilize the proven fusion techniques.

The hex-seal technique consists of fusing preselected boules of glass into a monolith to achieve the required thickness and diameter of a mirror blank. Subsequently, a meniscus blank can be accomplished by heating the plato blank and sagging it over a spherical refractory mold. This ability to fusion seal the glass offers the advantage of blank repair should fracture occur during manufacturing. Following the slumping, the blank is annealed. Further processing of the blank, such as grinding and coating, will not change the material properties or induce permanent stresses requiring additional annealing.

1. Mirror Material

The material and fabrication technique of mirrors blanks are major considerations for the success of any telescope. The mirror material must be carefully selected since its properties control the actual shape of the mirror. To obtain precision optics, the material should possess: a low thermal expansion to minimize surface deformation over the operating range; a low mass; and a high specific stiffness to resist mechanical distortions. The material must be isotropic, dimensionally stable with time and capable of accepting a high-quality polished surface and reflective coating.

2. Properties of Fused Silica and ULE

Corning manufactures two isotropic glasses with low coefficient of thermal expansion and excellent temporal stability, that are well suited for mirror substrates, fused silica Code 7940 at 0.5 ppm/°C $(10^{-8}/°C)$ and ULE Code 7971 with a 0 ppb/°C $(10^{-9}/°C)$ CTE. These glasses possess a low density and a high stiffness and are capable of being finished to a super-polished specular surface (Table I).

The Coefficient of Thermal Expansion (CTE) is a measure of the magnitude of the

* Paper presented at the Symposium on the JNLT and Related Engineering Developments, Tokyo, November 29–December 2, 1988.

TABLE I

Material properties

Property (units)	Symbol	ULE®	Fused silica
CTE (ppm/°C)	α	0[a]	0.52[a]
Density (g cm^{-3})	ρ	2.205	2.202
Elastic modulus (GPa)	E	67.6	73.1
Specific stiffness (m)	E/ρ	3.12×10^8	3.38×10^8
Smoothness (Å)	r.m.s.	<0.5	<0.5
Softening temperature (°C)	–	1490	1585

dimensional change of the mirror substrate subjected to temperatures changes. Due to the low CTE of these glasses, they will exhibit insignificant dimensional changes when compared to a material like aluminum. For a given temperature change, aluminum's dimensional change is approximately 40 times that of fused silica and 1000 times that of ULE, Figure 1 (Jacobs, 1987). Neither glass exhibits thermal expansion hysteresis. Therefore, a ULE or fused silica mirror substrate will follow the same expansion path upon heating or cooling and will return to the exact original dimensions.

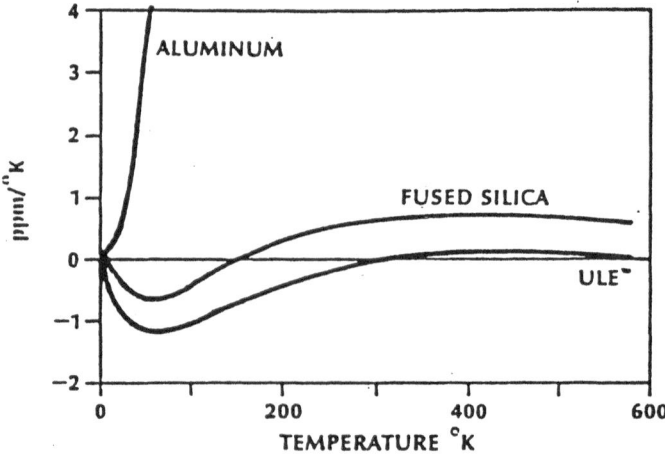

Fog. 1.   CTE comparison.

CTE inhomogeneity within the substrate can affect the optics of the system. Homogeneity measurements of glass samples can be accomplished by dilatometry, interferometry or by photoelastic sandwich seal techniques. Thorough testing of the actual substrate is prohibitive due to the time required and destructive nature of the testing. Past measurements of fused silica have indicated that a CTE range of 5 ppb/°C is achievable. With ULE, which is 92.5% silica and 7.5% titania, a rapid non-destructive determination of the CTE is possible. Due to a linear relationship of ultrasonic velocity, titania content and expansion coefficient, within a matter of minutes the CTE can be determined on any piece of ULE. Specifically, for ULE the material properties of CTE and elastic modulus are directly proportional to the level of titania in the glass. Secondly, there exists a fundamental relationship in any glass between the elastic modulus and sonic velocity (Hagy, 1973)

$$V = 103.37(E(1 + v)/\rho(1 + v)(1 - 2v))^{1/2} \times 10^{-8},$$

where V is the ultrasonic velocity, in μs; ρ, density, $g\,cm^{-3}$; E, elastic modulus; v, Poisson's ratio.

Therefore, as the titania level in the simple two-component glass varies, the CTE and elastic modulus will vary proportionally. The small changes of elastic modulus can be measured by the ultrasonic velocity, so a direct relationship of ultrasonic velocity to CTE exists. This correlation between ultrasonic velocity and CTE has been well documented (Hagy and Shirkey, 1975) and can be utilized to non-destructively measure ULE products:

$$\alpha = 59.670(V - 0.226405).$$

The correlation between elastic modulus and sonic velocity exists in multicomponent glasses but the unique relationship between a single oxide, elastic modulus and CTE does not exist, so this non-destructive method will not apply to all glasses. The CTE of a ULE mirror blank can be guaranteed due the relative ease of measurement.

3. Mirror Substrate Fabrication

Corning's method of manufacturing synthetic fused silica and ULE by the vapour-phase or flame hydrolysis process is quite different from conventional glass-melting processes. The glasses are produced by feeding the vapour of the oxygen and a tetrachloride mixture together with natural gas into burners over a rotating table at approximately 1700 °C. The chemical reaction or hydrolysis takes place in the flame forming sub-micron-sized silicate particles that collect and fuse into a large dense disc, or boule, approximately 1.5 m diameter by 12 cm thick.

$$CH_4 + 2O_2 = CO_2 + 2H_2O,$$

$$SiCl_4 + 2H_2O = SiO_2 + 4HCl \quad \text{fused silica},$$

$$SiCl_4 + TiCl_4 + H_2O = TiSiO_4 + HCl \quad \text{ULE}.$$

High glass homogeneity is maintained through strict process control and by starting with highly purified chemicals. Fused silica was first considered as a mirror substrate material for the Palomar Primary in the late-1920s.

Corning has well developed large mirror blank fabrication techniques for both ULE and fused silica. Due to the similarities of the properties of the two glasses, the processes for fabricating mirror blanks are essentially the same. These glasses offer manufacturing flexibility because pieces can be readily fused without changing the material properties.

A large thin solid like the 7.5 m diameter JNLT blank can be produced by the hex-seal process whih as early as the 1960s has successfully produced large solid such as the 2.7 m McDonald Observatory blank and the 4 m Canadian blank. Since 1970, Corning's production has primarily been in the area of lightweight mirrors for space applications. The highlight of this production was the 2.4 m Hubble Space Telescope primary blank fabricated from ULE. These lightweight substrates require successful fusion sealing without losing blank shape, a technology more demanding than fusing solids.

Corning has fabricated a number of thin meniscus blanks with an 80 : 1 diameter to thickness ratio by the hex-seal process such as a 3 m diameter with 35 mm thickness blank. When the techniques of the vertical fusion sealing process proved successful it was known that blanks of any diameter, even 10 m and larger, could be fabricated by scaling up the equipment.

Recently a new type of solid mirror has been proposed by Larry Barr of the NOAO (Barr et al., 1988) that will utilize the same proven techniques of fusing horizontal and vertical seals (Figure 2). A ULE mirror that is liquid cooled would minimize dimensional changes and temperature gradients between the mirror blank and the surrounding air to improve the telescope seeing ability. Currently Corning is manufacturing an experimental piece to fine tune this process of fusing the hexagonal pieces while maintaining open channels.

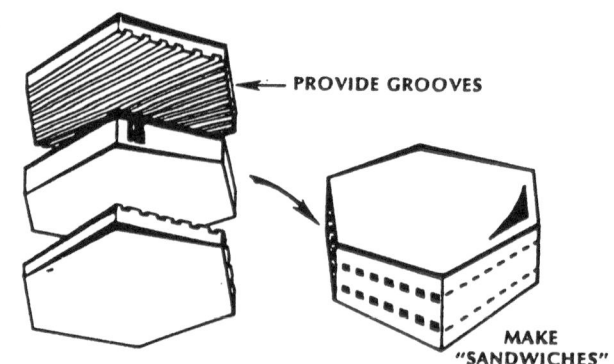

Fig. 2.   Mirror blanks with channels.

4. Proposed JNLT Fabrication Method

The hex-seal process utilizes fabrication techniques that enhances design freedom. The process for a large solid meniscus, such as the JNLT blank would begin with inspection and classification as each boule is produced. This allows glass selection flexibility to optimize the location of each boule within the final blank. The highest-quality glass is selected for the optical surface.

A sufficient number of boules would be stacked and fusion sealed. The stack height would be twice the final thickness for production efficiency, with the quality layer located in the center of the stack. The stacks would then be sliced, producing two stacks of the appropriate thickness with a guaranteed quality layer on top of each. These stacks would then be sliced into hexagonal shaped pieces, $1\frac{1}{4}$ m flat to flat. The surfaces and edges would then be carefully ground to facilitate a tight fitting mosaic assembly (Figure 3). The 34 hexagons stacks required for the JNLT would be assembled in the furnace to the required diameter along with partial pieces to fill in the periphery.

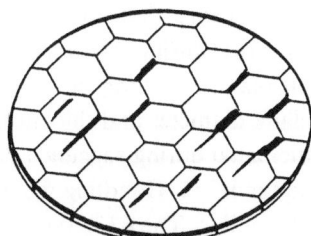

Fig. 3.   Hex-seal assembly.

The advantage of the non-destructive measurement technique of ULE enables optimization of the CTE profile within the blank. A dimensionally stable substrate can be engineered to ensure a uniform average axial CTE across the blank and to minimize the residual strain for long-term optical stability. The CTE average of each of the full-thickness hexagons of the JNLT blank would be guaranteed to be within a range of $\pm 5 \times 10^{-9}/^{\circ}C$.

The furnace would be heated to over 1500 $^{\circ}$Cn to vertically fusion-seal all the hexagons into a monolithick plano blank. The resultant fusion seal will be continuous, clear and detectable only by the presence of very small bubbles, typicall 0.1 to 0.5 mm in diameter. The existence of the small bubbles has been proven not to be detrimental to optical finishers. Testing indicates that glass rod samples containing fusion seals have strength equivalent to rods containing no fusion seals. After cooling, the top surface of the sealed blank would be ground flat using the furnace turntable as a grinding turntable. The blank would be turned over, edge ground to the desired diameter and then the second surface would be ground to final thickness.

To obtain the correct radius curvature, the meniscus blank would again be removed from the furnace while a convex refractory mold is fabricated and ground on the turntable. The piano blank si placed on top of the form and the furnace is carefully heated to sag the blank over the refractory shape. On subsequent cooling, the blank will be annealed. The furnace that would be used for the 8 m-class mirror blanks is the same furnace that was used for previous large solid substrates. This turntable was designed to support a 42 ton blank and since the proposed 7.5 m blank would be approximately 23 ton the turntable support is more than adequate. The turntable would be expanded from the current 5.4 to 8.9 m diameter.

An important feature of a glass mirror blank is that the possibility of repair exists. Should a blank be fractured in its fabrication, repair could be accomplished by reassembling the clean fractured pieces and reheating to fuse the affected area. Due to the high-service temperatures of both glasses, further high temperature coating processes will not change the blank's properties or dimensions. Local stresses, due to finishing or grinding the actuator pockets, can be removed by acid treatment so no additional heat treatment is required.

Stresses in the blank during fabrication and handling are naturally of concern. These stresses are tmeporary stresses due to the weight of the blank setting on support points. Corning designs all processes to ensure that tensile stresses do not exceed 1000ψ $(0.70\ \text{kgf mm}^{-2})$. The maximum stress during the entire hex-seal and sag manufacturing process occurs when the plank blank is setting on the sag form (Figure 4). The sag stresses are a function of the blank diameter and thickness and the diameter of the sag form. The maximum stress anticipated during sagging a 7.5 m blank such as the JNLT substrate would be 630ψ $(0.4\ \text{kgf mm}^{-2})$. Handling would be done by a basic 3-point support, at $120°$ apart, at a 63% radius. The 63% radius results in a zero stress at the center and periphery of the blank with a maximum stress at the support pads. The maximum stress due to 3-pont support would be less than 200ψ $(0.14\ \text{kgf mm}^{-2})$. The residual stress due to frozen-in temperature gradients would always be less than 120ψ $(0.08\ \text{kgf mm}^{-2})$. As these are first-order approximations a complete stress analysis would be finalized at the initiation of the program. The analysis is essential for the design of the handling and shipping container. Corning's expertise in glass technology is well known worldwide, and all necessary areas of Corning's facilities would be used for this program.

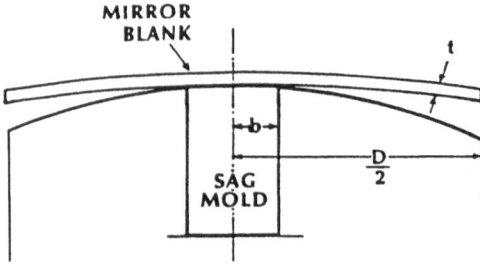

Fig. 4.   Sag form.

Manufacture of the JNLT blank would take approximately 3 years to produce the required glass and fabricate the blank. Transporting the blank from the Canton New York facility can be readily accomplished due to its close proximity to the St. Lawrence Seaway. The blank would travel approximately 20 miles (32 km) overland to reach the Seaway. Once the Seaway it can travel by boat to the Atlantic Ocean or navigate through the Great Lakes to Chicago. At Chicago it can continue on to the Mississippi River which connects to internal waterways within the United States as well as to the Gulf of Mexico.

5. Conclusion

In conclusion, the low expansion glasses, ULE and fused silica are excellent mirror materials and offer fabrication flexibility. Corning has thad the technology to fabricate 7.5 m substrates since 1965 and has successfully manufactured many large solids and thin blanks. Corning has world-renowned research and engineering facilities to successfully produce large mirror blanks and looks forward to the opportunity to be a part of the team to advance the science of astronomy.

References

Jacobs, S. F.: 1987, *Applied Optics and Optical Engineering,* Vol. 10, Academic Press, New York, Chapter 2. 'Dimensional Stability of Materials Useful in Optical Engineering', p. 71 (0–300 K data adapted from Jacobs).

Hagy, H. E.: 1973, *Appl. Optics* **12**, 1440.

Hagy, H. E. and Shirkey, W. D.: 1975, *Appl. Optics* **14**, 2099.

Barr, L. D., Beckers, J. M., Pearson, E. T., and Hobbs, T. W.: 1988, 'Reducing Mirror Seeing Problems in Meniscus Mirrors', *ESO Conference on Very Large Telescopes and their Instrumentation.*

LARGE MIRROR FIGURING AND TESTING*

K. BECKSTETTE, M. KÜCHEL, and E. HEYNACHER

Carl Zeiss O-Labor, Oberkochen, F.R.G.

(Received 8 February, 1989)

Abstract. We describe a method for figuring and testing large aspheric mirrors using a rectangular, flexible lap (the so-called 'Membrane Tool') and a vibration stabilized interferometer. The rear side of the lap is covered with computer controlled dynamic pressure actuators which determine the amount of material to be removed for surface error correction. This method has been developed in the laboratory and tested to some extent by figuring the $f/2.2$ primary of the 3.6 m ESO–NTT. We describe the ongoing developments and the manufacturing plan for 8 m-class mirrors.

1. Introduction

Many large telescope projects with 8 m-class primaries are under discussion or even have been started during the last years. One example is the $f/2.0$, 7.5 m diameter primary of the JNLT to which this meeting is devoted. In order to keep telescopes reasonably compact, these mirrors will have focal ratios between 1 and 2, considerably faster than any existing astronomical telescope. They present a serious technological challenge, because of their size, accuracy requirements and their asphericity.

The difficulty of polishing increases with asphericity, as the natural tendency of the polishing process is to produce spherical surfaces. The asphericity, the deviation of the surface from the closest sphere, increases dramatically with faster focal ratios: i.e.,

$$\mathrm{d}z \approx D(D/F)^3 .$$

The absolute value of $\mathrm{d}z$ depends on the fit parameter for the reference sphere. Table I lists asphericities for various mirrors, including the $f/1.0$, 500 mm diameter mirror currently being polished as part of the membrane tool process (MTP) development for the SOFIA $f/1.0$, 3 m primary.

TABLE I

Typical asphericities

Telescope	Focal ratio	Diameter (m)	$\mathrm{d}z$ (μm)	Relative factor
JNLT	2.0	7.5	580	2.7
VLT	1.8	8.2	870	4.2
NTT	2.2	3.6	210	1.0
SOFIA	1.0	3.0	1860	8.9
Test	1.0	0.5	310	1.5

* Paper presented at the Symposium on the JNLT and Related Engineering Developments, Tokyo, November 29–December 2, 1988.

Astrophysics and Space Science **160**: 207–214, 1989.
© 1989 *Kluwer Academic Publishers.*

The table shows that even the small $f/1.0$ test mirror has a higher asphericity than the fastest large astronomical mirror polished to date, the ESO–NTT primary.

Testing has always been an essential tool to the optician. For the new class of large telescope mirrors, testing problems arise due to the long focal length, asking for insensitivity of the test method to air turbulences and vibrations. The huge size requires higher lateral resolution to detect high-frequency errors.

2. Traditional Techniques for Making Fast Aspheres

The traditional process of making large aspheric mirrors for astronomical telescopes makes use of the fact that it is easily possible to polish spherical mirrors to very high precision with simple equipment. By use of this first step of lapping (or loose abrasive grinding) and polishing the best fitting sphere, has the big advantage, that a nearly perfect rotational symmetric workpiece is the starting pont for the aspheric figuring process.

Based on this symmetry, the iterative aspherisation process with medium size tools can be carried out relatively fast. The remaining non-rotational symmetric errors cause the use of smaller and smaller laps measuring $\frac{1}{100}$ or less of the mirror area and are the reason for time-consuming local fine correction. Additionally these small tools tend to produce zones and ripples appearing on smaller and smaller spatial scales. The whole process depends heavily on measuring the actual surface shape after every correction step. Logical evolutions of this traditional technique are:

– *Large area compensated tools*: A large tool (approximately workpiece dimension) is moved across the mirror. Due to different tangential and sagittal radii, only small amplitude motions are allowed. Local corrections are then performed by area weighted pitch trimming (petal tool). Of course trimming has often to be changed according to the actual surface profile. Therefore, tedious and time-consuming work is necessary acquiring, in addition, a high degree of experience.

– *The computer-controlled polisher* (*CCP*), a system which is well known in literature and has been used successfully for small to moderate-size optics (Jones, 1982). Here a very small lap with small polishing strokes is guided on its patch across the work. The removal pattern is realized by dwelling more or less on a spot depending on whether it is high or low. Good results concerning acute edges and the development of r.m.s.-values were reported.

3. Special Problems for 8 m-Class Mirrors

The traditional method can be pushed to make somewhat larger or faster mirrors but the difficulty increases rapidly with increasing asphericity. The major problems for the conventional methods are:

– *Material removal*: Comparing for example a 3.5 m, $f/3.7$ mirror to a 7.5 m, $f/2.0$ mirror, the volume of the material to be removed by polishing when starting the aspherisation from the sphere, increases by nearly two orders of magnitude. Even if the relative diameter for the tool used would be the same, the aspherisation process would last ten times longer.

 – *Tool handling and preparation*: For the manufacturing of 3.5 m mirrors, laps of 3.5 m diameter have been used. It seems not possible to handle and prepare 8 m diameter laps in a reasonable time and with appropriate safety.

 – *Local error correction*: Starting with aspericities of less than 50 μm, the traditional techniques could rely on producing the aspheric deformation by polishing and on saving the rotational symmetry during the aspherisation process. Increasing the deviation by more than an order of magnitude and keeping into account, that more flexible laps have to be used, would also increase the remaining non-rotational symmetric errors by more than a factor of 10! This means that the iterative method of local fine correction with small tools would be very time consuming and perhaps will not converge at all.

These considerations led us to the conclusion that a method for making large fast primaries should fulfill the following requirements:

 – generation of the asphere by diamond wheel grinding;
 – process designed to remove relative large errors of the grinding step on the aspheric surface; i.e., a lapping not a polishing process!
 – large tool area for volume removal;
 – to be used for figuring and fine correction without modification;
 – relative stiff tool to prevent ripple production;
 – fast tool preparation;
 – fast and environment insensitive metrology useful during lapping and polishing.

The following chapters contain a detailed discussion of two main components for successfully polishing a large mirror, starting from the diamond wheel ground state: the tool and the metrology.

4. The Membrane Tool Process

The fundamental assumption of all removal predicting theories is given by Preston's law (Preston, 1927) which states that the wear per unit time is proportional to relative velocity and tool pressure. The basic idea in our process of membrane polishing is to split the polishing process into its relevant parts, i.e., relative motion and local pressure (Figure 1).

The tool consists of two major parts: a relative thin membrane which carries the polishing pitch and performs the relative motion between tool and workpiece and a set of actuators at the tool's rear side which apply the necessary removal pressure. This pressure is dynamically controlled by a computer. This lap works in principle like an arrangement of many small size laps, used, i.e., on the CCP, working in parallel. The actual removal of each *subtool* is controlled by the pressure applied through the actuator in contrast to the dwell time approach of the CCP. The membrane is designed flexible enough to accommodate the desired variations in curvature of the asphere and stiff enough to provide a smooth print-through function for the actuators.

There are some important advantages in this membrane tool approach compared to the discussed techniques:

 – The tool can actively remove errors with low and medium spatial scales. The

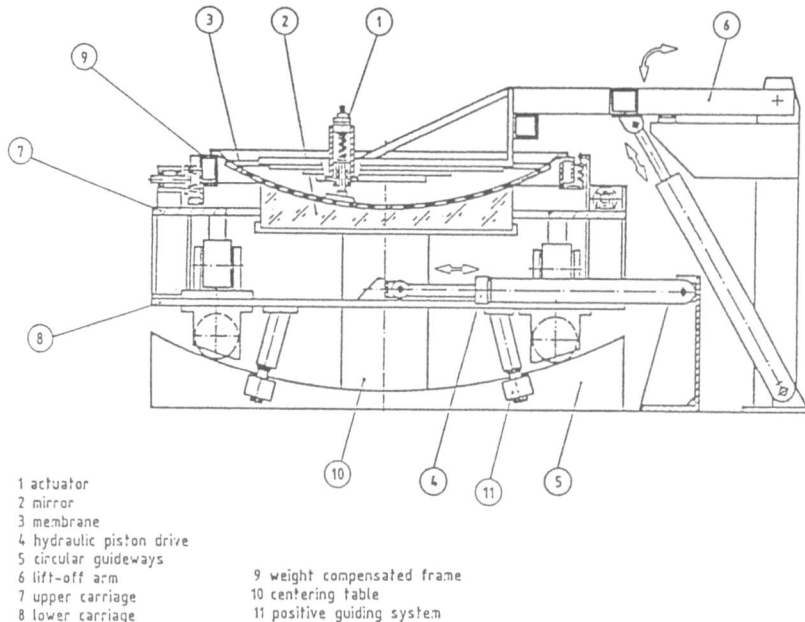

1 actuator
2 mirror
3 membrane
4 hydraulic piston drive
5 circular guideways
6 lift-off arm   9 weight compensated frame
7 upper carriage   10 centering table
8 lower carriage   11 positive guiding system

Fig. 1.   Principle of the membrane polisher.

smallest spatial scale error which can actively be reduced is only determined by the
actuator size.

– The tool does not rely on its own shape or the shape of the mirror to remove the
low-frequency errors. This reduces mirror support problems during manufacturing and
also opens the possibility to design a tool flexible enough to accommodate the variations
in curvature of the asphere.

– As the tool covers (at least in one dimension, see below) the whole surface of the
work it can apply bending moments at the edges of the *subtools* which prevents the
inherent edge problems of small laps and the production of ripples.

– The tool area can easily be made large, producing a good volume removal to
accommodate the large surface errors produced by diamond wheel grinding.

– To change the removal function from one iteration step to the other, no tool
preparation time is needed. The metrology data can, after some mathematical modifications, directly be fed into the control computer.

– Producing a *hole* in the surface during fine correction with a small area lap can be
a big problem, as the whole surface of the mirror has to be *set back* with the small lap,
a very time-consuming and error prone process. That is why the optician and even the
CCP tend to remove during one iteration step less material than predicted. The membrane tool, working all the time on the whole surface area overcomes this problem; the
convergence factor can be increased.

Figure 2 shows a setup of this tool especially adapted to the configuration for
manufacturing a large primary. The tool has a rectangular shape. The relative motion

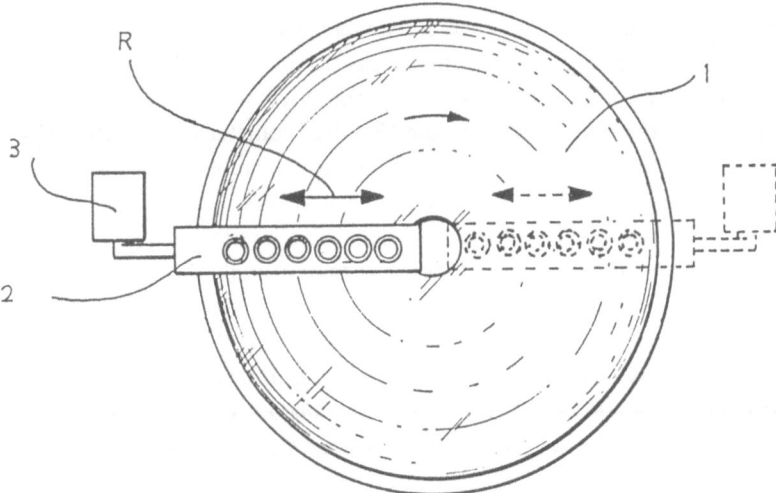

Fig. 2.   Membrane Tool for rotational symmetric parts: (1) rotating mirror, (2) oscillating membrane tool, (3) tool drive.

between tool and mirror is performed by rotating the mirror and radially stroking the tool.

5. Testing as an Integral Tool

As figuring of large aspheres has and will be performed in an iterative approach where the results of metrology determine the next polishing step, testing has to be addressed as an integral part of the fabrication process. The convergence of the process is not only driven by the predictability of the tool removal but also by metrological accuracy. Two subsequent test results are used to predict the tool behaviour for the next polishing step. Thus high accuracy is required for a controlled and fast converging process. Metrology closes the feedback loop in a closed-loop controlled process.

Additionally, as discussed above, when starting from the diamond wheel ground asphere, the test method has to accept a rough surface with relative large errors. Regarding the large dimensions of the test tower, two additional requirements on metrology arise when manufacturing a large mirror:

– insensitivity against air turbulence;
– insensitivity against vibration.

All these different aspects were covered when lapping and polishing the ESO–NTT. So we report here the test methods used and discuss the enhancements required to manufacture a very large mirror.

5.1. IR-INTERFEROMETRY FOR THE ESO–NTT PRIMARY

When figuring the ESO–NTT primary with lapping, IR-interferometry at a wavelength of 10.6 μm (CO_2) was used. In contrast to the manufacturing plan for an 8 m-class mirror where we will start from a CNC ground asphere, the NTT primary was figured with lapping, starting from the sphere. The figuring was performed in many cycles of measurement and lapping, until the final aspheric shape of the mirror was reached. Null lenses where used to adapt the plane wave from the Twyman–Green interferometer to the aspheric surface. The detector in the IR-interferometer was a pyro-electric vidicon. Taking into account the MTF of the vidicon, approximately 2000 points are resolved on the mirror. Phase-shifting interferometry is used to obtain a measurement precision despite the large wavelength of the used light. The interferometer was controlled by a micro-computer which also performs the on-line data analysis. The pointwise reproducibility of the measured mirror surface was approximately 250 nm, that is $\lambda/40$ for $\lambda = 10.6\ \mu m$.

IR-interferometry is insensitive to air turbulences and vibrations. An enhancement of the existing setup would be to use a CCD with a higher resolution than the pyro-electric vidicon. Data for a *closed-loop* control of the membrane tool is available in the analysis computer.

5.2. VIS-INTERFEROMETRY FOR THE ESO–NTT PRIMARY

By use of a phase-shifting interferometer equipped with a high-resolution CCD-camera, working at He–Ne wavelength of 632.8 nm, the influence of air turbulence in the test tower can be reduced to harmless level by averaging data sets. Thus the speed of the individual measurement becomes important. For the 8 m-class mirrors the same simple and cheap approach can be used. But it has to be kept in mind, that the test tower length will grow by approximately a factor of 2 and that the air volume will increase by nearly a factor of 10! If the air turbulence effects are proportional to the air volume the number of the individual measurements has to be 100 times higher. For the manufacturing of a very large mirror, measurement speed will be increased by more than this factor.

Polishing the ESO–NTT primary, vibrations of the test tower are compensated by a 3-axis stabilization of the interferometer. The axial motion and the tilt of the mirror with respect to the interferometer is measured and fed to a control loop which performs the active stabilization by tip-tilt controlling a mirror in the beam path (Figure 3). Thus frequencies up to 1000 Hz can be eliminated. This allows measurements during the day, with running machines, air conditioning, etc. The stabilization system also introduces the required phase shifts for the data acquisition under control of a microcomputer. This microcomputer later performs the calculations for the wavefront at all the data points on the mirror, at present approximately 40 000 points. When manufacturing a very large mirror, it will be possible to increase this number by a factor of 20. The surface of an 8 m mirror can be measured at a sampling spacing of 8 mm by 8 mm.

Important for the reproducibility of the measurement is the number of individual wavefronts to be averaged. A good compromise between measurements time (see above) and precision, about 50 individual wavefronts were averaged, leading to a local uncertainty of the surface shape of approximately 4 nm.

One great advantage of computerized interferometry is the fact that the measured wavefronts are directly available for all kinds of mathematical analysis. In the case of the ESO–NTT primary it was important to reach the best possible 'intrinsic quality',

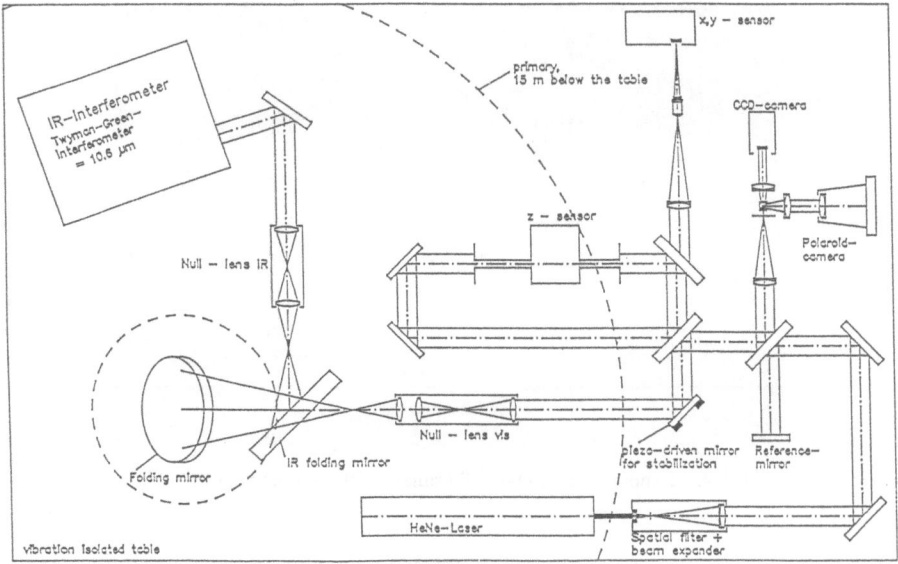

Fig. 3.   Schematic of the vibration stabilized IR- and VIS-interferometer used for ESO–NTT.

which was defined through the geometric spot size E_{80} after the subtraction of various error terms from the wavefront. For analysis of the polishing process these error terms were routinely subtracted and the spot size calculated.

6. Development Steps Towards 8 m Diameter

The Membrane Tool Process is under ongoing development at Carl Zeiss. Starting with a setup of five linear arranged actuators over a five by five array, working on a plano surface, the membrane tool has been developed to an extent where good results have been shown on a 500 mm $f/1$ spherical mirror.

Two aspects of the Membrane Tool Process have been used, figuring the ESO–NTT 3.6 m $f/2.2$ primary:

– rectangular, non-rotating tools;

– computer controlled pressure actuator with dynamic pressure variation.

By use of these tools in cooperation with the above-mentioned metrology, it was possible to reach the final result of an 'intrinsic quality' $E_{80}^{*} = 0.096''$ with a complete computer controlled process. This intrinsic quality is the geometric spot size after subtraction of low-order Seidel terms: focus, decentering coma, third-order spherical aberration, third-order astigmatism, triangular coma, quadratic astigmatism.

A complete version of the Membrane Tool will be used during the mirror fabrication hardware study for the Stratospheric Observatory For Infrared Astronomy (SOFIA), a 3 m $f/1$ telescope to be flown on a Boeing 747. During this study which will be finished in mid-1989, a 500 mm $f/1$ Paraboloid with 50 mm thickness is manufactured, using the above discussed Membrane Tool Process.

Fig. 4.   Lapping the ESO–NTT primary with two rectangular tools.

7. Summary

Membrane Tool Polishing is a method especially designed to figure very large mirrors with low focal ratios in a short time. It has been extensively tested experimentally, principally verified polishing the ESO–NTT primary and will be fully developed for SOFIA. It shows a good convergence factor and produces superior optical quality. The estimated lead time for a 8 m-class mirror, starting from the aspheric ground surface is less than 2 years.

The measurement technology for manufacturing very large mirrors is available and tested. Further improvements with the trend towards shorter times for the data analysis and acquisition, more data points, and consequently towards a still higher precision are under an ongoing development.

References

Jones, R. A.: 1982, *Appl. Optics* **21**, 562.
Preston, F. W.: 1927, *J. Soc. Glass Techn.* **9**, 42.

WONG – Can the membrane lap handle non-axial-symmetrical or local error? And how?

BECKSTETTE – The membrane lap is designed to work on non-axial-symmetric errors. For these we used already a 500 mm full aperture size lap. Local errors can be handled by the computer-controlled actuator concept.

BECKERS – How fast an 8 m mirror can you polish with your membrane technique?

BECKSTETTE – We calculated the necessary membrane parameters for 8 m $f/1.0$ mirrors and also for non-symmetrical $f/0.5$–$f/1.0$ anamorphic mirrors. Both can be worked on without problems.

KODAIRA – How long a time will be needed for polishing and figuring of an 8 m-class mirror?

BECKSTETTE – Starting from a diamond-wheel ground surface of approximately 0.5 mm P/V, we expect approximately 1.5 years. It has to be tested, if this can be reduced to 1 year.

LARGE MIRROR POLISHING AND FIGURING*

W. SCOTT SMITH

Contraves Goerz Corporation, Pittsburg, U.S.A.

(Received 8 February, 1989)

Abstract. The fabrication of the next class large aperture telesopes will require techniques which must be extrapolated from current experience. The areas of substrate support, optical testing, and figuring strategy needs to be modified or extended to handle the 7 to 8 m diameter aspheric primary mirrors. This paper discusses the application of advanced optical testing techniques to the large optical paths and the extension of existing aspheric figuring techniques to assure a smooth surface.

1. Introduction

Two requirements must be met for the successful completion of a large aspheric optical component. First, an accurate test method must be available with computer reduction software which yields a faithful representation of the optical surface. Secondly, the support of the optic under test must be repeatable and able to be modeled. Ideally the optic should be tested in the same configuration as it will be mounted for use.

Once these two requirements are met, the basic goal of optical figuring remains unchanged: polish on the high spots. The actual procedure used to optically figure a mirror varies among contractors. At Contraves Goerz we have developed a variation on a theme that has successfully produced several smooth $f/1.5$, 1.5 m diameter paraboloids.

2. Optical Testing

The optical testing for both in-process and final acceptance will utilize a center of curvature null lens and a real-time interferometer. Due to the long optical pathlength, 32 m in the case of an 8 m diameter $f/2$, careful consideration must be taken of the testing environment. Ground vibrations will be accentuated by the long distances. Microthermal turbulence problems will increase due to the increased air volume in the optical path.

To reduce the effects of microthermal turbulence, a standard commercial heterodyne interferometer with a 100×100 CCD array is used to average several individual interferometric tests. Under the assumption that the thermal turbulence is random in nature, then the distortion in the test data is reduced by the square root of the number of *interferograms* in the average. CGC has developed software that allows for the transfer of the full 100×100 data set to an auxiliary computer for averaging and analysis.

---

* Paper presented at the Symposium on the JNLT and Related Engineering Developments, Tokyo, November 29–December 2, 1988.

Experiments over a 30 m vertical path indicate that this approach is viable for the reduction of thermal turbulence effects for large aperture primaries.

The heterodyne interferometers use four sets of data to produce one phase data set. The time required to gather enough data for one data set is about 30 ms. The peak in the ground-vibration spectra occurs around 10 and 30 Hz, caused by machinery, transport equipment, etc. The effects of vibration can be the complete loss of data gathering capability or may become evident from the degradation of the data. The wavefront will typically take on a sawtooth appearance. In the past, this situation was resolved by testing at times when the vibrations were lower, in the middle of the night. Now computer code has been written that will remove the vibration-induced wavefront errors.

Over the past two decades CGC has put an emphasis on the development of computer software for the display of optical wavefront obtianed from interferometric measurements. The initial efforts analyzed the wavefronts in terms of Zernike polynomials. This method was chosen since it allowed the easy removal of residual alignment errors from the test data. These polynomials are global in nature and as a result do not give a faithful representation of localized figure errors. The computer programs presently used rely on a low-order fit to Zernike polynomials for the removal of test-alignment errors. The wavefront matrix is calculated by a local interpolation method. Rather than relying on the global figure, each matrix point is calculated utilizing its nearest 20 data points.

The output of the computer reduction program is a matrix, variable in size depending on the optical component and its stage in the figuring process. Smaller optics use a 31×31 matrix, larger optics use a 121×121 matrix. With the increase in computer speeds this matrix size will be increased to 500×500 as the focal plane arrays in the interferometers are increased in density. Within the time frame of the availability of mirror substrates 512×512 CCD arrays will be commercially available in heterodyne interferometers. For an 8 m diameter mirror this will yield a 16 mm data sampling over the aperture. This will be more than adequate for assurance that higher spatial frequency errors are not present on the mirror.

The development of these computer programs is responsible, to a large degree, for the reduction in the optical figuring time over the past few years.

3. Optical Figuring

With the advent of adequate optical data reduction software, new approaches were taken to optical figuring. Computer-controlled polishing techniques were developed that generally utilized small polighing laps, 0.1 lto 0.2 m. This lap usually drives in a planetary or orbital motion and is scanned over the optical surface. The time spent over a point on the surface is then proportional to the measured wavefront error at that point. Several problems arise from this approach. The science of polishing is not that well understood. The polishing characteristics may change over the time duration of the run. The determination of the work function of the polishing lap is required to properly overlap the orbits. Finally, and most serious, the use of small polishing laps causes an increase in the high spatial frequency wavefront errors.

Since the business at CGC is sually one of a kind optics, computer-controlled machines are not cost effective. Instead the opticians have developed a technique which yields rapid convergence to the final specified optical figure. A plotter is used to produce full-scale contour maps of the optical surface. The contours are then transferred to the optical surface one at a time. Starting with the highest contour, each is worked in succession. The final polishing cycle of a run covers a large portion of the surface. For the larger components, the smallest polishing laps utilized are 0.35 m in diameter. By use of a combination of moderately flexible laps and soft polishing pitch, the lap maintains contact with moderately aspheric surfaces. In this case, moderately aspheric means an $f/1.5$, 1.5 m diameter primary. For this example the smallest lap used on this mirror was $\frac{1}{5}$ of the diameter. The main advantages to this method come from the faster glass removal from use of a larger polishing lap; the elimination of the high-spatial frequencies which result from the use of small polishing laps; and the basic simplicity of the procedure. On the most recent large primary it was the opticians goal to reduce the residual figure error on the surface by 50% on each polishing and testing cycle. In practice he achieved a 30% average wavefront reduction per cycle. It is interesting to note that the development of these procedures was done by the opticians in coordination with help from the engineering department for software development.

4. Substrate Support

Two types of support systems have been used by CGC for large mirror fabrication support. For the 1.5 m diameter optics 36 support ponts are used. The hydraulic support utilized 10 cm diameter diaphragms and were arranged at positions determined by a finite element analysis, FEA. The 36 supports were divided into three systems of 12 units piped together. The three systems were separately pressurized to correctly tilt the substrate and space it so that the individual diaphragms were in an unstressed position. With the substrate properly positioned each of the units was isolated by a valve, becoming an individual hard support point.

The second type of support used the same positions as the hydraulic system, but were cast blocks of pitch 10 cm square by 2.5 cm thick. The high viscosity of the pitch provided the support while the flow was fast enough to allow the mirror to achieve an unstressed condition over a few days. The testing was done with the mirror supported on edge, with a horizontal optical axis, so that any support-induced figure errors would have been measured. Both methods proved to be adequate, however, the hydraulic system did develop a leak occasionally.

5. Results

Over the past three years CGC has produced five 1.5 m diameter primary mirrors. In no case was the residual wavefront error limited by either the test or the fabrication capability. The decision to accept the optical was based only on the specification. Two interesting results occurred. The learning curve (Figure 1) shows the increased efficiency

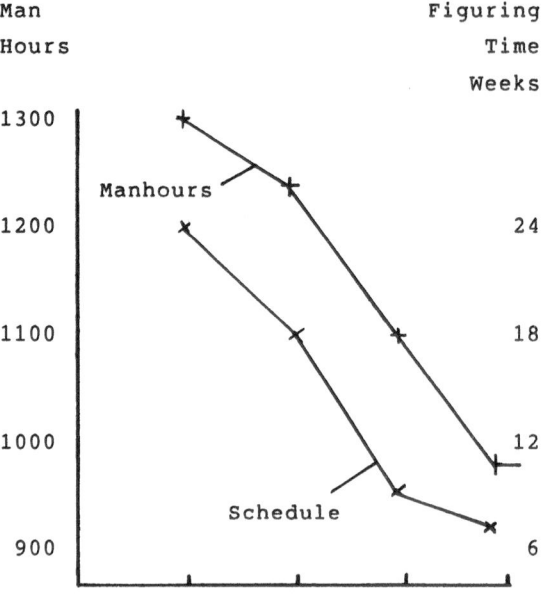

Fig. 1.   History of 1.5 m diameter mirrors.

which results both from the classic learning curve along with the improvements in the testing over this time span. Figure 2 is a contour plot of the most recent mirror. It is also turns out to have the best wavefront. This is not the limit of technology at CGC but rather a mirror that is better than the specification. The surface is accurate to 0.013 waves aims at 0.633 µm wavelength or 0.008 µm r.m.s. Geometrically 81% of the wavefront falls within a 0.25″ diameter.

6. Summary

The technology exists for the fabrication of the 8 m-class mirrors. The testing over long optical pathlengths has been demonstrated. The focal plane arrays of the appropriate density are available for heteordyne interferometers and will be implemented in time for testing these mirrors.

A computer-controlled machine will be necessary for the generating, grinding, and polishing. Machines of this capacity exist for generating. Modifications to these machines for grinding and polishing have been proposed. Computer-controlled grinding and polishing machines exist in the 5 m capacity range and the extension to 8 m would be straightforward.

The remaining problem of a mechanically weaker substrate and the resulting more difficult support problem can be handled by present finite-element analysis procedures. For the final testing of the 8 m class mirrors the test support structure must emulate the final mounting in the telescope.

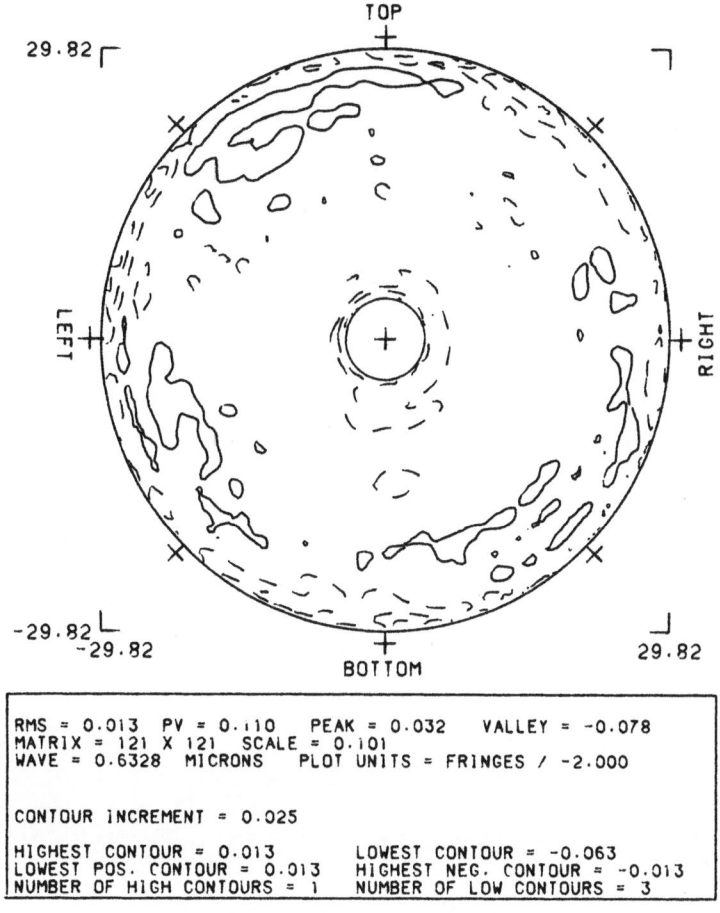

Fig. 2.   Contour plot of 1.5 m diameter mirror.

BECKSTETTE – What is your experience in using CCP with small laps? Do you see edge problems when using 0.5 m laps on an 8 m diameter mirror?

SMITH – We are not currently using CCP, but are using the computer to direct the optician. On our last 1.5 m aperture primary the smallest polishing lap was a solid 0.35 m diameter.

The edge is the steepest slope and largest wavefront error. Polishing was stopped because we met the specifications, we were not yet limited by fabrication procedure or testing.

KNOWLEDGE ENGINEERING APPLICATION IN THE PROBLEM OF SUPPORTING A LARGE MIRROR*

MASAAKI WATANABE, YOUICHI AGEISHI, and KEN NAKAMURA

AdIn Research Inc., Tokyo, Japan

(Received 2 December, 1988)

Abstract. We propose an application of knowledge engineering in the problem of active supporting for a large mirror. To reduce calculating load and to shorten cycle time, we divide the system hierarchically into a managing central controller and many autonomous local controllers, and use reasoning instead of dynamical analysis. The reasoning is based upon pattern matchings between observed error pattern and test patterns in the knowledge base.

1. Introduction

The mirror support problem for a large mirror involves many difficulties, i.e., large number of supporting points, high accuracy of supporting force, and invisibility of surface deformation of mirror during observation. When we deal with external disturbance, we must discriminate it from supporting errors in a moment, which makes the problem far difficult.

As an approach to solve such a problem, we propose a knowledge engineering application in the mirror support problem.

Keypoints of the proposed system are the hierarchical division of control system and use of reasoning instead of dynamical analysis. Reasoning is performed through pattern matching between observed error patterns of supporting forces and test patterns in knowledge base.

2. Hierarchical Division of System

In the present study we divide the system hierarchicaly into a managing central controller and 39 autonomous local controllers (Figure 1). Each local controller governs 10 actuators (Figure 2). Our aims are as follows:

(1) It makes surface deformation easy to sense as a pattern.

(2) It disperses and reduces calculating load and shortens cycle time.

(3) The sensitvity of single load cell is improved by statistics.

The local controllers operate autonomously. They distribute their responsible supporting forces assigned by the central controller to their actuators, and report the sum of achieved forces to the central controller.

---

* Paper presented at the Symposium on the JNLT and Related Engineering Developments, Tokyo, November 29–December 2, 1988.

Astrophysics and Space Science **160**: 221–224, 1989.
© 1989 *Kluwer Academic Publishers.*

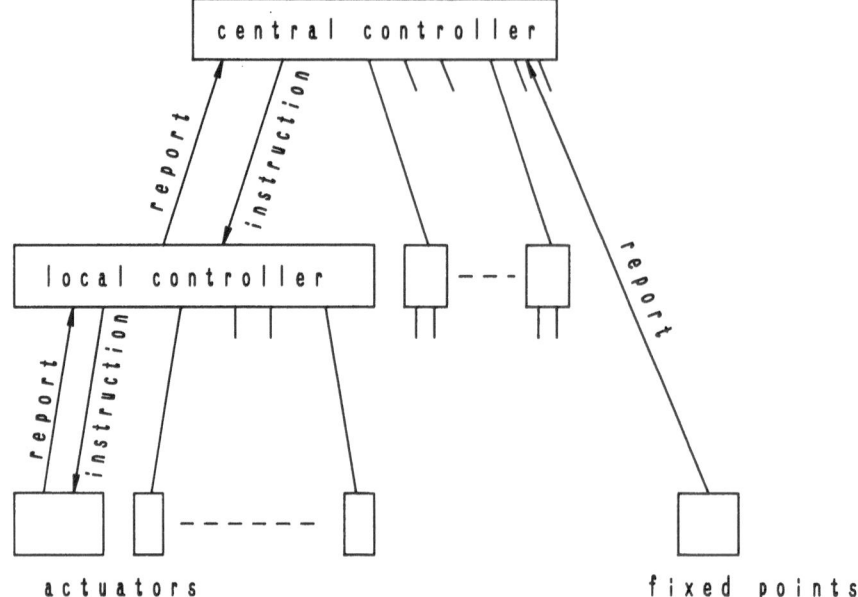

Fig. 1.   Hierarchical control system consists of a central controller, 39 local controllers, and 390 actuators.
Three actuators are fixed to settle the mirror.

Fig. 2.   Every ten actuators form a group and are governed by the local controller. Surface deformation
within one group is negligible because of local rigidity of mirror.

The central controller compares observed error pattern with test patterns in knowledge base. Then it discriminates external disturbance from supporting errors by reasoning. Finally it determines next supporting forces.

3. Pattern Matching and Reasoning

The knowledge base contains the static ideal supporting forces, the basic deformation patterns and their correcting force patterns. All patterns are related to each other to form a knowledge network.

The first pattern matching is carried out with the most popular pattern group (Figure 3). After that, the coincidence of the last matching determines the next pattern group successively. The pattern matching terminates when the coincidence reaches a certain threshold or the matching count reaches the limit.

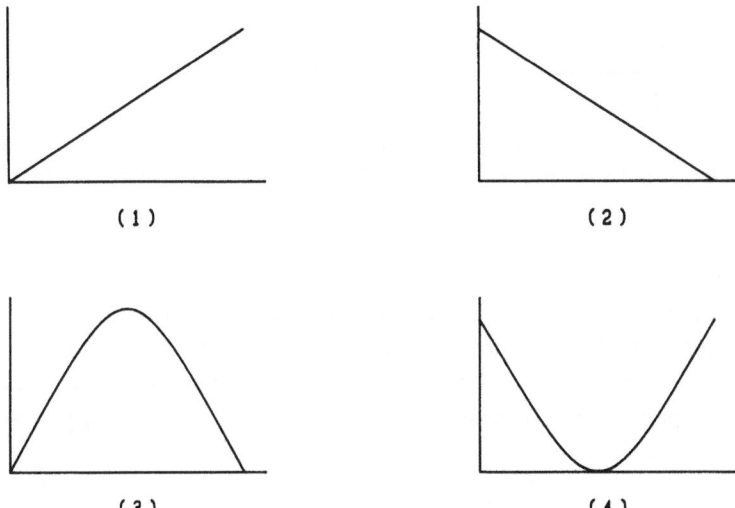

(1)   (2)

(3)   (4)

Fig. 3. Smooth and simple basic deformation patterns. They are referred at the first time in pattern matching. Sufficient number of patterns may be less than 30, because the amplitudes of higher-order vibrations are expected to be small.

The supporting forces are determines by mixing the correct force pattern and the negative feedback with appropriate rates according to the coincidence (Figure 4). As a special case, when the coincidence is lower than the threshold for any test pattern, no correctin is performed and the control operates as damping.

The knowledge network is improved by learning. Both the linkage and the reference order are changed. The cycle time of the system, therefore, may be shortened progressively through practical operation.

4. Summary and Discussion

It is presented that the fast active supporting system of a large mirror is realized by hierarchical construction of control system and reasoning through pattern matching.

This system aims at the fastness of control rather than the completeness. This strategy may cut off a greater part of deformation in each moment and put the mirror back to its ideal shape in a short time.

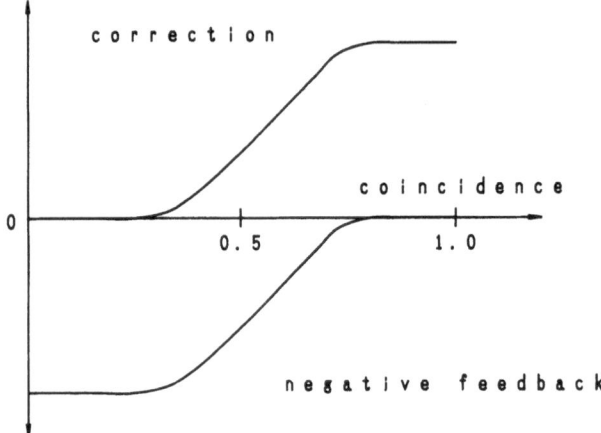

Fig. 4.   Next supporting force is determined by mixing the correcting force and the negative feedback according to the coincidence. The shape of mixing rate curve must be adjusted during trial operation.

This system is most useful when the cycle time is set shorter than the vibrating period of the mirror because the surface deformation is supressed as soon as it starts. To realize such a quick control, it is necessary to measure supporting forces with high accuracy and at small intervals. The high precision load cell is indispensable. There is a possibility to construct such a load cell by combination of the traditional load cell and the direct measuring of spring length of each actuator to monitor rapid change of supporting force.

A MEANS FOR IMPROVING TELESCOPE DRIVES*

P. R. GILLINGHAM

Anglo-Australian Observatory, Australia

(Received 10 January, 1989)

Abstract. In seeking to minimize local seeing, some large telescopes are being planned with less wind protection than has been traditional and some sites with very good seeing appear to suffer significantly from ground-transmitted vibration. A great care is needed in designing new telescopes to ensure that vibration does not become a dominant cause of image degradation. A supplementary drive system is proposed which could improve on past tracking performances even in such adverse conditions. The system may also reduce the costs of very large telescopes by easing the demands on the stiffness and accuracy of their mountings and drives.

1. The Need for Improvement

The necessity to minimise local degradation of seeing is leading to large telescopes being planned with less protection from wind than in the past. At some sites there is also the danger that significant vibration will be transmitted through the ground, either from far away or from whatever wind protection does surround the telescope. Stiff alt-azimuth mountings with heavy gear or roller drives partly combat the wind effects but may increase the susceptibility to high-frequency ground vibration. In any case, constructing mountings and drives with low compliance and extreme smoothness is difficult and very costly.

2. Small Effort Required to Sustain Telescope Oscillation

If one relied only on the very low damping inherent in a metal structure, the effort required to maintain an unacceptable oscillation of the optical axis would be extremely small. Consider a body with rotational inertia of 1000 tonne m², having a natural frequency of 2 Hz and material hysteresis equivalent to viscous damping 0.01 times critical (structural steel has less than this). The amplitude of the torque to keep the body oscillating through ± 0.5 arc sec at its natural frequency is 7.7 Nm. For a 10 m radius of action of the force, the corresponding force amplitude is only 0.77 N (78 g or 2.8 oz). The power absorbed by this oscillating body is less than 1 mW!

Thus unacceptable telescope oscillation can very easily be excited. Fortunately, this also indicates that it should be possible to counter such oscillation with subtle means.

---

* Paper presented at the Symposium on JNLT and Related Engineering Developments, Tokyo, November 29–December 2, 1988.

Astrophysics and Space Science **160**: 225–230, 1989.
© 1989 *Kluwer Academic Publishers.*

2. A Promising Solution

It is proposed that conventional drives around the altitude and azimuth axes be supplemented by inertial drives, linked to the main optics as directly as possible. For a telescope 'tube' comprising primary mirror and secondary mirror cells joined by Serrurier or similar trusses to a 'centre section', the inertial drive could act on the centre section near the attachment points for the trusses.

Figure 1 shows how inertial drives might be implemented mechanically for rotation around the altitude axis and the orthogonal 'train' axis. Weights are attached to the ends

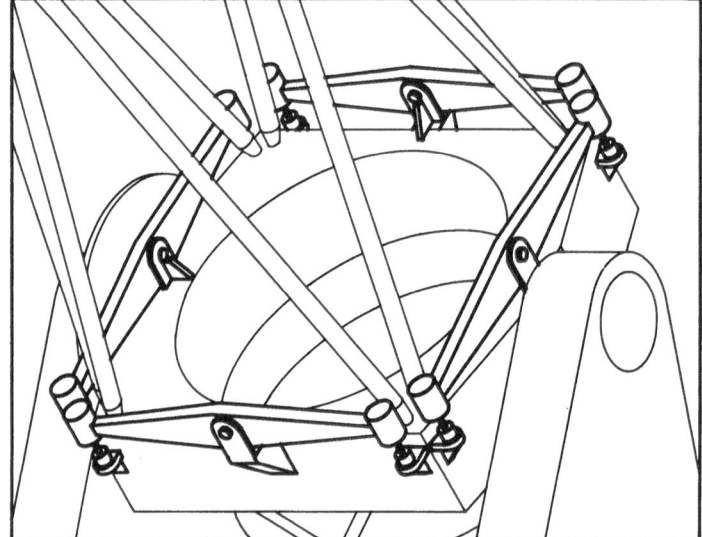

Fig. 1.   A possible mechanical layout for inertial drives around two orthogonal axes.

of pivoted levers ('seesaws'). Each weight is connected to the centre section through an electro-magnetic driver. To minimise torsional vibration of the centre section, two seesaws can be used for each drive axis.

4. Simulation of Inertial Drive

To explore the practicality of using an inertial drive, a simplified model, as shown in Figure 2(a), has been simulated in a Fortran program. Figure 2(b) shows the model with angular rotation represented by linear motion, so the various elements can be differentiated. The model represents rotation around only one axis, but the same principles apply to drives around both altitude and azimuth axes.

Each millisecond, the torques applied by the shafts, dampers, and motors were calculated and the resulting motions used to update the calculations for the next millisecond. As an example, the total torque on wheel 3, representing the optics assem-

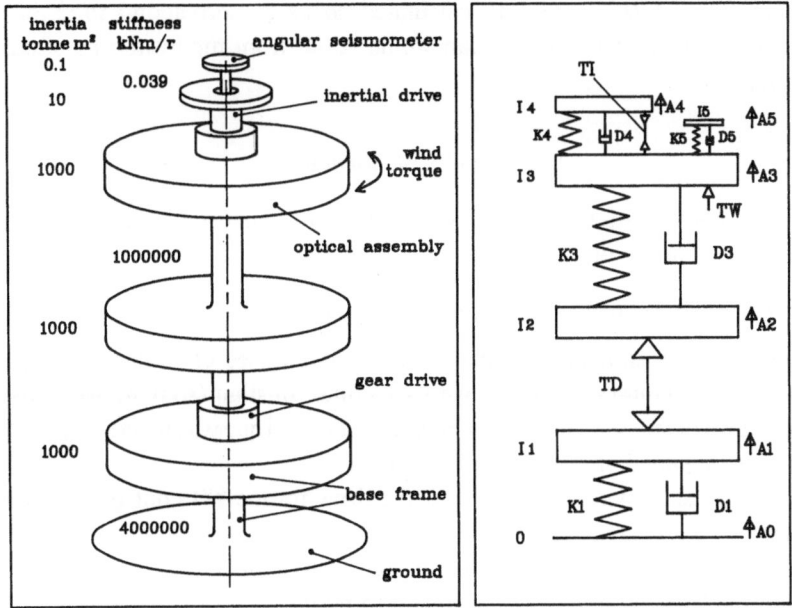

Fig. 2.   (a) Rotational model. (b) Model redrawn with rotation represented by linear motion.

bly, is given by

$$T3 = -K4(A3 - A4) - D4(V3 - V4) - K3(A3 - A2) - D3(V3 - V2) -$$
$$- K5(A3 - A5) - D5(V3 - V5) + TW + TI,$$

where K is a torsional stiffness, D a damping constant, A an angular displacement, and V an angular velocity, and TW is the torque due to wind. The torque TI applied by the Inertial Servo in trying to maintain a fixed orientation of the optics assembly 3, may, as an example, be given by

$$TI = -KI(A3 - A5) - DI(V3 - V5),$$

where the angular error inferred for 3 is from the motion of 3 relative to 5, which is the inertial mass of an angular seismometer.

TD in Figure 2(b) represents the torque applied by a conventional gear drive. If only an inner gear loop was implemented, TD would be a function only of the differences in angle and angular velocities between wheels 1 and 2. The efficacy in stabilizing the optics of a servo applying torque at only this point is limited by the compliances in its links to the optics and to the ground. The inertial drive avoids these limitations.

4.1. STANDARD CONFIGURATION USED IN SIMULATION

For the purpose of the simulation reported here, standard values of the system parameters were adopted. The rotational inertias and stiffnesses were as shown in Figure 2(a); note that the inertia of the drive seesaw was $\frac{1}{100}$ that of the optics assembly. The damping coefficients $D1$ and $D3$ were both calculated so as to correspond to $\frac{1}{20}$ of critical damping (i.e., substantially higher than would result from the hysteresis of structural steel alone).

5. Results of Simulation

Several servo configurations and several types of disturbance have been simulated; only a few examples can be discussed here.

5.1. ATTEMPTING TO HOLD OPTICS STATIONARY IN WIND

Wind buffeting was simulated by generating a time series of torques from gaussian distributed random numbers, filtered with a 1 s time constant (so that, over most of the frequency range of interest, the r.m.s. amplitude of the torque was inversely proportional to frequency). Figure 3 shows the performance:

(a) with only an 'inner gear loop': i.e., the drive torque TD controlled to minimise $A2 - A1$;

(b) with an 'outer gear loop': TD controlled to minimise $A3$;

(c) with the 'inertial drive': TI controlled to minimise $A3$ (in addition to inner and outer gear loops).

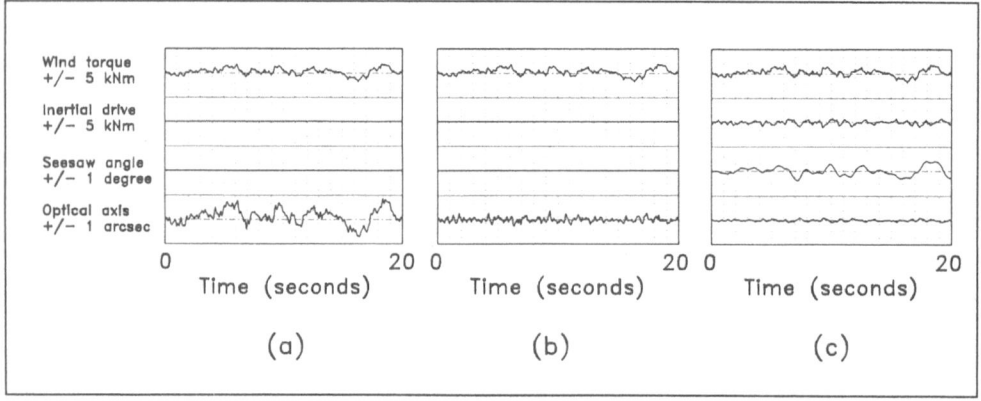

Fig. 3.   Simulated performance, reacting to wind.

In cases (a) and (b), the gain and damping were adjusted to minimise the r.m.s. error. In (c), the gain was restricted rather arbitrarily to keep the seesaw angular range within bounds. The results show that the inertial drive can greatly improve performance with reasonable seesaw angles.

5.2. USE OF AN ANGULAR SEISMOMETER TO CONTROL THE INERTIAL DRIVE

In general it is not possible to measure $A3$ absolutely, as assumed above. A proposed method of sensing errors in $A3$, when a suitable guide star is unavailable, is to use an 'angular seismometer' and measure angle $A3 - A5$ (see Figure 2(b)). (The inertia, stiffness, and damping of the angular seismometer were chosen to give a period of 10 s with critical damping.) The 'error signal' $A3 - A5$ was used to control the inertial drive loop in the next example (see Figure 4), while the outer gear loop was controlled to minimise the seesaw angle. With the same 'filtered wind' disturbance, an even better result was achieved than above (higher gain could be used without excessive seesaw angles).

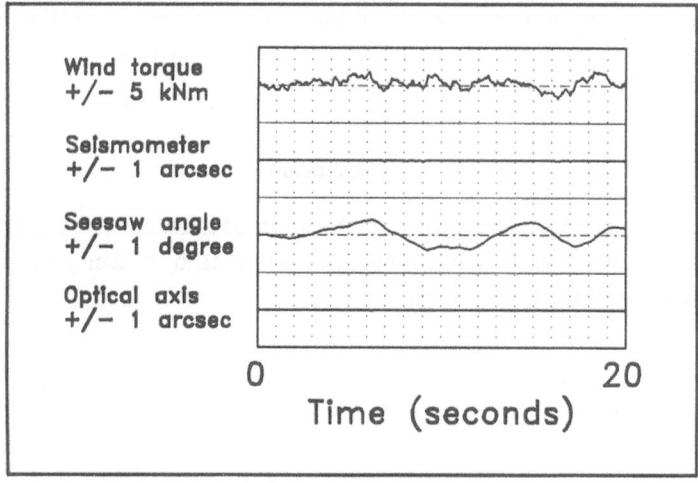

Fig. 4.   Performance with seismometer control.

5.3. USE WITH AUTOGUIDING

Seeing motion of a star was simulated (from the same sequence of filtered gaussian noise as for the wind torque) and used as the demand for $A3$. Figure 5 shows the performance, (a) with the outer gear loop and (b) with the addition of the inertial loop. Again much better performance is available with the inertial drive.

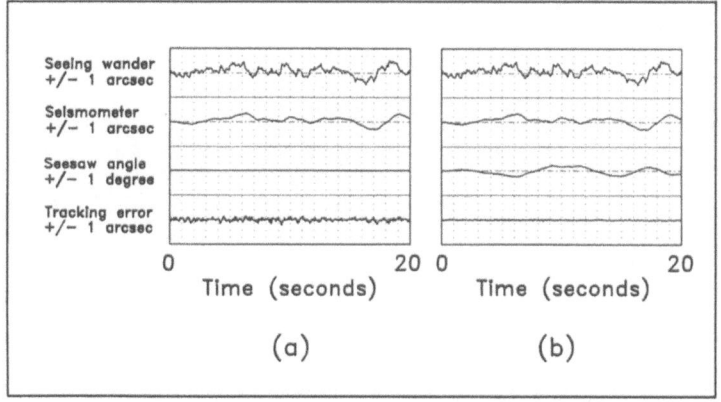

Fig. 5.   Performance with auto-guiding.

5.4. Coping with ground vibration

Simulating vibration input from the ground – making $A0$ in Figure 2(b) a function of time – also showed the advantage of the inertial drive, with the important further point that the performance is even better if the base frame and drive stiffnesses are *reduced*. This suggests that, at least on some sites having problems with ground-transmitted vibration, deliberate vibration isolation, e.g., with air springs, may be advantageous. It also demonstrates that mounting a telescope on a tall tower, for better elevation above seeing disturbances, need not be excessively expensive since, if one relied on inertial drives for vibration control, the tower could be quite compliant.

6. Conclusion

Inertial drives appear capable of giving much improved telescope dynamic performance together with big cost savings in building large telescopes, by reducing the need for very stiff structures and ultra-smooth gear drives.

THE DESIGN OF THE CENTRAL-AXIS REFLECTOR (ZAS)[*]

Hermann Hügenell

C.F. Angstenberger, Consultants

Bismarckstraße 116, D- 6700 Ludwigshafen am Rhein

West Germany

(Received 28 February, 1989)

ABSTRACT

The Central-Axis Reflector, the design principle which is presented below is a segmented-mirror telescope.The inventions relate
mainly to the optical system and to the tracking apparatus.
A large number of small individual mirror bodies, ground off-axis
(hexagonal/polygonal) produce one primary mirror with closed circular aperture when joined together.
The overall design of the tracking apperatus results directly -
and thus without unnecessary adornment - from the two planes of
motion which have been reduced to a minimum but which are required for tracking of the telescope.

INTRODUCTION

In order to find a way of achieving large lightgathering areas, attempts have now been made,
chiefly in the USA, to combine several ajoined individual elements to produce one primary mirror
with a large diameter, the Multiple Mirror Telescope (MMT). This principle is now been implemented in the "Keck Telescope" on Mauna Kea, Hawaii. As a Segmented Mirror Telescope(SMT) it is the
first design of this kind.

1. Manufactoring method for the primary mirror of the ZAS

(by way of example of a 20 m primary mirror)

If a large diameter of a segmented primary mirror as a reflector is to be obtained, it must be
ensured that both the manufacture of the mirror bodies and grinding and polishing the mirrors and
their mass (dead weight) remain within feasible limits. On the basis of this considerations, the
dimensions of the individual mirror segments allowing for all static requirements, must be designed so that they remain feasible and manageable.

The subdivision shown in Fig. 1 was thus selected for the 20 m primary mirror. This area sub -
division shows the maximus longitudinal extend of a mirror body (inside the primary mirror) to be
1.35 m. Fig. 1 also shows that this configuration of the primary mirror comprises 11 different
mirror forms, various quantities of which are required the lowest quantity being six.

Fig.1

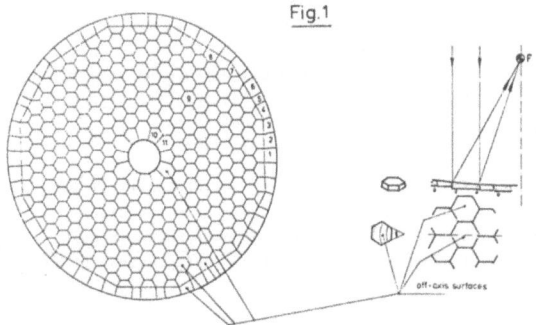

Tab. 1

Mirror bodies with No.	No. of mirror bodies when manufacturing the blanks	
1	6	
2	12	
3	12	
4	12	
5	12	
6	6	366 mirror blanks
7	18	
8	30	
9	246	
10	6	
11	6	

* Paper presented at the Symposium on the JNLT and Related Engineering Developments, Tokyo, November 29–December 2, 1988.

Astrophysics and Space Science **160**: 231–239, 1989.
© 1989 *Kluwer Academic Publishers.*

Fig. 2

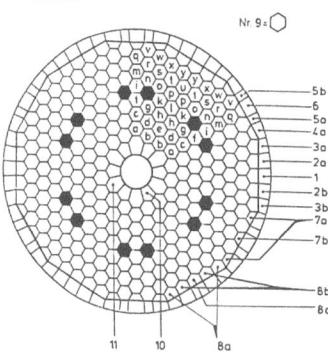

Fig. 3a

Tab. 2

Number of mirror body	Frequency of identical operations when preforming, grinding and polishing the mirror blanks, Laser hologram inspection
1	6 times
2a	6
2b	6
3a	6
3b	6
4a	6
4b	6
5a	6
5b	6
6	6
7a	12
7b	6
8a	12
8b	12
8c	6
9a	6
9b	12
9c	6
9d	12
9e	6
9f	12
9g	12
9h	12
9i	6
9j	12
9k	12
9l	6
9m	6
9n	12
9o	12
9p	12
9q	6
9r	12
9s	12
9t	12
9u	6
9v	12
9w	12
9x	12
9y	12
10	6
11	6

366

Fig. 3b

Photographs: Heraeus Quarzschmelze

Fig. 2 shows the positional arrangement of the mirror segment 9j within the primary mirror and demonstrates that this, as shown in table 2 is manufactured 12 times with the same machining operations for technical preforming as a blank for grinding and polishing and for surface inspection using a laser hologram.

For colour reproduction of these figures see colour section.

Table 2 shows the frequency of identical machining operations when preforming the blanks by the manufacturer and those for grinding and polishing and inspection. The data shown in both tables (1,2) have a positive effect on the economic feasibility study for manufacturing the finished primary mirror.

2. Structure of a mirror body made of quartz

Fig. 3 shows the structure of a mirror body made of quartz for a circular mirror. The copper plate and the base plate, approximately 6 mm in thickness are absolutely congruent and are joined homogeneously to the prefabricated honeycombed structure which is arranged between both plates in

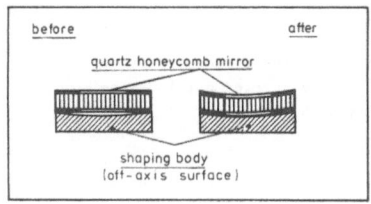

Fig. 4

the smelting furnace in a separate operation and
at a specific temperature (see manufacturing process of Heraeus-Quarzschmelze, Hanau) in such a way
that the finished mirror body can be inserted,with
a weight reduction of over 80 %. The same manufac -
turing principle can also be used for the multi -
form-complex segments of the ZAS.

The mirror body has a thickness of approximately
15 cm is then technically preformed in such a way
that only fractions of a millimeter need to be re -
moved during the subseqeunt grinding process. This preforming method which substantially reduce grinding and polishing work permits astonishing short manufacturing times for the overall
primary mirror and thus has a positive influence on economic factors accordingly. This is be -
cause only approximately 700 kg of quartz need to be removed by grinding from the overall 20 m
mirror. (By comparison: on the 5 m Palomar mirror, it was necessary to remove apprx. 5 tonnes
of glass when grinding). All necessary mirror blanks, of whatever form can be prefabricated
with the method shown in Fig. 4. Of their surfaces to be polished, as off-axis sections, only
extremely slight layers need to be ground off and polished. Every mirror segment preformed in
this way can be supplied by the manufacturer (Heraeus-Quarzschmelze) with surface tolerances
of 1 mm.

3. Structure of a mirror body made of Zerodur

Honeycombing with Zerodur is also possible. But at present it cannot be preformed like it
is possible with quartz. And with respect of quartz-preforming it could be more labor-inten-
sive to produce those segments. A proper structural study must be done with both materials.
Of course there is the need of temperature control on quartz but the situation is not critical particularly in ho-
neycombed structures.

Fig. 5. Light weight Zerodur
mirror made by Schott Glass-
werke, Mainz. For colour
reporduction of this figure
see colour section.

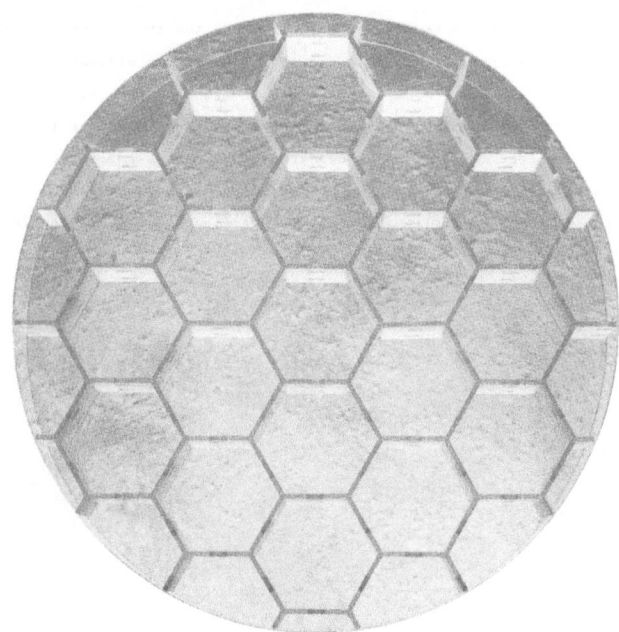

4. The grinding and polishing method

In order to permit each off-axis section, with whatever shape, to be finished it was necessary to develope a grinding and polishing method which can be carried out today on existing, polar coordinate- controlled grinding and polishing machines.

Fig. 6 shows the principle of this method and the positive guidance (3 degrees of freedom) of the grinding tools. This grinding method on which the guide plinth (1) and on which the bearing block (3) with the mirror body (4) mounted on it is moved by means of displacement in accordance with its positioning within the primary mirror finishes the mirror surfaces which lie well below the values required for astronomy ($\pm$ 30 nm).

5. Surface inspection by means of Laser hologram

After finishing of the mirror surfaces, the mirror segments are moved on a runner to the

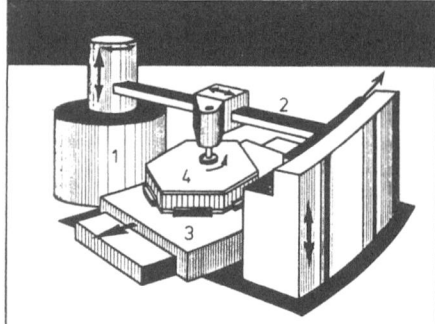

Laser hologram tester which is attached directly to the grinding machine. This tester inspects the mirror surfaces in order to establish whether they complied with the required tolerances and, if necessary, the mirror surfaces are remachined on the polishing unit.
This dispenses with the need for the expensive, high testing tower and the interruptions in work required for this.

Fig. 6

6. Mounting and adjustment of the individual mirror bodies

All mirror segments made of quartz, taken together, have an overall mass of approx.15 tonnes. (By comparison: Mount Palomar = 15 tonnes, 8 m element ESO-VLT = 15 tonnes). An extremely sturdy and stable mounting platform is required for precise and adjustment of these masses in order to also guarantee tracking of the entire telescope in the sky. In this respect, the Keck Telescope which is being constructed on Hawaii poses the same mounting problems. Enquiries established that the study results are already available for the 10 m mirror. These results will substantially shorten the investigations before the 20 m mirror of the ZAS.

Machine-controlled interferometer. Small workpieces are measured over their entire area. Larger workpieces are observed and measured in scanning mode.

Fig.7

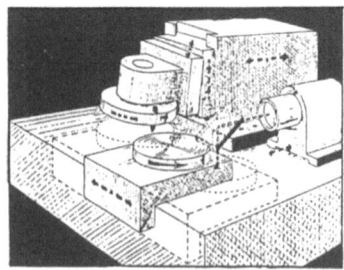

In addition, using modern (such as carbon fibre-reinforced plastics) will have a very favourable effect upon technical measures related to interior fittings as illustrated by way
of example of one possible method:

Pinned to its bearing plate, each lower segment will already be aligned mechanically to
within fractions of a millimeter at the focal point. Fig. 8 shows the bearing plate of a
hexagonal segment with the hydraulically controllable retaining pins which can align and
controll the mirror body in 3 planes of motion.

This bearing plate should be able to be moved at least 50 cm against the direction of
the incident light in order to guarantee rapid fitting and removal of a mirror segment,
e.g. for remetallization.

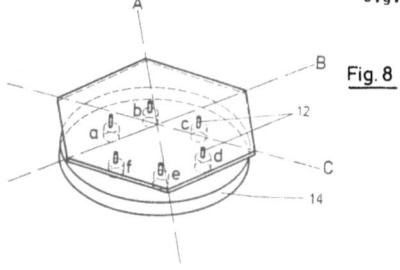

Fig. 8

A, B and C are the tilting
planes,
12 = retaining pins for the
mirror bodies
13 = bearing plate
14 = mounting platform
15 = pressure valve

Fig.9

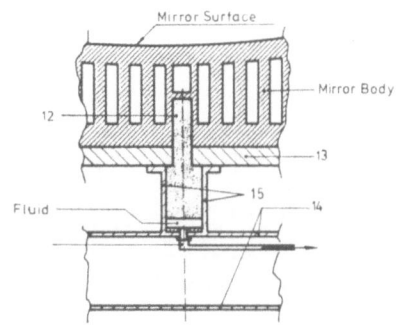

Fig. 9 shows a section through the bearing
system. From an optical point of view, the
lack of one mirror body within the 20 m primary mirror has no effect. This means: it is
possible to make optimum use of the telescope.
Surfaces which are to be remetallized may be
removed step by step and recoated accordingly
without disturbing operation of the telescope.
Remetallization is performed in a high-vacuum
chamber which should be located near to the
mirror segments which have a maximum size of
approx. 1.35 m; the ZAS can be considered to
be an economic solution.

7. Sensors, measuring and control system, precise adjustment

When all mirror bodies are mounted on the mirror platform, they are already aligned, as
previously mentioned, to within an accuracy of fractions of a millimeter to the focal point.
A sensor system with an ultra-fine measurement resolution and which is cabable of recording
the position changes of each mirror body in the lowest nanometer range is required in order
to achieve the necessary, minute deviation tolerances in the focal point of approx. + 5 nm.
Meeting this extremely difficult technical engineering requirement was a difficult nut to
chrack on the Keck Telescope. During that project, attempts were initially made to measure
the positional change using piezoelectric position controllerswhich project into the reflection field and to compensate for this positional change mechanically using a control element.
This concept has now been dropped since the intrinsic thermal radiation of the infrared signal to be received from outer space would have entailed too much of a disturbance.

An alternative, several systems developed for the Keck principle, which were mounted be-
neath the thin mirror or are even, if mirror thickness was appropriate, mounted in the mirror body, permitting focal point tolerances of + 10 nm. A tolerance of only 30 nm for low
frequencies is required for astronomy purposes. The thickness of the mirror bodies of the
ZAS is eminently for such a sensor since as early as the manufacturing stage for the mirror
blanks, preparatory measures can be taken to prevent influences to the inherent stability
of the mirror.

Thus in conjunction with a central monitoring and control unit, all 366 individual mirrors can be precisely adjusted to such an accuracy that the required focal point tolerances
are reached. The technical control components of the adjusting apparatus have also been in-
investigated in order to meet the stringent requirements of the deviation tolerances which
lie in the region of nanometers. The fact that this state of art must, of course , be in -
vestigated in order to establish whether it can be used for the 20 m mirror is understand -
able, solely of the fact that such a large mirror has never yet been built. However, to
conclude, we must state that the investigations to be con -
ducted which relate to adjustment comprise only applica -
tionable and adaptional investigations of a technology which
has already been developed. Thus, the solutions illustrated

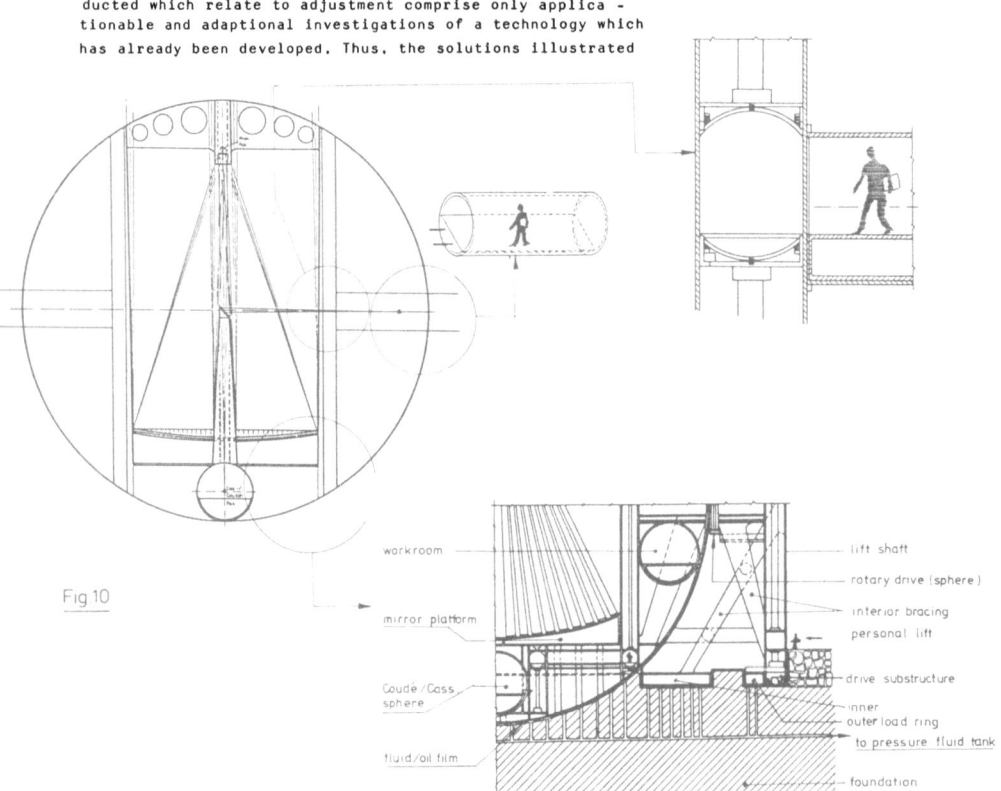

Fig 10

are feasible proposals. Variants of the mounting system on the basis of state-of-the-art technology are conceivable.

8. Optical path and image processing

    The dimensions of the tube are derived from the aperture ratio . In the case of the 20m
ZAS, this value would be 1/1.5. Of course , it is possible to achieve even smaller aperture
ratios(e.g. 1/1) with the manufacturing method for the primary mirror developed for the ZAS.
This would, admittedly, shorten the overall length of the tube which was to date absolutely
necessary, with plastic telescope construction, in order to reduce the moving masses. How -
ever,on the ZAS whose tube is to be incorporated in a sphere and on which the axis of rota-
tation acts in the center of the tube length in the altazimuth mounting, it was possible to
retain this aperture ratio of 1/1.5. This does,of course also effect the shaping of the individual mirror bodies. This means that the extent of the edge camber of thus designed primary mirror remains feasible (approx. 83,3 cm axis - edge).

As can be seen in Fig. 10, the recording room for the Cassegrain/Coude focus in form of a
sphere, is located the mirror platform. Access can be gained to it in the axis of rotation
of the inner platform which moves on a fluid/oil film via lifts and a catwalk. The sphere
(44 m diameter) which incloses the tube offers adequate space with its enclosed volume of
almost 45.000 m^3 to accomodate such workroom in the sphere without disturbing optical image

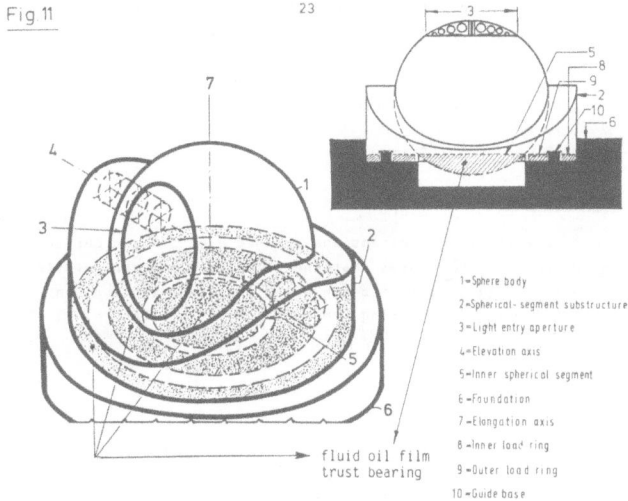

Fig 11

1 = Sphere body
2 = Spherical- segment substructure
3 = Light entry aperture
4 = Elevation axis
5 = Inner spherical segment
6 = Foundation
7 = Elongation axis
8 = Inner load ring
9 = Outer load ring
10 = Guide base

fluid oil film
trust bearing

recording at all.
Fig. 10 also shows the optical path of the incident light on the basis of its marginal rays.
The central whole in the primary mirror, with a diameter of approx. 2.5 m, is filled by a
retaining tube for the tilted mirror of the Nasmith focus, the tube becoming narrover in the
direction of the focal point. This permit the system to be converted to the Cassegrain/Coude
focus within as short a time as possible. For this purpose, the light which is reflected at
the secondary mirror is routed through the retaining tube into the inside of the approx. 6 m
sphere from where it passes on directly to the recording instruments.

Modern image processing equiptment such as larger CCD- cameras, electronic image evaluation
systems, remote picture configuration technology as a controlling function, tv- cameras and
a great deal more are obligatory.

9. The great eye, a future-oriented tracking apparatus

Fig.11 shows the great eye, the 44 m sphere with a tube aperture slightly greater than
20 m and the spherical-segment substructure in which the sphere, together with all interior
fixtures, rotates on a fluid/oil film about the unstressed suspension axis.
The spherical-segment substructure also rotates, floating on a fluid/oil film, with the
sphere which it guides positivly, about the vertical axis of the telescope. All masses are
thus elevated and require only very low rotary drive forces. The telescope can thus be aligned to any position in the sky with this dual-
plane contol system.

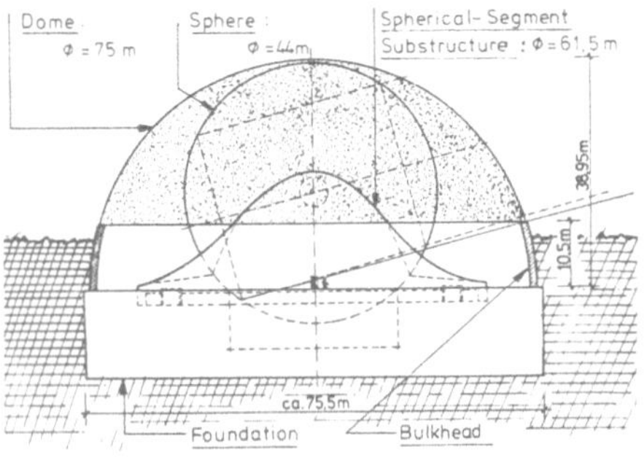

It is also shown how the spherical-segment
substructure got its shape. It can be said
that: the overall design of the tracking apparatus results directly - and thus without
unnecessary adornment- from the two planes
of motion which have been reduced to a minimum but which are required for tracking of
the telescope.

Fig.12

10. Protection against atmospheric influences

Such a system requires a protective dome with optimum dimensions spanned over it in order to
protect it against atmospheric influences. Fig.12 shows that the geometrical characteristics
of the telescope permit a hemispherical protective structure on which the lower section con -
sisting of a reinforced concrete wall being the loadbearing component of the hemisphere, is

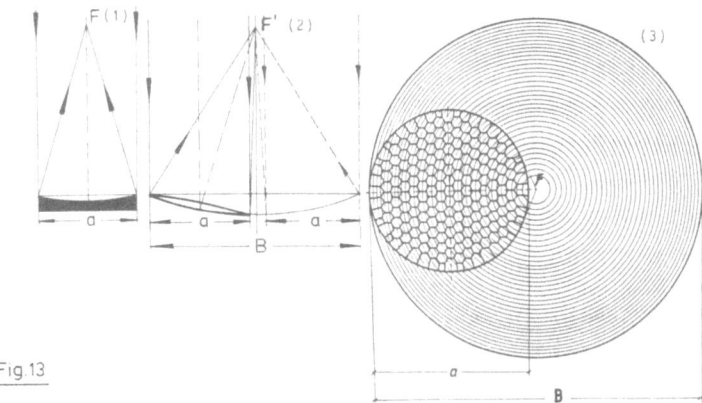

Fig.13

designed such that the rotating dome can be fitted directly on it. The dome, made of steel,
titanium or carbon fibre reinforced plastic, for instance, may be coiled on a self-supporting
framework and would have an overall height of approx. 28.5 m. This would mean that the dead
weight of such a large dome could be reduced substantially. The opening gap which would have
to be made substantially, with a clear width of approx. 21 m can be sealed off to provide a
watertight seal using various methods. This may be achieved, for instance, by spherical part-shells which can be traversed on the inside or on the ouzside.

Fig.14

— Primary Focus

— Lift

Entry

Sectional view through the finished
Central-Axis Inclined
Reflecting Telescope'

Secondary Focii

—Primary Mirror
(off-axis)

11. One variant of the ZAS: The Central-Axis Inclined Reflector

   Fig.13   shows that, on the basis of the abovedescribed principle of geometrical area
subdivision of a 20 m mirror it is now also possible to create a very large mirror which
represents an orbicular section of an even larger, hypothetical large paraboloid/ hyper-
boloid.
The surface of the mirror body with diameter a and the focal point F, ground in rotationally fashion shown in diagram (1) is altered in such a way that it corresponds to an orbicular section (off-axis)of the hypothetical large mirror with diameter B (2) the surface

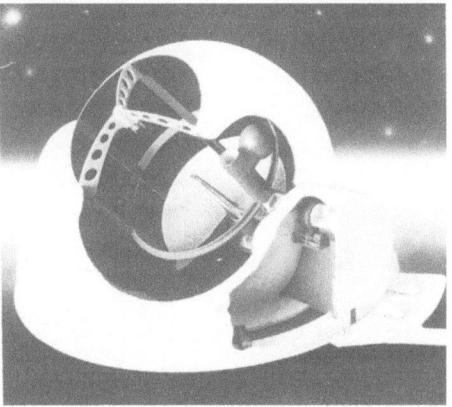

Photograph of
the ZAS -model
at the Tokyo -
Symposium on
Large Telescopes
1988

of which is shaped in
rotationally symmetric
fashion. Diagram (3)
shows the surface cha-
racteristics of our or-
bicular section with
diameter a. The primary
mirror with this shape
can now also be produced
using the grinding and
polishing method which is
described above.

   Fig.14   shows a sectional view through the finished "Central-Axis Inclined Reflector"
with non-occluding light entry aperture, primary focus lying outside the incident light cone
and the other focus variants shown here. For colour reproduction of this figure see colour section.

NEW PLANS AND ACHIEVEMENTS WITH EXISTING TELESCOPES

CHARACTERISTICS OF MAUNA KEA RELATING TO THE JAPAN NATIONAL LARGE TELESCOPE*

D. N. B. HALL

University of Hawaii, Honolulu, U.S.A.

(Received 26 March, 1989)

Abstract. The characteristics of the Mauna Kea site are reviewed. An extensive site survey by NOAO during 1984–1985 showed that the average seeing was in the 0.4–0.6″ range and that r_0 was approximately 30 cm. An image quality study at the UH 88-inch telescope in 1987 showed the free atmosphere seeing was 0.5″ and the boundary layer contributed less than 0.25″.

This paper reviews the characteristics of Mauna Kea, particularly as they relate to the Japanese National Large Telescope (JNLT), giving particular emphasis to the image characteristics as we now understand them. The tropical inversion, normally located at an elevation of 2.4 to 3.0 km, traps cloud and moisture below it. We now believe that isolated island peaks probably provide the best sites we know of today for optical and infrared astronomy. There are two such sites in the Northern Hemisphere: Hawaii and Canary Islands. There appear to be no comparable island sites in the Southern Hemisphere. The best candidate, Reunion in the Indian Ocean, does not have mountains high enough to get into clear air enough of the time.

The typical astronomical characteristics of Mauna Kea are summarized below:
- elevation 4.2 km,
- pressure 616 mb,
- latitude $+20°$,
- dark sky,
- $\sim 50\%$ photometric nights,
- $\sim 70\%$ useful nights,
- low H_2O and IR emissivity,
- excellent image quality.

The image quality is essentially that of the free atmosphere, indicating the isolated summit does not appreciably perturb the airflow in the upper atmosphere. The site is protected by strong lightning ordinances by the County of Hawaii, including a revision enacted this year, which will ensure that there will be no deterioration in the foreseeable future, i.e., well into the next century. University of Hawaii's (UH) policies for the Mauna Kea Science Reserve also prohibit new radio frequency transmitters within the Science Reserve and will lead to phasing out of existing transmitters in the next few years.

* Paper presented at the Symposium on the JNLT and Related Engineering Developments, Tokyo, November 29–December 2, 1988.

Astrophysics and Space Science **160**: 243–248, 1989.
© 1989 *Kluwer Academic Publishers.*

The most significant astronomical characteristic of Mauna Kea is undoubtedly the superb image quality. Several years ago, the University of Arizona, Smithsonian Astrophysical Observatory, and the UH collaborated with the National Optical Astronomy Observatories (NOAO) to carry out exhaustive characterization of several sites. This included extensive monitoring of a number of key parameters on Mauna Kea over the year from December 1984 to late-1985. In particular, the survey carried out differential image motion measurements from a 30 cm telescope and produced results which indicated that the median r_0 on Mauna Kea was of the order of 30 cm corresponding to a medium image full-width–half-maximum of around 0.4″ (Figure 1). A scaling of this

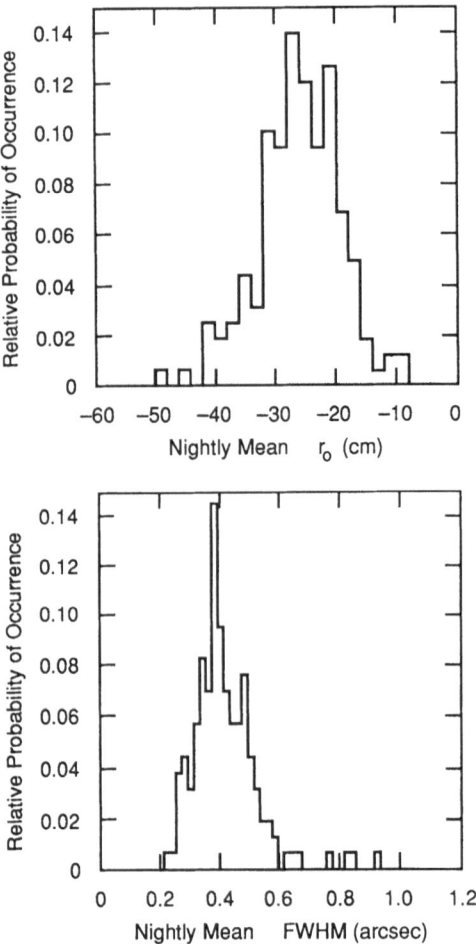

Fig. 1.   Nightly mean derived values for full-width–half-maximum and r_0 for the NOAO site survey.

to telescope diameters in the 2 to 4 m range indicated that the normal images for existing telescopes on Mauna Kea should nearly always be in the 0.4–0.6″ range.

This result was very surprising and raised some questions as to the validity of the measurement technique. In particular, it raised the question as to why the existing telescopes on Mauna Kea were not already routinely achieving that level of image quality.

Since the NOAO survey, UH and Canada–France–Hawaii Telescope (CFHT) staff have been attempting to resolve these questions. At the UH 88-inch telescope there have been a number of thermal changes. CFHT has made similar modifications which Bob McLaren discusses in his presentation. The major changes at the University of Hawaii 88-inch were to block off heat paths into the dome and to install fans which, during the day, circulated the air around the dome from off the refrigerated floor. These simple measures reduced the thermal gradient in the dome between the floor and the upper levels of the dome by about an order of magnitude. Since then, observers have remarked on a very substantial improvement in image quality. In order to assess the improvement at the 2.2 m telescope, in November 1987, we carried out an intensive two-week campaign to really understand all aspects of the image quality and we are planning a similar campaign in June 1989 to obtain comparable results under similar conditions.

The 1987 campaign at the UH 88-inch, consisted of a variety of techniques, including:

(1) the University of Nice's SCIDAR to monitor seeing in the free atmosphere above the summit of Mauna Kea;

(2) acoustic sounders to probe microthermal structure in the first few hundred meters of the boundary layer;

(3) 30 m towers instrumented with anemometers and microthermal sensors;

(4) direct CCD imaging at the UH 2.2 m. During the CCD imaging, great care was taken to focus correctly and to eliminate all sources of thermal disturbances;   (5) installation of a shearing interferometer at the 0.6 m telescope; and

(6) data from Hilo radiosonde ascents which are launched within 30 km of the summit.

During the campaign the seeing was measured at early int he night, around the middle of the night, and late in the night. The image quality was very consistently around 0.7″. The measured values are shown in Figure 2 together with the corresponding high-altitude contributions of the seeing as measured by the SCIDAR. Figure 2 shows that it remains very stable most of the time, although there are occasional periods when the high-altitude seeing blows up and dominates completely. Those periods were later found to be associated with periods of high wind shear in the upper atmosphere.

The reader is referred to the paper by Azouit et al. (1988) for details of the analysis of the June 1987 campaign. The key result that the high altitude image quality as shown here is the dominant contributor to the observed image size. The boundary layer associated with Mauna Kea was much smaller, 0.25″ or less, throughout the two-week period. If substantial image degradation (0.4″) associated with the $f/35$ infrared secondary is corrected, then the 2.2 m telescope should be limited by the free atmosphere seeing. We were not able to detect any image degradation due to the local dome/telescope atmospheric environment.

If the known image degradation associated with the 88-inch telescope $f/39$ secondary are taken into account, then the results are entirely consistent with the NOAO survey results (Figure 3).

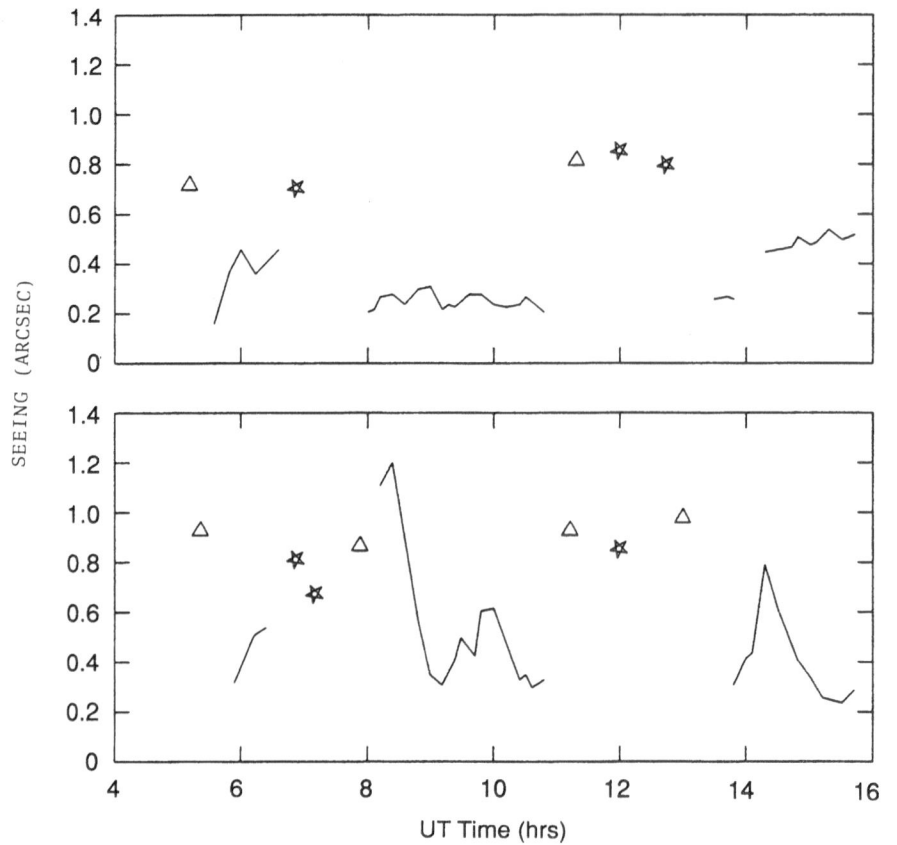

Fig. 2.   Comparison on two nights of the high altitude seeing variation (solid line) with the measured seeing
(SCIDAR = triangles; CCD = stars).

The basic conclusions from the June 1987 site campaign over the two-week period
were:

(1) The raw full-width–half-maximum image diameter over the two-week period was
0.73" before any correction;

(2) the free atmosphere throughout the campaign had a median value of 0.5", and
remained close to that value except for occasional blowing-up of the image associated
with transient shear layers in the upper atmosphere;

(3) there was substantial contribution associated with the telescope's $f/35$ secondary;

(4) the boundary layer contributed less than 0.25" over the two-week campaign; and

(5) telescope- and dome-seeing effects were not detectable at the 0.25" level.

Over the two-week period there was no correlation of the image quality with the
temperature difference between the primary mirror and the air, although during the
period the primary mirror ranged as far as 3 °C out of equilibrium in both directions.
We have been able to test for significant correlations down to the 0.4" level. This agrees
with similar experience at CFHT and does raise some questions about currently
accepted results, based primarily on laboratory experiments that there is a strong,

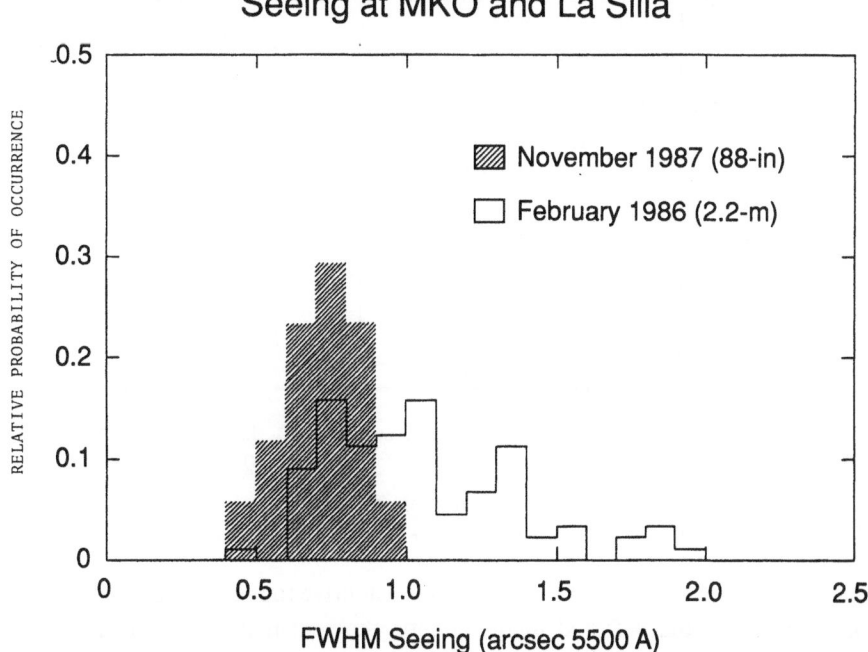

Fig. 3.   Comparison of seeing at Mauna Kea (shaded) and La Silla (unshaded) using the University of Hawaii 88-inch Telescope at Mauna Kea and the 2.2 m telescope at La Silla.

probably linear, relationship between the air/mirror temperature difference and the image quality through the telescope.

Thus we now appear to have reduced dome and telescope seeing effects at both the UH 88-inch and at CFHT, to where they are not a limiting factor. At the UH 88-inch we have problems ith our optics and tracking which are being corrected. All the indications are that, once that is done, we should achieve a median image quality of 0.5″ and best images of 0.3″.

JNLT is the first large telescope being designed for Mauna Kea since these results were established. The significant implications for the JNLT are illustrated in Figure 4 which gives the figure of merit, D/θ as a function of telescope diameter D for various image sizes. D/θ is essentially a measure of the limiting magnitude for point sources fainter than the sky background. A number of telescopes are identified on the figure. The key feature is the sensitivity of the JNLT where combinations of the 7.5 m aperture and superb images give huge gains in sensitivity to point sources. In the case of the extremely good seeing, the JNLT will achieve sensitivities very comparable to the best space facilities.

As we heard in the Introduction, and in later talks, the infrared is of vital interest for the JNLT. For random turbulence, the seeing should only improve as $\lambda^{-0.2}$, implying 2.2 μ median-image quality of 0.4″ with best imaging down to around 0.25″. The parameter r_0 scales as λ to a median value of 1.5 m and a 10% value of 2.2 m at 2.2 μ, an appreciable fraction of the JNLT mirror diameter. With a median r_0 of 10 m at 10 μ, the JNLT will normally operate at its diffraction limit of around 0.3″. It is clear that

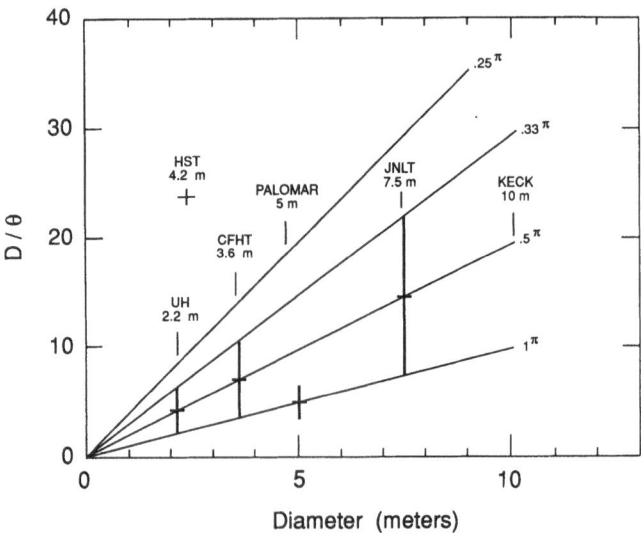

Fig. 4.   Figure of merit D/θ for representative large telescopes.

the imaging requirements in the 2 to 5 μ region will drive the requirements on telescope performance. With this infrared image quality, the sensitivity of the JNLT to point sources is greatly enhanced and in fact approaches cryogenic telescopes in space, simply because of the very large improvements in angular resolution and the corresponding reduction in background. In the infrared the JNLT will provide major gains over the performance of the best existing 3 to 4 m-class infrared telescopes.

At visible wavelengths the case is often made that the gain with aperture of a 7 to 10 m-class telescope can be offset by integrating correspondingly longer (3 to 6 times) on a 4 m-class telescope. However, dark time is not available in unlimited amount and there is a maximum of about 30 nights of dark time in a year to get long integrations on a particular object. Allowing for typical weather and six-hour integrations each night, there really is an upper limt of around 100 hours of integration time on a particular source. It may seem that a hundred hours is a much too high estimate to take for integration time. However, a number of UH programs with the Mauna Kea telescopes are based on ten-hour integrations. At this level, it is possible to do visible spectroscopy at CFHT down to magnitude limits of around 24, and IR imaging on UKIRT to K magnitudes somewhat fainter than 20. The 7.5 m aperture and superb image quality of the JNLT will allow observations to substantially fainter limiting magnitudes than these, opening many exciting observational opportunities at the limits of current technology.

References

Azouit, M., Cowie, L., Erasmus, A., Lugten, J., Roddier, C., Roddier, F., Songaila, A., and Vernin, J.: 1988, *A Description of Results from the November 1987 Mauna Kea Site Campaign* (in prep.).

KODAIRA – How have you determined the partial contribution from the telescope optics, 0.4" ?
HALL – The image degradation is dominated by errors in the $f/35$ infrared chopping secondary which was used to achieve the correct image scale. We are obtaining a new $f/35$ secondary.

OPERATIONAL AND TECHNOLOGICAL DEVELOPMENTS FOR UKIRT*

T. J. LEE

Royal Observatory, Edinburgh, Scotland

(Received 10 January, 1989)

Abstract. There are a number of parallels between the UKIRT and the JNLT both of which are major astronomical facilities of nations which do not have premier observing sites in their own country. Some elements of experience with UKIRT relevant to the JNLT are described. These include matters related to personnel and to instruments.

1. Introduction

There are a number of parallels between UKIRT and the JNLT. This paper will present two topics which have contributed to the success of UKIRT as a National Telescope for a wide variety of users and of astronomical programmes. These two topics are quite different although their aim, to enable effective astronomy to be carried out at infrared wavelengths, is the same. First, I will discuss some of the human aspects of telescope operations and then the instrumentation philosophy with examples of the present state-of-the-art.

Observatory experience within Britain up until the mid-1970's was with the telescope, operational support, engineering support, instrument development and astronomical research activities all in the same place. Clearly this is difficult to implement on any high mountain site and extremely so, at 4200 m, the altitude to which one must go for the full range of ground-based observations. Thus UKIRT, a wholly British telescope on foreign soil, was for us a new venture. The international aspects in the form of an agreement between the SERC and the University of Hawaii have been straightforward and positive.

2. Human Aspects

Human beings are indispensable to all but the most simple of the operating functions and dominate the research and technical development. Success is dependent on attracting, retaining and motivating high quality staff. The further one is away from normal living environments the more one is constrained in the choice of people one can involve in the various activities related to the telescope. Astronomers have an urban bias but form only about 20% of the staff of a modern observatory, this bias can be expected to be stronger for he other 80% of the employees such as engineers, technicians, and administrators. In modern industrial societies a good home life has become an expec-

* Paper presented at the Symposium on the JNLT and Related Engineering Developments, Tokyo, November 29–December 2, 1988.

Astrophysics and Space Science **160**: 249–253, 1989.
© 1989 *Kluwer Academic Publishers.*

tation and this must be considered in the management of work. Specialists require the stimulation of the more challenging parts of their discipline to work at their best. These factors must be considered carefully in engineering and operational plans.

Activities connected with UKIRT have been distributed as shown in Table I. On the summit, the most hostile and remote of the environment functions are kept to the

TABLE I

		MK OBSERVATORY
		CFHTelescope
ROYAL OBSERVATORY	JOINT ASTRONOMY CENTRE	UKIRTelescope
EDINBURGH	HQ, Hilo	JCMTelescope
		CSObservatory
		Keck Telescope
		Others
Policy and Direction	Management and Organisation	Operations
Long range planning	Software Development	Maintenance
User Liaison	Instrument updating	Astronomy support
New Instruments and systems	Repair of subsystems	First Line repair
Service observing	Engineering support	Instrument changes
Research	Astronomy support	Service observing
	Research	

minimum required for good observations to be made. Technology is used to cut down the amount of work required here. Importantly it is recognised that no staff member spends all of his work time at the telescope but part of it is spent at the sea level base. Appropriate work for all staff exists at the base.

Location of the Sea Level Base in the town of Hilo was deliberate to obtain optimal support from staff. More than 50% of the poulation of the Big Island live within 30 min drive of the base facility. For the majority of people Hilo gives greatest chance of access to health care, education, entertainment and employment for family members as well as housing at reasonable cost, also importantly in finding support outside the home in case of absences on the mountain, a necessary part of telescope life.

The separation of work between the sea level base in Hilo and the home base in Edinburgh is interesting. In practice home base technological staff are rarely engaged in the shorter term activities even of an urgent nature and thus the program for new developments has their full attention and it is possible to execute a well-managed and highly successful instrumentation programme with a low level of disruption. A continuing programme of developments to exploit new technologies is a key to doing leading research work in astronomy.

In summary, the disruption of work between summit, home base, and sea level base is important and the location of the sea level base crucial.

3. Technological Developments

For UKIRT the operational work takes place on the Big Island both on Mauna Kea summit and at the Joint Astronomy Centre headquarters in Hilo, and also in Scotland at the Royal Observatory, Edinburgh where we deal with such aspects as service observing and planning for the future in conjunction with our community of users. This also includes instrumentation plans, and procuring or building the instruments. It is important to remember that infrared technology is much less mature than visible technology so that in order to have state-of-the-art performance one must have a strong and continuing programme of new developments. We look for factors of ten improvements of overall performance on 4- or 5-yr time-scale.

To exploit any instrumentation it must be able to be used with relative ease by the average observational astronomer, otherwise a team of instrumentation experts is required at all times and this is not a realistic way of operating. Therefore, for UKIRT we adopted a plan for Common User instruments. This requires that the instruments are capable of automatic operation through a computer and that the instruments are well engineerd and reliable so that their performance is stable. Performance must also be competitive. The cost and effort to build such instruments lies somewhere between that of a graduate student project and a space instrumentation project. More specifically the equivalent cost is 10^8 to 10^9 yen.

Until about two years ago each of our instruments had just one or at best about a dozen detectors, this meant that data rates were relatively slow and it was difficult to make observations of large extended areas of the sky. In 1986 two-dimensional infrared detector arrays became available, this has caused very big changes. Fortunately we had been able to anticipate this and were in a position to build an infrared camera complete with the data acquisition system and image display and processing software so that we could make a complete instrument to exploit these two-dimensional arrays which have about 3500 detectors. Our first infrared camera, IRCAM (McLean et al., 1986) has been in action on the telescope for well over half the nights in the past year. This camera is just the first of a whole series of possible infrared instruments. Table II shows a range of examples of cameras and spectrographs which one might build to cover the range of short, mid, and long wavelengths. We have built two infrared cameras which can be used at different plate scales and have the ability to make polarimetric images and spectral line images by using filters or Fabry–Pérot interferometers. To do more extensive spectroscopy we have designed a grating spectrometer which has all of its optics cooled to cryogenic temperatures and the first version of this spectrometer will be undergoing tests in the laboratory a year from now. This spectrometer called CGS4 (Atad-Ettedgui and Mountain, 1988) will use a 58×62 iodium antimonide array and initially will be used at resolving powers up to about a 1000 for long slit spectroscopy. Later we can use an echelle to work at resolving powers up to about 20 000 and when larger arrays of 256×256 elements become available we can cover a greater spectral range.

Images which have been obtained with the IRCAM on UKIRT have been published (McLean, 1987a). Selected examples are shown.

TABLE II

CAMERAS		
IRCAM 1 2.5 < λ < 5.4 InSb 58×62 1.2 ARCSEC/PIX BROADBAND NARROW BAND FABRY PEROT POLARISER	IRCAM 2 2.5 < λ < 5.4 InSb 58×62 0.6 ARCSEC/PIX 2.4 ARCSEC/PIX BROADBAND NARROW BAND FABRY PEROT POLARISER	
"NEW IR CAMERA"		
SWIR 0.8 < λ < 2.5 CMT 256×256 .8, .4, .2, .1, .05 BROADBAND NARROWBAND FABRY PEROT	MWIR 2.5 < λ < 5.4 InSb 256×256 ARCSEC/MM NARROWBAND FABRY PEROT	LWIR λ > 8 Si:XX 56×56 1.2, .6, .3 NARROWBAND FABRY PEROT
SPECTROGRAPHS		
SWIRGAS CMT 256×256 or InSb 256×256 RESOLVING POWER $3×10^2$ or $2×10^4$	CGS4 InSb 58×62 RESOLVING POWER 200 to 1000	LIRGAS Si:XX 58×62 Si:XX OTHER 10^2 TO 10^4

In the future, infrared cameras can be designed and built to take advantage of larger arrays, and those which work in the intermediate infrared between 8 and 30 μ. At these wavelengths the high thermal background strongly influences the type of instrument which can be built. In the design of all of our future instruments we look at the suitability of the design for use on an 8 m telescope and we are also looking at configurations which would allow the use of adaptive optics which we believe are necessary to get the full performance of very large telescopes especially at wavelengths below the diffraction limit. This is presented more fully by Lee and Wade (1988) and Lee *et al.* (1988).

For many reasons, some of them given by Professor Hayakawa in his introductory talk, we expect that infrared observations will play a major part in the use of telescopes with apertures of 8 m above. These include the energy distribution of sources be they distant galaxies, active galaxies, regions of star formation, or dying stars. At the longer wavelengths one gains in resolution as well as in collecting area of the telescope and in the infrared one is looking through fewer atmospheric turbulence cells, the consequences of which were discussed by Professor Hall. Larger telescopes bring potentially greater gains to the infrared than they do to the visible.

References

Atad-Ettedgui, E. and Mountain, C. M.: 1988, *SPIE Proc.* **916**, 27.
Lee, T. J. and Wade, R.: 1988, 'The Edinburgh Infrared Instrumentation Package for a VLT', *ESO Conference on Very Large Telescopes and their Instrumentation,* Garching, 21–24 March, 1988 (in press).
Lee, T. J., McLean, I. S., and Weade, R.: 1988, 'A Camera for Infrared Astronomy and Its Performance on the 3.8 m UKIRT', *4th International Conference on Infrared Physics, Infrared Physics* (in press).
McLean, I. S.: 1987a, 'Results with the UKIRT Infared Camera', *Infrared Astronomy with Arrays,* Hilo, 24–26 March, 1987.
McLean, I. S.: 1987b, *SPIE Proc.* **287**, 138.
McLean, I. S.: 1988, *Sky Telesc.* **75**, No. 3, 254.
McLean, I. S., Chuter, T. C., McCaughrean, M. J., and Rayner, J. T.: 1986, *Proc. SPIE* **627**, 430.

BARR – Do you feel that chopping secondaries are still important for large telescopes?

LEE – Probably we will not need to chop to the background even in the thermal IR. However, we may need a secondary with tip tilt for image stabilization.

OKUDA – Could you tell me your experience of the effect of OH airglow emission in observations of near-IR, e.g., K-band? It is the main limitation of the observations?

LEE – OH airglow is the limiting background at $\lambda \leq 2.5 \, \mu$. There is a gap between 2.25 μ, the long-wave limit of the $\Delta v = 2$ bands and 2.6 μ where the $\Delta v = 1$ bands commence. Current IR detectors do not have the OH fringing problems seen in CCDs at shorter wavelengths.

OKAMURA – Could you tell us the integration time of the beautiful IR image of M51, or a typical integration time of the IR camera for the observation of external galaxies?

LEE – Five to 15 min per frame. We need to mosaic a number of frames for larger galaxies because our array is just 58 × 62.

LENA – The infrared plans for the European VLT may not be as grim as you described them, we are currently considering to give a high priority on the IR use of the first 8 m telescope to ensure the ground-based follow-up of the Infrared Space Observatory mission in 1995.

LEE – ISO is an important driver for our ground-based instrumentation programme for UKIRT.

RECENT DEVELOPMENTS AT CANADA–FRANCE–HAWAII TELESCOPE*

ROBERT McLAREN

Canada–France–Hawaii Telescope Corporation, Kamuela, Hawaii, U.S.A.

(Received 10 January, 1989)

Abstract. Recent developments at CFHT are described with particular emphasis in the following areas: image quality, techniques for high-resolution imaging, computers, and communications.

1. Image Quality

Figure 1 presents seeing statistics ⟨FWHM⟩ based on observer estimates from shortly after the telescope was commissioned until the present. It must be kept in mind that these data are quite heterogeneous in their origin, coming from four different foci and based on TV guide camera images, spectrograph slit images, CCD frames, etc. Nonetheless, it is clear that rapid improvement occurred in the early days as gross problems in

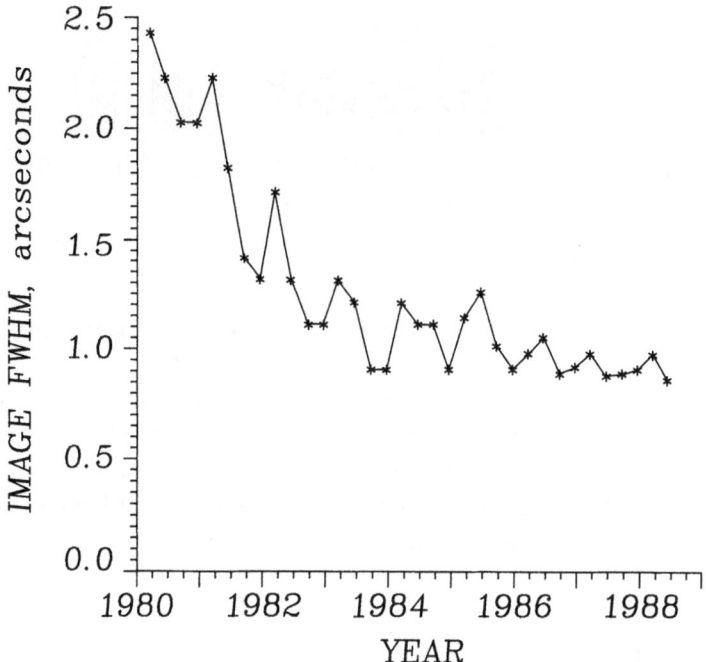

Fig. 1. Quarterly averages of the image quality ⟨FWHM⟩ at CFHT based on observer estimates for the period 1980–1988.

* Paper presented at the Symposium on the JNLT and Related Engineering Developments, Tokyo, November 29–December 2, 1988.

the optics were corrected and major sources of dome seeing eliminated. In recent years the seeing has stabilized around average values of 0″8 in the summer months and 0″9 in the winter. Cooling of the telescope hydraulic oil, which eliminated the largest single heat source (12 kW) in the dome (June 1987), has not produced a dramatic change in the average seeing, although it appears that the poorer than average seeing which used to occur in certain directions has now disappeared.

Figure 2 is based on a much smaller but more homogeneous and reliable set of image quality measurements. The period covered is April through September 1988, and only

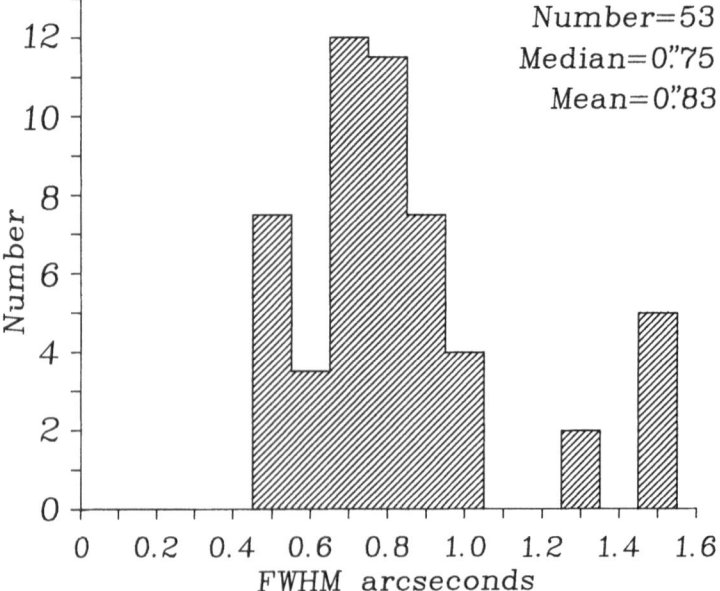

Fig. 2. Frequency distribution for image quality obtained on direct CCD images during the six-month period April–September 1988.

results from direct CCD images have been included. The 53 measurements have a median of 0″75 and a mean of 0″83.

Recent studies conducted on Mauna Kea (see contribution by D. N. B. Hall to this Symposium) strongly suggest that the average intrinsic seeing is $\langle FWHM \rangle \sim 0″5$. The question then arises as to what accounts for the difference between the image quality actually obtained and that provided by the site. The various contributions to image quality can be represented in the following way:

Image quality $\sim 0″85$,

= Optical quality
+ Support and alignment $\sim 0″65$,
+ Guiding
+ Dome seeing
+ Site $\sim 0″50$.

From what has been said above, it is apparent that the first four items on this list are contributing $\sim 0\rlap{.}''65$ on average to the observed image size. At CFHT, we have embarked on a program to separately evaluate and minimize each of these contributions. Our measurement techniques and what we have learned so far can be summarized as follows.

– Optical Quality

The original acceptance tests for both the primary and secondary mirrors showed that they are capable of producing images with FWHM = $0\rlap{.}''15$. We have no reason to suspect that this has changed, and, therefore, we do not believe that the intrinsic quality of the mirrors is a significant source of image degradation.

– Support and Alignment

CFHT has developed a CCD-based Shack–Hartmann tester. This device gives us the capability to quickly obtain an accurate and quantitative measurement of the optical performance. The Hartmann tester measures only the optics – i.e., seeing does not affect the results. We are also investigating the curvature sensing device recently proposed by Roddier (1988) as an alternative diagnostic for optical support and alignment.

At present, the most significant effect that we see is residual astigmatism in the primary mirror. The problem is perceptible at zenith and grows with increasing zenith distance. During 1989, we will carry out an extensive retuning of the radial support system, and we expect that this will correct the problem.

– Guiding

The deleterious effect of poor guiding is well known to all observers. Quantitative evaluation is easy to perform by comparing short and long exposures.

CFHT's autoguider is based on a digital guide star image obtained from a *leaky memory* attached to the guide ISIT camera. The TCS computer calculates the center of the guide image and uses the difference between the measured center and the desired position as an error signal to correct the telescope tracking rate.

The autoguider usually performs well, producing long-exposure images as small as $0\rlap{.}''6$. Nonetheless, it does introduce some image quality degradation as illustrated by the results from the DAO VHR camera presented below. Thus we are trying to improve the autoguider algorithm and the telescope control servo-system to achieve better performance.

– Dome Seeing

We will be using a shearing interferometer (Roddier, 1976) to measure the combined site plus dome seeing. The shearing interferometer, since it operates in the pupil plane, provides results which are independent of telescope optical performance, and thus it complements the Shack–Hartmann device. The plan is to use the interferometer during a seeing campaign (see contribution by D. N. B. Hall to this Symposium) and to extract the dome-seeing contribution by comparing the observed site seeing with that obtained simultaneously at the telescope pupil.

We already have in place an extensive system for continuously monitoring the dome thermal environment comprising several dozen temperature sensors situated at various points on the telescope and throughout the dome. The temperature readings are logged automatically by a computer. It is our intention to augment this system with probes to measure the atmospheric temperature structure constant C_t^2 along the light path.

2. Techniques for High-Resolution Imaging

Two different techniques are being pursued to achieve spatial resolution significantly greater than that normally obtained. The DAO Very-High-Resolution (VHR) camera developed by Robert McClure (Dominion Astrophysical Observatory, DAO) and René Racine (Université de Montréal) in collaboration with CFHT staff is shown in Figure 3.

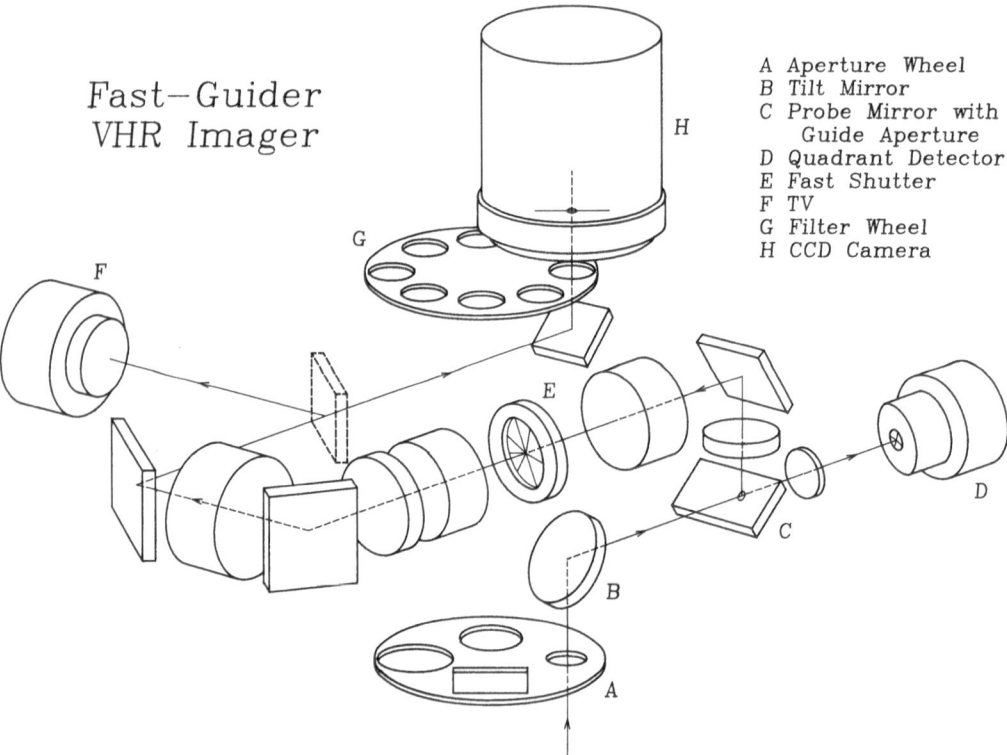

Fast–Guider
VHR Imager

A Aperture Wheel
B Tilt Mirror
C Probe Mirror with
  Guide Aperture
D Quadrant Detector
E Fast Shutter
F TV
G Filter Wheel
H CCD Camera

Fig. 3. The DAO VHR camera provides high-resolution imaging by combining a tilt mirror for rapid guiding with a fast shutter.

An image of the sky is formed first on the probe mirror C and subsequently on the CCD camera by means of transfer optics. Light from a guide star ($m \leq 17.5$) passes through a hole (1.''2 diameter) in the probe and falls on a quadrant detector. A servo drives the tilt mirror B (speed up to 500 Hz) to keep the guide star centered in the quadrant. A fast (10 Hz) shutter E is provided to block the light during intervals of poor seeing as judged by the total amount of light reaching the quadrant detector. The performance of the camera is illustrated in Figure 4, which shows several images of the center of the globular cluster M15 obtained under various conditions. Images as small as 0".36 can be achieved. So far we have not used the fast shutter.

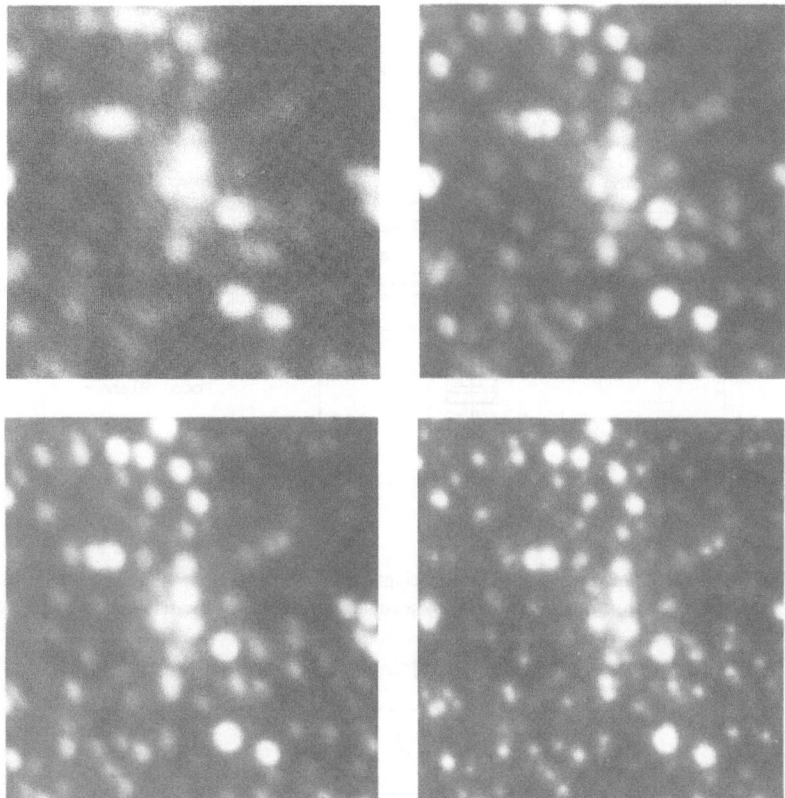

Fig. 4.   CCD images of the core of M15 illustrating the performance of the DAO VHR camera. *Upper left*: A 'normal' image taken without the camera illustrating the median image quality of 0".8 FWHM. *Upper right*: Image obtained with the VHR camera but with the tilt mirror guided manually by observing the signal from the quadrant detector on an $X - Y$ oscilloscope display. FWHM = 0".6. *Lower left*: VHR camera with the tilt mirror controlled by the servo. FWHM = 0".5. Some astigmatism can be seen in this image and in the previous one. *Lower right*: Same as previous but with telescope stopped down to 1.2 m aperture. FWHM = 0".36. This large improvement results from the combination of decreased residual aberrations and increased effectiveness of fast guiding for the smaller aperture.

The second VHR imager is a collaborative effort between a French team (G. Lelièvre and R. Foy, Obs. de Paris; J.-L. Nieto, Obs. Pic-du-Midi-Pyrénées; Jean Arnaud, currently at CFHT) and the CFHT staff. The Segmented Pupil Imager (SPI) is shown in Figure 5. An image of the telescope pupil is formed on the diagonal elliptical mirror. This mirror is segmented into eight sectors, and each sector is tilted so that an array of eight separate images of the field is produced on the photocathode of the photon-counting camera CP-40. The photo-events are recorded on magnetic tape, and the

R. McLAREN

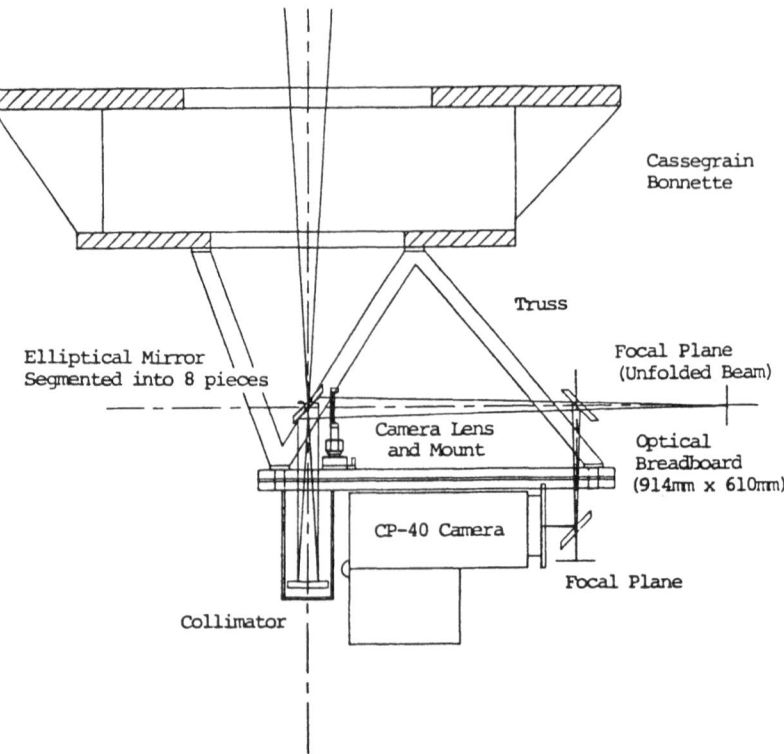

Fig. 5.   The segmented pupil imager (SPI) forms eight separate images of an object on a photon-counting
camera. Each image is produced by a separate segment of the telescope pupil.

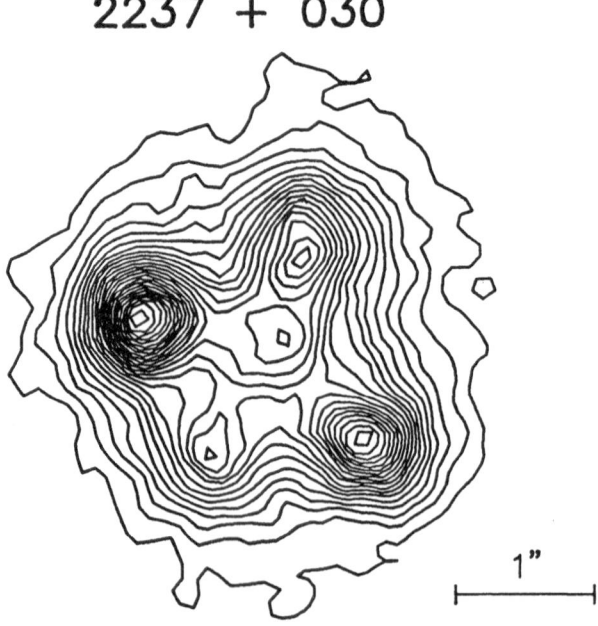

Fig. 6.   An image of the gravitational lens object 2237 + 030 obtained with the SPI camera. The integration
time for the individual images was ~ 1 s and the total integration time was ~ 15 min. The resolution is
$\sim 0\overset{''}{.}40$ FWHM.

resolution enhancement is performed *post-facto* by a computer program in the following manner. A guide object is identified and photo-events are accumulated long enough to establish a reliable centroid (typically 1 s or less). The individual images are shifted to make the centroids coincide and then co-added. In effect, the SPI camera corrects the wavefront tilt separately for each of the eight sub-pupils. Because the analysis is done after the fact, various algorithms can be tried, including selection if desired, without sacrificing any data *a priori*. An example of what can be achieved is illustrated in Figure 6, which shows the gravitational lens 2237 + 030 at a spatial resolution of 0″.40.

3. Computers and Communications

CFHT has adopted the Hewlett–Packard 9000-series computer as its new standard for instrument control and data acquisition. The operating system is UNIX with HP's real-time extension. Standards for communications are RS-232, HP-IB (IEEE 488) and Ethernet. The user interface is based on X windows, and IRAF is used as well for some applications.

For data reduction we use Sun 3/60 workstations plus a Sun 4/280 processor for more demanding applications. There are local area networks (LAN's) at both the Waimea base facility and at the observatory linking the various computers together. The two LAN's are linked by a 9.6 kbps dedicated telephone circuit, which will be upgraded to DS1 capacity (1.544 Mbps) early in 1989. Connection to Internet, Bitnet, and SPAN is via a 9.6 kbps leased line to the University of Hawaii campus at Manoa. Figure 7 illustrates our current computer complement and network topology.

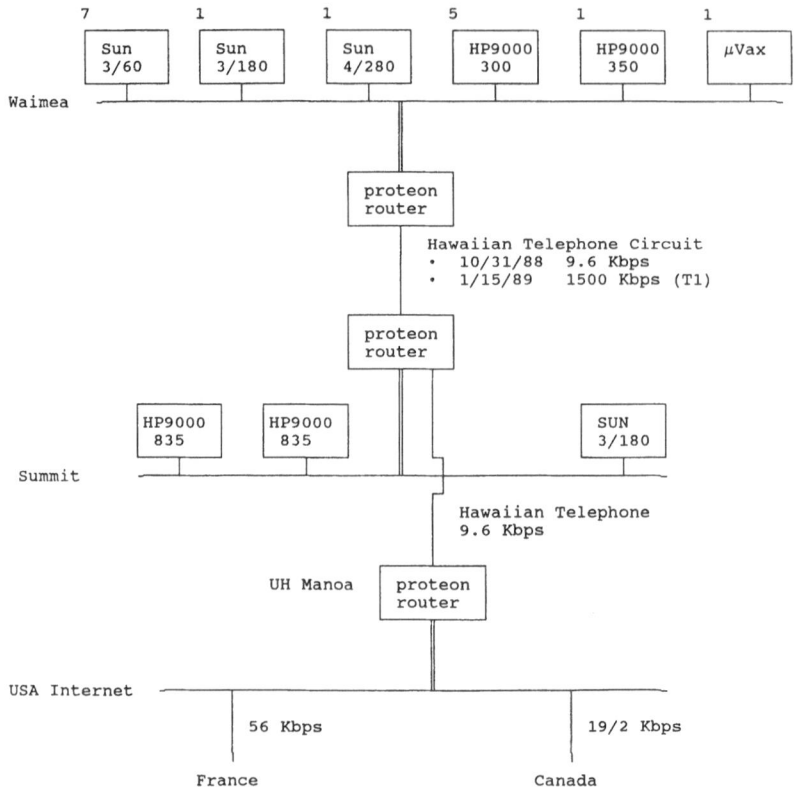

Fig. 7. The present computer configuration and network topology at CFHT.

References

Roddier, C.: 1976, *J. Opt. Soc. Am.* **66**, 478.
Roddier, F.: 1988, *Appl. Optics* **27**, 1223.

LAING – For the single-mirror image sharpening device, do you know what fraction of the improvement comes from removing telescope errors, and how much from atmospheric correction?

McLAREN – Almost all of our experience to date is with the camera used at the full aperture of the telescope, in which case the improvement comes from guiding out the motion of the image caused by the atmosphere. Typically we achieve FWHM = $0''.55$ with the camera, when conventional guiding would give FWHM = $0''.65$. We had expected a bit greater improvement than this, and the fact that we don't get it probably results from residual optical imperfections. On one occasion, we stopped the telescope down to 1-m aperture and achieved FWHM = $0''.40$ under similar intrinsic seeing conditions using the fast guider. The improvement from $0''.55$ o $0''.40$ results from a combination of reducing D/r_0 and reducing telescope aberrations. At present, I cannot give a quantitative breakdown for these two components.

COOPERATION ASPECTS OF INTERNATIONAL OBSERVATORIES*

SAEKO S. HAYASHI

Joint Astronomy Centre, Hilo, U.S.A.

(Received 10 January, 1989)

Abstract. From the user's point of view, it is essential to promote collaborations to make full use of large telescopes, by covering the wide range of the wavelengths. To do that efficiently, it is necessary to work with other facilities, often abroad. Other talks in this symposium are concerned about the spatial resolution or the coverage at one wavelength or one to two octaves in wavelength. Though the above is important, that is not enough to attach the mysteries in the Universe, since a considerable amount of energy is emitted in the infrared, and mass exists in molecular gas form. In this paper, two examples of international, multi-wavelength work are shown which turned out to be beneficial for the world-wide community. One is the Japan–U.K. arrangement mainly between the radio and infrared facilities, and the other is the arrangement around James Clerk Maxwell Telescope (JCMT), challenging a new regime of the observational astronomy.

1. Japan–U.K. Collaboration

This alliance started in 1983 (peaceful one compared to the military alliance between the U.K. and Japan that started in 1902) between the observational astronomers in Japan, U.K., and Hawaii for the study of the compact objects. The motivation was to make the effective use of the big facilities at the infrared (UKIRT: United Kingdom Infra-Red Telescope, 3.8 m) at Mauna Kea, Hawaii, and at mm-wave (NRO: Nobeyama Radio Observatory, Nagano, Japan, 45 m; the interferometer was not yet commissioned). These areas were growing rapidly at that time. In practice, the cooperation has been done by joint proposals and exchange of the astronomers. It has been a good exercise for Japanese astronomers to be exposed to such an international arrangement, particularly instructive for the young astronomers.

At first stage, the group at Kyoto University sent an infrared polarimeter to UKIRT as a common user instrument. The first major impact from this instrument was the discovery of the large polarization in the vicinity of the protostellar disks (the initial data was taken at UH 2.2 m) (Nagata *et al.*, 1983). This tool, though very simple, has turned out to be a very successful one in studying the geometries of the dark clouds and the disks in relation to the magnetic field, which at that time emerged as a important subject (Hough *et al.*, 1986; Moore *et al.*, 1988; Nagata *et al.*, 1987; Sato *et al.*, 1988; Tamura *et al.*, 1987, 1988; Yamashita *et al.*, 1987a, b, 1988).

On the same year, a simultaneous observation of Mkn 421, a BL Lac object, was coordinated. To identify the radiation mechanism of this conspicuous object, the following facilities joined with the best spectral coverage available; X-ray satellite

---

* Paper presented at the Symposium on the JNLT and Related Engineering Developments, Tokyo, November 29–December 2, 1988.

Tenma, UV satellite IUE, ground-based optical telescopes at Mt. Lemon, Okayama, Dodaira, IR telescopes at Mauna Kea (UKIRT) and Agematsu, and radio telescopes at Michigan and Nobeyama (Makino *et al.*, 1987). This strategy of combining the space and ground, X-ray to radio, passed on to the currently ongoing programs for the quasars using the Ginga (X-ray satellite), UKIRT, and JCMT.

In general, the interests among the collaborators are the elementary processes and the global structure of the star formation. For the beginning, the interaction among the astronomers were just exchanges of information, like preprints. Then, suddenly they realized that they found good collaborators and ended up with observing together and discussing unceasingly. The targets are the active galactic nuclei including the galactic center, the active star-forming regions, dynbamically shocked regions, and the photo-dissociation regions.

Extensive studies have been made toward Orion as a representative of dynamically excited, active star-forming region (e.g., Hough *et al.*, 1986; Hasegawa *et al.*, 1987; Burton *et al.*, 1988; Hayashi *et al.*, 1985), NGC 2023 as a representative of a radiatvely excited region (e.g., Gatley and Kaifu, 1987; Gatley *et al.*, 1987; Tanaka *et al.*, 1988a, b), M17 for a large-scale dynamic of star-forming region (e.g., Gatley and Kaifu, 1987; Rainey *et al.*, 1987) (Figure 1), and IC 443 as an example of SNR (White *et al.*, 1987).

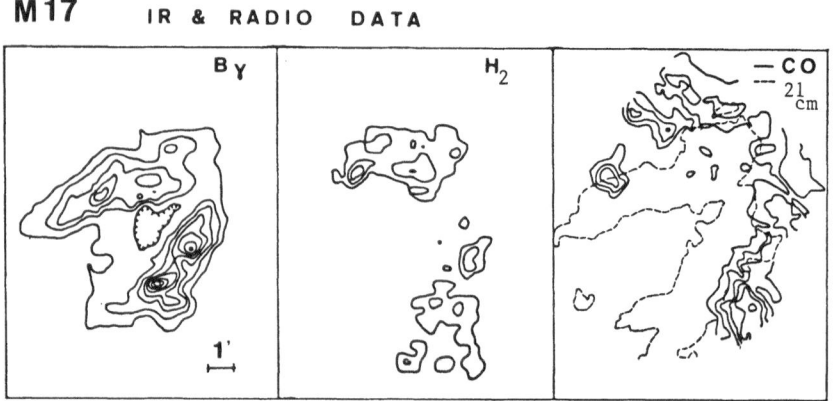

Fig. 1.   Dynamical structure of M17 nebula, a site of star cluster formation. The young stars are in the *bottom left*, and the ionized gas shown in Br γ (*left*) as well as the 21 cm radio continuum (*right*, dashed) is an expanding H II region. The molecular hydrogen (*middle*) shows the shock front outside of this H II region. The CO molecular line data shows that molecular cloud is pushed by the expanding H II region and forms an expanding molecular shell. In all, these pictures delineate a blister H II region. Figures based on Gatley and Kaifu (1987).

The coexistence of shocked molecular hydrogen and CO molecule was clearly shown, proving that the shocked region is not slab-like, but rather clumpy. Also the molecular abundances and the excitation were discussed. The extensive and intensive studies of the molecular hydrogen emission lead to the astrophysics of the fluorescent emission and the quantitative measurement of the ortho-para ratio of this basic molecule.

Discussions with the scientists outside of astronomy was also useful, e.g., the chemist or physicist of the spectroscopy, doing model calculations or laboratory experiments (e.g., Takayanagi *et al.*, 1987).

The astronomers around Nobeyana radio telescope benefitted from the idea of the imaging by fast scanning, without sticking to the minor detail of the individual pixel. The idea of frequency switching was transferred to the infrared observations. The setup of the Fabry–Pérot was modified to allow the on-line and off-line technique. Also the idea of the velocity resolved mappings (just natural for the radio astronomers) was implemented to UKIRT. These are powerful methods in observing the details of the diffuse and extended emission.

2. Progress at JCMT through Supra-National Collaboration

JCMT, a 15 m radio telescope at Mauna Kea started its operation for the common use in September 1987. The constructing and commissioning of instrumentation, and the use of the telescope time of this telescope has been arranged with supra-national collaboration, which is for the world-wide use. The astronomers and engineers are mostly from the three host countries: United Kingdom, Holland, and Canada; and are merged together to challenge this new observational field. One example of this joint effort is the surface adjustment of JCMT, the task which is essential to achieve enough efficiency of the main reflector.

The back structure of the main reflector was constructed in Holland which follows the homologous design that works against the gravitational deformation. The light-weight panels and the motorized adjusters were fabricated in the U.K.; the accuracy of the individual panels ranges 10–15 μm r.m.s.

In order to adjust a large number of panels into a smooth parabola, it is necessary to have a precise and handy measurement system, which is rather difficult for this scale size. The radio holography method developed in the U.K. and U.S.A. has its origin (at least its naming) in the optical method. This requires an extra antenna (small one) as the reference to do the interferometry. This method is used in Texas 5 m, Nobeyana 45 m, Effelsberg 100 m, etc.

The new concept of the phase retrieval, so-called phaseless holography, has been developed by a French group and applied to IRAM 30 m telescope. This method does not need a reference antenna, and needs two sets of amplitude measurement. A British group applied this method to JCMT, of which initial setup was done by a laser-ranging machine. The measurement and the adjustment cycle at Mauna Kea for JCMT has been carried out by the U.K.–Dutch–Hawaii–Japan league. With about a dozen of iterations since spring of 1987 after adopting this phaseless holography, the surface has been improved. At present, this phaseless metrology tells that the surface is better than 30 μm r.m.s., and this good number was confirmed to be less or around 35 μm by an actual astronomical observation; the planet measurement was done by a German–U.S.A. group, which brought their own high-frequency instrument (UCB/MPE heterodyne receiver) (Harris, 1988; Webster, 1988). Note that the elevation of these two measurements were quite different, which means that the homology design of JCMT functions very well in cancelling the gravitational deformation at lower elevation.

The struggle of the adjusting large reflectors is shown in Table I. Note that these three largest telescopes at each frequency have similar characters; large number of panels,

TABLE I

Surface adjustment of the radio telescopes

Item/tel.	NRO 45 m	IRAM 30 m	JCMT 15 m
Method of measurement	conventional holography	conventional holography, phase retrieval	phase retrieval
Source	geostationary satellite	transmitter water/SiO maser	ground-based signal source
Frequency	19.45 GHz	22/86 GHz	94 GHz
Panels	600	210	276
Accuracy	60 μm r.m.s.	26 μm r.m.s.	12 μm r.m.s.
Adjusters	700 @ corners	870	828
Motor step	10 μm		3 μm
Grid	64×64, 128×128	32×32	65×65, 97×97
Resolution	1.7 m–43 cm	1.4 m	25 or 17 cm
Achieved	140 μm	60 μm [a]	< 38 μm [b]
λ/D (")	12 (115 GHz)	9 (239 GHz)	5–12 (810–345 GHz)
ε/D (μm m^{-1})	3.1	2.0	< 2.5

[a] Measured as 69 μm, after corrections 59 μm.
[b] Measured as 26 μm, planet measurement < 38 (35) μm.

motorized adjusters, radio holographic or similar measurement, and the resultant error versus the observation frequency. In a way, they are close to the limit with the current design or current site on the ground.

The consequence of this task, combined with the considerable improvement of the telescope pointing, is that JCMT really works at the sub-mm range. During the more than one year of commissioned use, the continuum observations have shown the dust emission around the protostars as well as the late-type stars, and the line observations probed into the hot and dense molecular cores (e.g., Hayashi, 1988).

We believe that the current system at JCMT, both the measurement and the adjustment, is valid for improving the surface even more. Compared to other telescopes, this one has a big advantage of being enclosed so that the temperature gradient in the telescope structure can be reduced. When the number of something like 25 μm comes to be a reality, this facility will be efficient in the short sub-mm of the far-infrared region where the energy and the temperature of the radiation mechanism is quite different from the current understandings based on mm-wave and near-IR data. To attack this brand-new field, it becomes far more essential for the astronomers to work with the people from other observational fields.

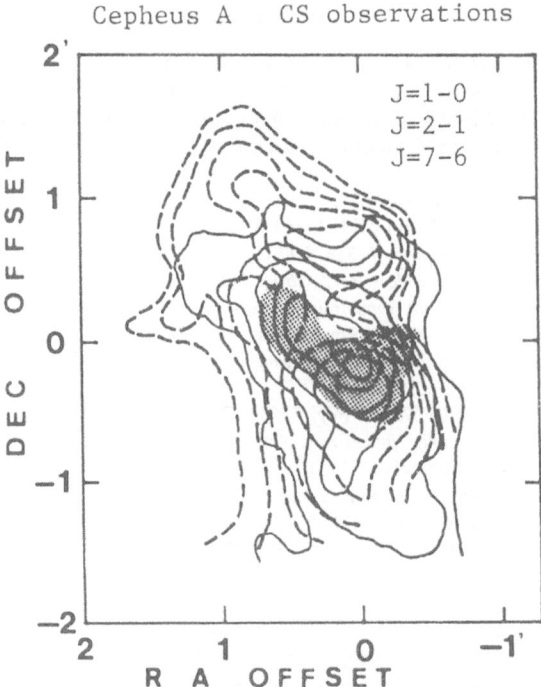

Fig. 2. Hierarchy in Cepheus A star-forming region. The multi-transition studies of CS molecule reveals on onion-like interior of the protostellar disk. Low excited CS ($J = 1$-0: dashed line) is extended and show a cold massive envelope. Toward higher transition, it probes deeper into the disk ($J = 2$-1: thin line). The inner most region shown in CS 7-6 (hatched area) is the immediate vicinity of a protostar and represents hot and dense region; 100–1000 times denser thanthe outer envelope of the CS $J = 1$-0 emission (e.g., Hayashi, 1988).

3. Statement

The key to the successful collaboration is to have common scientific targets. The conversed sentence is also true. For progress in astronomy, it is essential to work together with facilities of other wavelengths that are often abroad. The activity, the collaboration, and the competition in astronomy is no longer limited to the Atlantic side of the Earth, but extended to the Pacific side; across the date-line where the majority of tomorrow's people live.

Acknowledgements

The Joint Astronomy Centre is operated by the Royal Observatory, Edinburgh on behalf of the Science and Engineering Research Council of the United Kingdom, the Nederlandse Organisatie voor Zuiver Wetenschappelijk Onderzoek, and the National Research Council of Canada. Nobeyama Radio Observatory, a branch of the National Astronomical Observatory, is a cosmic radio observing facility open for outside users. Many thanks to my collaborators everywhere on this Earth (at this moment).

References

Burton, M. G. *et al.*: 1988, *Monthly Notices Roy. Astron. Soc.* **235**, 161.
Dent, W. R. F. *et al.*: 1985, *Monthly Notices Roy. Astron. Soc.* **217**, 217.
Gatley, I. and Kaifu, N.: 1987, in M. S. Vardya and S. P. Tarafdar (eds.), 'Astrochemistry', *IAU Symp.* **120**, 153.
Gatley, I.: 1988, in R. Pudritz and M. Fich (eds.), *Galactic and Extragalactic Star Formation*, D. Reidel Publ. Co., Dordrecht, Holland.
Gatley, I. *et al.*: 1987, *Astrophys. J.* **318**, L73.
Harris, A. I. *et al.*: 1987, Int. J. IR and MM Waves **8**, 857.
Hasegawa, A. I. *et al.*: 1987, *Astrophys. J.* **318**, L77.
Hayashi, M. *et al.*: 1985, *Monthly Notices Roy. Astron. Soc.* **215**, 31p.
Hayashi, S. S.: 1988, in *International Symposium of Sub-mm and mm Astronomy*, held in Kona, Hawaii.
Hough, J. H. *et al.*: 1986, *Monthly Notices Roy. Astron. Soc.* **222**, 629.
Hough, J. H. *et al.*: 1988, *Monthly Notices Roy. Astron. Soc.* **230**, 107.
Makino F. *et al.*: 1987, *Astrophys. J.* **313**, 662.
McLean, I. S. *et al.*: 1987, *Monthly Notices Roy. Astron. Soc.* **225**, 393.
Moore, T. J. T. *et al.*: 1988, *Monthly Notices Roy. Astron. Soc.* **234**, 95.
Nagata, T. *et al.*: 1983, *Astron. Astrophys.* **119**, L1.
Nagata, T. *et al.*: 1988, *Monthly Notices Roy. Astron. Soc.* **227**, 543.
Rainey, R. *et al.*: 1987, *Astron. Astrophys.* **171**, 252.
Richardson, K. J. *et al.*: 1988, *Astron. Astrophys.* **198**, 237.
Sato, S. *et al.*: 1988, *Monthly Notices Roy. Astron. Soc.* **230**, 321.
Takayanagi, K. *et al.*: 1987, *Astrophys. J.* **318**, L81.
Tamura, M. *et al.*: 19*Monthly Notices Roy. Astron. Soc.* **224**, 413.
Tamura, M. *et al.*: 1988, *Monthly Notices Roy. Astron. Soc.* **231**, 445.
Tanaka, M. *et al.*: 1988, *Monthly Notices Roy. Astron. Soc.* **231**, 445.
Tanaka, M. *et al.*: 1989, *Astrophys. J.* (in press).
Webster, A. S.: 1988, in *International Symposium of Sub-mm and mm Astronomy*, held in Kona, Hawaii.
White, G. J. *et al.*: 1987, *Astron. Astrophys.* **173**, 337.
Yamashita, T. *et al.*: 1987a, *Publ. Astron. Soc Japan* **39**, 809.
Yamashita, T. *et al.*: 1987b, *Astron. Astrophys.* **177**, 258.
Yamashita, T. *et al.*: 1989, *Astrophys. J.* (in press).

THE PERFORMANCE OF THE APACHE POINT OBSERVATORY 3.5 M TELESCOPE*

EDWARD J. MANNERY, WALTER A. SIEGMUND, and CHARLES L. HULL

University of Washington, Washington, U.S.A.

(Received 18 January, 1989)

Abstract. The Astrophysical Research Consortium 3.5 m telescope facility on Apache Point (2800 m above sea level) near the National Solar Observatory in southern New Mexico is nearing completion. The telescope mount has been installed and testing and fabrication of remaining subassemblies are underway. The $f/1.75$ lightweight honeycomb primary mirror was cast April 1988 by the Steward Observatory Mirror Laboratory and is currently being figured.

The 3.5 m optical telescope is an altitude over azimuth mechanical structure with Ritchey–Chrétien optics. The lightweight (1800 kg) mirror leads to a mount weighting only 41 000 kg; readily available rolling element bearings are used to achieve the necessary performance at low cost and without the heat dissipation of externally pressurized types. Drive torques are applied by DC servo-driven capstans. These are coupled by friction to large diameter drive disks on each axis. No gears are used. Position feedback comes from low cost incremental encoders, also capstan coupled.

We have recently completed a series of measurements of the telescope mount. These measurements show that the telescope is very stiff; the lowest natural frequencies are about 7.2 Hz. Initial tracking performance is good and the mount shows high resistance to wind-induced vibration. Our experience during acceptance testing suggests that routine power spectral analysis of drive motor torque and other parameters could be an important tool in the early detection of failures.

1. Introduction

Apache Point Observatory is a new 3.5 m telescope facility located at an elevation of 2800 m above sea level, in the desert of southwestern United States 18 km SE of Alamogordo, New Mexico. It is owned and operated by Astrophysical Research Consortium (ARC). Member universities are University of Chicago, University of Washington, New Mexico State University, Princeton University, and Washington State University.

The telescope is intended to be a general purpose imaging and spectrographic telescope. To exploit the best seeing expected at the site, the design goal is to produce images better than 0".4 fullk width at half maximum under these conditions. Certain types of scientific programs and modes of operation are difficult to accommodate at the national facilities. The telescope and operation environment are expected to be particularly well suited for many of these programs – for example, for synoptic observations, surveys, remote observations, and flexible response (Balick *et al.*, 1988; Owen *et al.*, 1988). Flexible response is needed to take advantage of transient events such as supernovae and unusual atmospheric conditions such as periods of exceptionally good image quality.

* Paper presented at the Symposium on the JNLT and Related Engineering Developments, Tokyo, November 29–December 2, 1988.

The telescope project design phase started in 1983. Major contracts were awarded in 1986 and early 1987. The telescope and buildings were essentially completed in November 1987 and are awaiting the installation of the primary mirror. The primary was cast in April 1988 at the University of Arizona and is currently being figured. It is expected to be installed early in 1990. The project cost including instruments and staff salaries but not the primary mirror blank was $10 million.

The $f/1.75$ primary mirror is a 3.5 m lightweight borosilicate blank with a mass of only 1800 kg (Siegmund *et al.*, 1986). This is the same as a meniscus blank with a thickness of 9 cm. The primary was cast by the Steward Observatory Mirror Laboratory at the University of Arizona (details are given in an article by Angel in these proceedings).

The secondary and tertiary mirrors are lightweight borosilicate hot gas fusion blanks and were fabricated by Hextek, Inc., Tucson, Arizona. They are 780 mm in diameter and have masses of 40 kg each.

The telescope enclosure rotates with the telescope in azimuth on a track located 5.5 m above the ground. The telescope chamber co-rotating floor is at 8.5 m and gives excellent access to instruments mounted at the Nasmyth foci, 1.3 m above this floor. Openings in the rear corners of the enclosure improve air flow around the telescope (Siegmund and Comfort, 1986). The upper enclosure frame consists of weight efficient *I*-beams for minimum thermal mass.

An enclosed corridor, which is used as an air exhaust duct, connects the telescope enclosure and the operations building which contains the computers and electronics shop. This corridor serves as a pathway for people and equipment; this is particularly useful during the winter when up to 1 m of snow can accumulate. Cables connecting the telescope and instruments to computers in the operations building are placed in a cable tray located on one wall of this corridor.

2. The Telescope

The telescope mount is an altitude over azimuth design (Mannery *et al.*, 1986). The total moving mass is 41 000 kg. A preloaded pair of spherical roller bearings are used on each side of the altitude structure. The main load carrying azimuth bearing is a spherical roller bearing. This bearing is located at the apex of large steel inverted cone. At the upper end of the cone is a large diameter steel disk which serves as a bearing, drive, and encoding surface. Four guide rollers, two of which are driven, contact this disk and define the azimuth rotation axis together with the bearing at the cone bottom.

To reduce wind torque on the telescope, the main telescope secondary truss tubes diameters were minimized and are only 100 mm in diameter. Even at this size, they contribute more drag than the secondary assembly. However, further reduction of truss diameters would result in unacceptable lateral stiffness. The square secondary frame is placed in compression by the secondary vanes. Vane tension is set so that the rotational mode of the secondary about the optical axis is above 10 Hz.

The mirror cell and the tube center section is a one-piece weld to reduce weight and cost. The mirror is removed from the telescope by an overhead bridge crane mounted near the ceiling of the telescope chamber.

Fig. 1.   The APO 3.5 m telescope. Visible above the floor are the forks and altitude structure (white). The main telescope secondary truss tubes are only 100 mm in diameter to minimize wind loading. The secondary counterweight will be replaced by the real secondary assembly late in 1989. The mirror cell and the tube center section is one piece to reduce weight and cost. The drive torque is transmitted to the altitude structure via the large diameter drive disk mount on the underside of the mirror cell. Openings in the rear corners of the enclosure improve air flow around the telescope.

The telescope is driven by rollers which transmit torque via friction. No gears are used. Each axis is very closely coupled to a large diameter circular disk. The altitude disk is only a portion of a full disk. Each disk is driven by a 10 cm diameter roller. This gives a reduction of about 35 : 1. The drive roller is part of a three-stage roller speed reducer. The overall ratio from the motor to the telescope is about 1200 : 1.

Sony Magnesensors give absolute encoding to about 0″.1 at 15 deg intervals. A roller attached to an incremental optical encoder is drivenby friction by each large disk. The encoders are made by Heidenhain, Inc. and give the mount a resolution of 0″.01.

3. Dynamic Performance

To measure the natural frequencies of the telescope structure and drive system, the telescope was excited by a mechanical impulse. It was struck repeatedly at roughly random intervals of 0.5 to 2 s with a 3 kg wooden block.

The primary sensor was a linear velocity transducer with a resolution of better than 1 µs made by Schaevitz, Inc. This device consists of a permanent magnet inside a pickup coil. The transducer was mounted between the telescope and the observing chamber floor. It was sampled at 100 Hz. Samples were 20 to 40 s long. These data were transformed to power spectra using the fast Fourier transform algorithm. Data were also obtained from the axis increment encoders.

For the azimuth structure response, the impulse was applied to one fork and directed to maximize the impulse torque about the azimuth axis. The azimuth motor shafts were

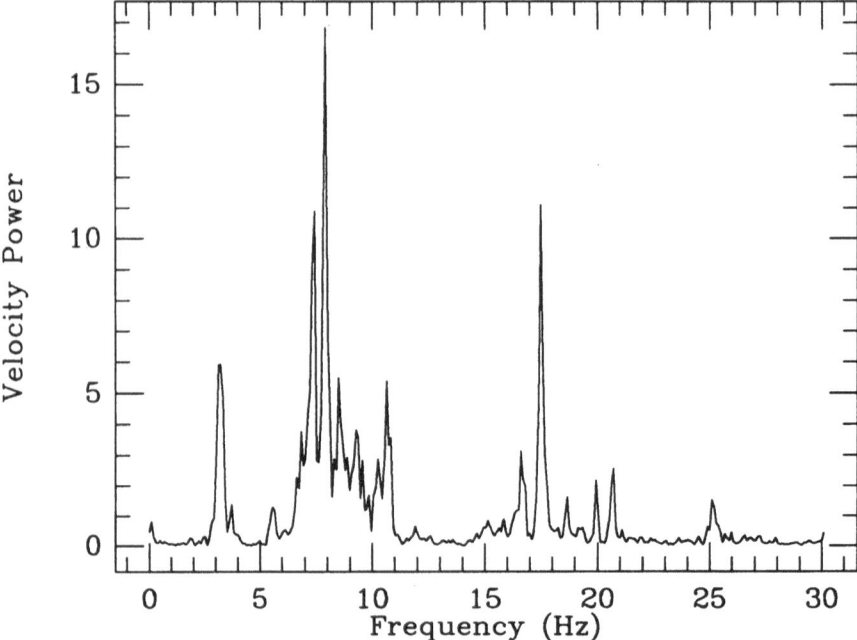

Fig. 2.   The azimuth axis power spectrum: the azimuth rocking vibrational mode due to lateral compliance of the azimuth bearings is at 7.2 Hz. The azimuth locked rotor resonance frequency due to the compliance of the azimuth drive is at 7.8 Hz. The other features seem to be vibration modes of the steel weldings except for the peak at 3 Hz which is the rocking mode of the building (the transducer was connected between the telescope and the building).

prevented from rotating. The locked rotor resonance frequency was 7.8 Hz. For the altitude structure response, the impulse was applied to the edge of the steel box structure surrounding the primary mirror to maximize the impulse torque about the altitude axis. The altitude motor shaft was prevented from rotating. The locked rotor resonance frequency was 11.7 Hz.

Other frequencies of interest include the azimuth overtuning frequency due to the compliance of the azimuth bearings at 7.2 Hz. The peak at 5.5 Hz has tentatively been identified as the pier-rocking mode.

Separately, we have investigated the dynamic behaviour of the secondary truss structure. The measured natural frequencies are all above 14 Hz.

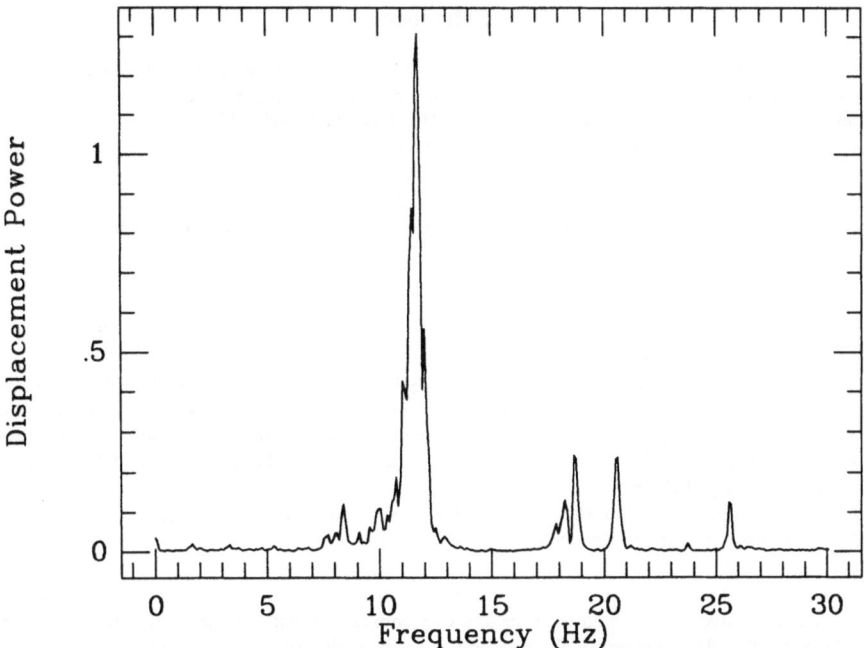

Fig. 3.   The altitude axis power spectrum: the main feature is the altitude locked rotor resonance frequency due to the compliance of the altitude drive at 11.7 Hz. The other features seem to be vibration modes of the steel weldments.

These high natural frequencies imply that the telescope is very stiff. For example, typical locked rotor resonance frequencies of all but the most recent large telescopes are below 5 Hz. High values of stiffness in the telescope and drive gives good resistance to wind-induced vibration. In addition, high stiffness permits high servo loop gains and low tracking error.

4. Conclusions

The proposed Columbus, Magellan, and National Optical Astronomy Observatory telescopes have larger diameter drive, encoding and bearing surfaces coupled to the optics via very direct load paths. The ARC telescope shares many of these design features; the resulting excellent dynamic performance and low mass helps to confirm the promise of this new approach.

Our experience with dynamic power spectra suggests that this is a powerful tool for monitoring telescope performance and for early detection of failures. During acceptance testing at the factory, we discovered a strong peak in the power spectra of altitude motor torque while tracking at a constant rate. The identification of the peak with the motor shaft rotation frequency localized the fault in the motor or speed reducer, and not in the azimuth bearings. Several months later, the speed reducer failed and the component at fault was identified.

References

Balick, B., Loewenstein, R., Siegmund, W., and York, D.: 1988, in L. B. Robinson (ed.), *Instrumentation for Ground-Based Optical Astronomy: Present and Future*, Springer-Verlag, New York.

Mannery, E. J., Siegmund, W. A., and Balick, B.: 1986, *Proc. SPIE* **628**, 397.

Owen, R., Siegmund, W., and Hull, C.: 1988, in L. B. Robinson (ed.), *Instrumentation for Ground-Based Optical Astronomy: Present and Future*, Springer-Verlag, New York.

Siegmund, W. A. and Comfort, C.: 1986, *Proc. SPIE* **628**, 369.

Siegmund, W. A., Mannery, E. J., Radochia, J., and Gillett, P. E.: 1986, *Proc. SPIE* **628**, 377.

MACK – I commend your ability to get high natural locked rotor frequencies of the telescopes but why have you got such a low pier frequency?

SIEGMUND – The pier design was based on a soil modulus determined from seismic refraction tests. It was predicted to have a natural frequency of 15 Hz. We are very axious to understand this discrepancy. It is encouraging, however, that in the azimuth axis impulse test, the amplitude of the pier rocking mode is small compared to the azimuth overturning mode at 7.2 Hz. I suspect that this is due to the larger inertia and better damping of the pier.

ENARD – The completion time of the Apache Point telescope was about 6 to 7 years which is consistent with that of other similar projects such as the NTT or the WHT.

This suggests that start lead times announced for some larger projects may not be very realistic.

LAING – I would like to ask about your friction-driven incremental encoder. Do you have any problems of slippage and have you checked their accuracy by comparison with the absolute reference points?

SIEGMUND – We plan to measure slippage and accuracy within the next six months.

SOME COMMENTS ON LARGE TELESCOPE ASTRONOMY IN BRITAIN AND AUSTRALIA*

RUSSELL CANNON

Anglo-Australian Observatory, Epping, Australia

(Received 10 January, 1989)

Abstract. Many of the major British telescopes are described in this volume by other speakers directly involved with those facilities, so this contribution will concentrate on aspects of the operation of the Anglo-Australian Observatory (AAO) which seem particularly relevant to the JNLT. Some requirements for the success of the new very large optical telescopes are also discussed.

1. Current British and Australian Ground-Based Facilities

The oldest of the modern optical and infrared telescopes run by British and Australian astronomers are the two telescopes of the AAO on Siding Spring Mountain in Australia: namely, the 3.9 m Anglo-Australian Telescope (AAT) which became fully operational in 1975, and the U.K. 1.2 m Schmidt Telescope (UKST) which entered service about a year earlier. It may seem odd to mention the UKST at a conference devoted to telescopes which will make even the AAT seem small, but it is the most powerful wide-field Schmidt-type telescope in the world and it is making a major contribution to the cosmological and galactic structure programmes which dominate the use of the AAT, and which will no doubt feature strongly in the programmes of the new 8 m-class telescopes. The UKST was formerly operated by the Royal Observatory, Edinburgh, but in June of this year the AAO became responsible for its operation and, like the AAT, it is now funded equally by the British and Australian governments. The very successful combination of a powerful survey telescope with one of the world's largest multiple-purpose optical telescopes illustrates the point that the new generation of 8 m telescopes should not be considered in isolation. For maximum effectiveness, astronomers using the largest telescopes will also need access to smaller facilities of various types.

On the British side, the other major telescopes are the new 4.2 m William Herschel Telescope and the 2.5 m Isaac Newton Telescope, both on La Palma, the 3.8 m UK Infrared Telescope (T. J. Lee, this issue) on Mauna Kea, and the 15 m James Clerk Maxwell Telescope (S. Hayashi, this issue) for sub-millimetre and millimetre-wave astronomy, also in Hawaii. The latter is included here since the boundaries between optical, infrared, and sub-millimetre astronomy are becoming increasingly blurred, especially in the context of 8 m-class telescopes.

On the Australian side, the Australian National University's 2.3 m alt-azimuth telescope has been operating now for about 3 years on Siding Spring Mountain. This

* Paper presented at the Symposium on the JNLT and Related Engineering Developments, Tokyo, November 29–December 2, 1988.

was a low-cost telescope incorporating several technological innovations relevant to the new generation of large telescopes, and like the AAT it is used for both optical and infrared work. The next largest telescope in Australia is the much older 1.88 m at Mount Stromlo, which is virtually identical to the Okayama 1.88 m in Japan, and there are a number of smaller telescopes operated by the ANU. By contrast, Britain has practically no domestic optical telescopes which are still used for research.

Two exciting new astronomical facilities should become fully operational in Australia within the next couple of years. The Sydney University Stellar Interferometer (SUSI) is a two-beam optical interferometer, not to be confused with the now-dismantled Stellar Intensity Interferometer of Hanbury-Brown, and is currently under construction near Narrabri in New South Wales. It is being built on the same site as the Australia Telescope (AT), a new radio aperture synthesis array. The AT in fact consists of two arrays of radio telescopes, some of which can also be operated independently. The short baseline array at Narrabri consists of six moveable 22 m antennae spread over a 6 km baseline, while the long baseline array includes another new 22 m antenna near Siding Spring Mountain, the famous Parkes radio telescope and several other antennae. The development of optical astronomy in both Britain and Australia has depended very strongly on the world-leading radio astronomy in those two countries; without the strong support of radio astronomers it is doubtful whether many of the recent optical telescopes would have been built, and the AT will surely have a major influence on the future development of astronomy in Australia.

2. Instrumentation on the AAT

Having excellent instrumentation is as important as building large telescopes. The AAT has a comprehensive suite of instruments acquired over many years, and new state-of-the-art instruments and detectors are continually being added. The list of currently available instruments contains some sixteen separate items. Even so, not all possible options are covered and the AAT like other 4 m-class telescopes has to some extent specialised in particular areas of astronomy, partly in response to the interests of its user communities and partly to take best advantage of its climatic and geographic situation. The AAT can thus serve as an example, not of the specific instrumentation required for a new 8 m-class telescope such as the JNLT, but of the level of instrumentation that will be required for a national telescope.

Throughout its life, the use of the AAT has been dominated by optical spectroscopy at the Cassegrain focus, covering a wide range of wavelength regions at many different dispersions. Recent enhancements have included spectropolarimetry and multi-object spectroscopy; the trend to survey spectroscopy, using both optical fibres and multi-slit aperture plates, is now very strong and this wins about a quarter of the observing time. Altogether, optical spectroscopy takes up some two-thirds of the total time on the AAT. Infrared astronomy, including photometry, spectroscopy, and polarimetry, takes up a further fifth of the time, leaving only about 10% in total for direct imaging, photometry and special applications such as speckle imaging.

One major new instrument commissioned earlier this year seems certain to win a large amount of observing time in future: the coudé echelle spectrograph (UCLES). This instrument, built by a team from University College London, allows AAO users for the first time to combine high dispersion with wide wavelength coverage using modern detectors and, hence, to do high-resolution spectroscopy on faint objects. The recent increased use of the coudé focus has shown how very effective that focal station can be; instruments can be set up and tested without interfering with the current observing programme, and the instruments themselves do not have to be built to fit into a small space or to withstand being moved and tilted at all angles. They are also free from vibration or temperature variations, and so can be much more stable than instruments at the Cassegrain or prime foci. The coudé focus, and the fixed Nasmyth foci of alt-azimuth telecopes, will surely see much more use in future.

All instruments need detectors; for optical work, the AAT relies almost exclusively on the Boksenberg Image Photon Counting System (IPCS), especially at short wavelengths, and on CCDs, and most instruments can be used with either. Much effort is currently going into assessing possible new panoramic electronic detectors.

3. Future Large Telescope Plans in Britain and Australia

Many astronomers in both Britain and Australia feel that they must have at least a share in an 8 m-class telescope if they are to remain competitive ten years from now. On the U.K. side, R. S. Ellis has described here the work of the Large Telescope Panel and B. Mack has introduced one possible new 8 m telescope on La Palma. Australia is proceeding more slowly, having just set up a review panel with a brief to report on options for the next decade or so. As a first step, A. W. Rodgers of the Australian National University is proposing a new site-testing campaign. Although Siding Spring is very good by comparison with any site in Britain, or within Japan, it does not have as good an astronomical climate as they very best sites in the world. However, there may well be substantially better sites within Australia, especially in terms of clear nights and photometric conditions. There are no very high mountains in Australia, but there are some very promising mountain ranges in semi-desert country near Broken Hill in NSW and in South Australia. These lie at about $-30°$ latitude, in the band which has very little rainfall in either summer or winter. One consideration in developing such a site would be the logistics of running an observatory in a very remote site; a great strength of the AAT is its location in a region which provides very amenable living conditions for technical staff and their families. The excellent maintenance and back-up facilities at the AAT in many ways compensate, and sometimes more than compensate, for the inferior climatic conditions compared with those at some observatories in less hospitable locations.

A major motivation in searching for a better site in Australia is to try to find a location which would be attractive to international partners, perhaps including Japan at some future date, in view of Australia's southern hemisphere location. Some non-astronomers, and even some astronomers, argue that since there are many stars and galaxies in both hemispheres, it is not necessary to have good facilities in both. However, this is a very narrow and unscientific point of view, and there are of course crucially important reasons for needing all-sky coverage. The most important concern studies of our own Milky Way Galaxy which is very asymmetric as seen from Earth, and of the distribution of nearby galaxies and clusters, while the common assumption that quasars at high redshifts are distributed isotropically *is* just an assumption and has to be checked observationally. X-ray and infrared satellite surveys find interesting objects over the whole sky. There are also many unique objects in the southern sky: the galactic centre; the Magellanic Clouds (including now SN 1987A); the nearest classical radio galaxies, Centaurus A and Fornax A; and the nearest and largest globular star clusters.

At the AAO itself, the main AAT plans for new instruments are for an infrared array camera for direct imaging and, later, spectroscopy; the development of new large format optical detectors; and the possible provision of a 2-deg wide field at the prime focus, to be used particularly for multi-object spectroscopy. Design studies for the three main components of this last exciting possibility are underway. It is already clear that corrector lens systems can be devised which will give such a wide field, and the optimum design is now being sought; laboratory experiments are being started to find ways to position up to 300 fibres automatically; and designs for spectrographs optimised for multi-fibre work are being investigated. The latter may come from the parallel development of the 'FLAIR' multi-object system on the UKST. There a prototype system is already in use and funds have been allocated for construction of an optimised system. Although the 1.2 m aperture of the UKST is rather small, its $6.5° \times 6.5°$ field is so large that it is ideal for surveys of galaxy redshifts or sparse stellar objects down to around magnitude 17; it complements the AAT systems very well, being ideal for targets which occur at the rate of only a few per square degree while the AAT is most efficient for targets which have about ten times higher projected surface density on the sky. A new spectrograph being built for FLAIR may also be used as the prototype for the AAT prime focus.

4. Some Considerations Concerning New Large Telescopes

4.1. SCIENTIFIC TRENDS AT THE AAO

Although it is impossible to predict the main thrusts of astronomy research more than ten years from now, it is obviously wrong to assume that they will be the same as during the past ten years and, therefore, in planning new telescopes there is some point in looking at recent trends in the use of the AAT. The first decade was dominated by extragalactic astronomy involving quasars and galaxies. Recently, multi-object spectroscopy has become very important, still mainly for observational cosmology but also for galactic structure and in the study of clusters of both stars and galaxies. The driver here has been the availability of the FOCAP and Autofib fibre systems on the AAT, in conjunction with the use of the automatic measuring machines COSMOS and APM in Britain to measure many UKST plates. These have generated large-scale catalogues, as well as finding rare exceptional objects. It seems likely that the UCLES high dispersion spectrograph will lead to renewed interest in stellar astronomy, especially in detailed abundance analyses and in the solar-stellar connection, i.e., the study of flares, spots, active chromospheres, and astro-seismology in stars other than the Sun. Many 'classical' problems in stellar astronomy were left in abeyance rather than solved twenty years ago, and the time is ripe for a fresh attack using the much more powerful instrumentation now available. Another strong trend is towards improved optical resolution by a variety of interferometric and image-sharpening techniques, and of course there is certain to be a great growth in infrared imaging and spectroscopy with array detectors.

The general lesson to be drawn is that the astronomical research is driven by the available instrumentation rather than the converse, and that big breakthroughs occur when there is a sudden very large enhancement in some instrumental capability.

4.2. THE POWER OF NEW LARGE TELESCOPES

The gains for many 'conventional' fields of optical astronomy in simply going from a 4 m to an 8 m telescope will be quite small, especially in comparison with the enormous gains achieved in the last decade, where CCDs have given a 100-fold improvement over photography and multi-object spectroscopic systems have given similar gains in those fields where they are applicable. In particular, for background-limited work on faint galaxies, an 8 m telescope will only go twice as faint as a 4 m in the same observing time. For many applications, an 8 m telescope is simply four times faster than a 4 m, and with linear detectors one can often (although certainly not always) trade aperture for exposure time. A corollary of this result is that a quarter-share in a 8 m-class telescope may be only as useful as a full share in a 4 m; although multi-institution or international cosortia have a great deail to recommend them both financially and scientifically, it may not be worth sacrificing a good domestic programme for a small share in a larger programme. The oft-quoted analogy with particle accelerators for fundamental physics is not necessarily a good one in the case of astronomical telescopes.

It is, therefore, imperative that the new 8 m-class telescopes be designed particularly to do those jobs where the full advantage of increased aperture will be realised; for optical astronomy, this means in particular for photon-limited work such as high dispersion spectroscopy on objects brighter than the sky background, or where fast time resolution is required. The biggest advances, however, will surely be in the infrared, where major advances in detector technology are expected and where existing telescopes begin to be diffraction-limited. Only by emphasizing these fields of research is the funding for the new telescopes likely to be secured.

It is also essential that the new very large telescopes be built very well, with superb optical, mechanical, and control systems, if they are to out-perform the best existing telescopes. For example, even a small degradation in optical imaging performance could easily make an 8 m telescope no better than a 4 m telescope for many key projects. The scientific cases for 8 m telescopes often assume that they will have even better performance than the best telescopes in use today, but this will not be achieved easily or cheaply, especially given the novel technological problems to be overcome.

4.3. SOME CRUCIAL REQUIREMENTS FOR SUCCESSFUL LARGE TELESCOPES

The following summary of requirements applies particularly to *national* telescopes, designed to be used by many astronomers for a wide range of projects. There are of course valid arguments for building simpler and cheaper telescopes, to carry out specific tasks or to be used by smaller communities of astronomers, and not all of the requirements necessarily apply in such cases.

– The 8 m telescopes must have excellent optical and mechanical performance, and they must be on excellent sites, or they will be no better than the best existing smaller telescopes for many purposes.

– They must have excellent instrumentation.

– That instrumentation must include detectors, powerful computing facilities, and efficient software, especially for data reduction.

– The telescope must be able to attract and retain high calibre highly motivated people, including scientific, technical, and support staff.

– Workable management structures are very important; those running telescopes must be given reasonable levels of authority and responsibility, and plans should not be continually changed; committees should advise rather than steer.

– There must be adequate ongoing budgets to maintain operations; getting the initial construction capital is only half the battle.

– The operating budgets must also have generous provision for new instrumentation, to enable the telescopes to remain competitive.

– Adequate financial provision must be made for visiting astronomers to travel to and work at telescopes in remote sites, often overseas.

It is apparent that the Japanese already appreciate many of these points and are doing exactly the right thing in going for high-performance very large telescope now. The JNLT is an extremely exciting project which promises to be one of the very best astronomical instruments in the world.

DAVIES – Don Hall and Terry Lee emphasized that these are very large gains to be made at IR wavelengths with 8 m telescopes (compared to 4 m). In addition, new scientific applications that depend on the resolution diameters rather than the collecting area are possible at 8 m. In fact, the scientific applications that you listed as high priority for the AAO, namely IR imaging and échelle spectroscopy, are areas where large gains are to be made by going to 8 m. Given the choice would you build a 4 or 8 m telescope?

CANNON – Certainly, given that simple choice, I would have no hesitation in building an 8 m telescope! I agree entirely with what you say: the points I was trying to make were (i) that we must base the scientific case for a new 8 m telescope on precisely those areas where we will get the biggest gains and do new science, and (ii) we must make the 8 m telecopes and their instruments good enough to ensure that the theoretical gains are realized in practice.

ANDO – It seems to me that your major instruments were developed in Britain. Do you have facilities to develop instruments in Australia? If not, could you explain the reason?

CANNON – It is true that many of our instruments were built in Britain, although we do have facilities to build our own instruments as well, and some have been built by other groups in Australia. The reason is that the AAO was established as an organization specifically to operate the 3.9 m AAT for users from the two countries. It was not intended to function as a complete independent observatory. We do, however, have a significant in-house instrumentation capability, because we need to be able to maintain the telescope and its many complex instruments.

SEKIGUCHI – What fraction of observing time is used for the multiple-object spectroscopy?

CANNON – I guess recently it has varied between about 10 and 20%.

LESSONS FOR NEW LARGE TELESCOPES FROM THE AAT*

P. R. GILLINGHAM

Anglo-Australian Observatory, Australia

(Received 10 January, 1989)

Abstract. When it began operating in 1975, the Anglo-Australian Telescope set new standards for pointing, tracking, and efficient observing. Since then, several large telescopes with more advanced control systems and on better sites have come into competition but the AAT retains the reputation of having the best overall observing efficiency.

A number of organizational factors in the design and construction phase and in the Anglo-Australian Observatory's operational years have contributed to the AAT's success. Careful consideration of these factors should help groups planning the construction and operation of new telescopes.

1. Introduction

The Anglo-Australian 3.9 m telescope, which began fully-scheduled operation in the middle of 1975, was one of the first large optical telescopes to be planned with full computer control. Its good mechanical, optical, and electronic engineering together with innovative programming gave it a performance which set new standards for pointing, tracking, and the efficiency of routine observing.

Now, in 1988, there are a number of comparable telescopes on better sites with more advanced control systems but the AAT's overall observing efficiency is probably still the best amongst telescopes on which observing time is so widely distributed.

Why was the AAT initially so successful and how has its reputation been maintained so well? Although the author, having been so long and so closely involved with the AAT, is in danger of not seeing the wood for the trees, it may help other observatories and particularly those planning new large telescopes, to describe several aspects of the AAT design and construction project and of the observatory management. Some technical lessons from the AAT experience are still very relevant but this paper will be confined to organizational matters.

2. Project Management and the Design Team

Figure 1 shows the functional links between the various bodies involved in the telescope design and construction. The Anglo-Australian Telescope Board (AATB) was (and remains) a six-member body, half British and half Australian with one senior representative from the appropriate government department of each nation and the other members leading scientists involved with astronomy. The AATB meets twice each year.

During design and construction of the AAT, the full time employees of the AATB

---

* Paper presented at the Symposium on the JNLT and Related Engineering Developments, Tokyo, November 29–December 2, 1988.

comprised a Project Manager, an Executive Officer, and a small group of engineers, draftsmen, programmers, and clerical assistants. This group, known as the AAT Project Office, was set up in 1968 and, at its largest, in 1971–1972, had a staff of 17. It was housed in an office building in Canberra. One astronomer from each nation acted throughout the project as a special adviser and a few small advisory committees composed of astronomers and engineers were set up during the project. The first Project Manager was a structural engineer who had worked on the Parkes radiotelescope project and the first two engineers were from Mt Stromlo Observatory and from the Royal Greenwich Observatory. Thus the Project Office started with a background of very relevant experience.

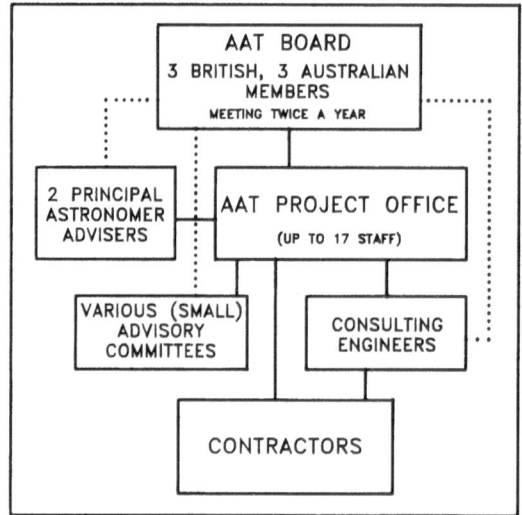

Fig. 1.   Organization of design and construction.

As Figure 1 indicates, day-to-day management of the project was very clearly the responsibility of the Project Office. The advisory committees, individual advisers, and consulting engineers had some direct input to the Board but they normally communicated with the Project Office.

3. Design Constraints and Help from Kitt Peak National Observatory

The agreement to build the AAT stipulated that it be based on the telescopes already being designed for Kitt Peak and Cerro Tololo. Thus, although the aims for overall performance were high, there was no necessity to break completely new ground, so efforts could be directed mainly towards refining the design. The AAT later diverged considerably from the Kitt Peak design – not one working drawing for the American telescopes was used for manufacture of AAT parts – but access to the already completed design studies and the continuing discussions and correspondence with the Kitt Peak design team were extremely important. Some studies initiated by the Project Office led to significant changes in the Kitt Peak design, so the benefits were not completely one way.

Because the Project Office had no 'in-house' research and development facilities, with detailed design and all construction contracted out, the major design decisions had to be conservative. For example, the primary mirror focal ratio was chosen to be $f/3.3$, rather than the $f/2.7$ adopted for its American predecessors, partly because commercial optical firms at that time were unwilling to guarantee as good a performance at $f/2.7$ as was desired.

Since the Project Office had no planned existence beyond the completion of the project, there was strong pressure to ensure that the telescope would work properly 'as built'. This was good as far as the telescope itself was concerned because it resulted in thorough engineering studies, with no temptation to accept a partial solution that might be 'fixed' after construction. For the auxiliary instrumentation, however, some drawbacks resulted. For instance, a few of the instruments meant to be available from the outset had to be frozen in their design too early and were virtually obsolete before they were used.

4. Contractual Matters and the Role of Consultants

The inter-government agreement specified that for all contracts, tenders would be invited world-wide, without favouring the partner nations. This resulted in a very broad distribution of contractors and consultants, with Canada, U.S.A., Britain, Switzerland, Japan, and Australia all making major contributions. In all the contracts, a favourable attitude of cooperation was maintained, with the contractors regarding the job as out of the ordinary and worthy of special efforts.

A direct consequence of keeping the project team small was the need to engage engineering consultants for several aspects of the work. Although this was expensive, if measured by the cost per man hour, the broad range of experience and expertise tapped in this way was very valuable. The Project Office, of course, kept closely in contact with the work of the consultants at all times.

5. Erection

Daily supervision of the erection of the building and dome was by staff of the Sydney civil engineering consultants who had been responsible for the details of their design, an engineer from this firm living on site for the duration of this and some later work. The consulting structural engineer who had been most concerned with the mounting contract stayed some months in Japan while the polar axis structure was assembled and tested then lived in Coonabarabran (about 30 km from the site) during the erection and initial testing of the telescope. Mechanical, electronic, and optical specialists from the British contractor for the optics and optical support systems spent several months on site.

6. Commissioning

Five of the Project Office staff lived for a year or more on site or in Coonabarabran during the commissioning of the telescope. They comprised a mechanical engineer (the author), supervising day-to-day commissioning, a computer scientist coordinating the digital hardware and software activities, a computer hardware specialist, and two computer programmers. About 6 additional staff were recruited to support the various activities. The astronomer who had been the Australian adviser throughout the project was appointed Commissioning Astronomer, with responsibility for coordinating the testing and tuning up of the telescope and auxiliary instruments. He made frequent visits, as did the Project Manager and other Project Office engineers still based in Canberra.

First light was in April 1974 but it was not until late October that the aluminising plant for the primary mirror was ready, so for six months functional testing of the telescope was possible with no pressure to make observations of direct astronomical interest. This was very valuable in encouraging thorough testing of the optical, mechanical, and electronic performance and allowing development of the pointing and tracking software with very little distraction. Several astronomers from the U.K. and Australia helped with the commissioning activities; in the case of those from the U.K., they generally stayed on Siding Spring mountain for about a month.

The first Director took up his post late in 1974 and the Anglo-Australian Observatory (AAO) was established, with its administrative base in Sydney, early in 1975. After the primary mirror was aluminised, time was increasingly devoted to astronomical observing, and fully scheduled observing began on 28 June, 1975.

The commissioning staff living at or near the site had continual access to the telescope so they were able to make fast progress (spending a great proportion of their time on the job). It was soon apparent that the telescope would live up to our high expectations and this greatly encouraged our efforts. Because the author lived only 200 m from the telescope and the programmers adopted the practice of working from early afternoon till early morning, full advantage was taken of opportunities for witnessing observers' problems and for testing and refining the telescope performance (e.g., when it was cloudy) until many months after fully scheduled observing started.

7. Observatory Management

Figure 2 shows in outline how the observatory is organized. The AAT Board controls the AAO as it previously controlled the Project Office. The Board has included some very distinguished and influential scientists and public servants at the highest level. Bi-national funding and the strength of this board, with responsibility for just the one telescope (until mid-1988, when the AAO took over operation of the UK Schmidt Telescope) have helped maintain a healthy budget despite the reductions made by both governments in their science expenditure. The fact that all AAO employees can meet and put their views directly to the people at the top of the management 'pyramid' has been good for staff morale.

The senior AAO staff member, the Director, is appointed for a limited time as are all but one of the astronomers. On the other hand, with a few exceptions, technicians, tradesmen, and clerical staff have indefinite appointments. The AAO staff was limited

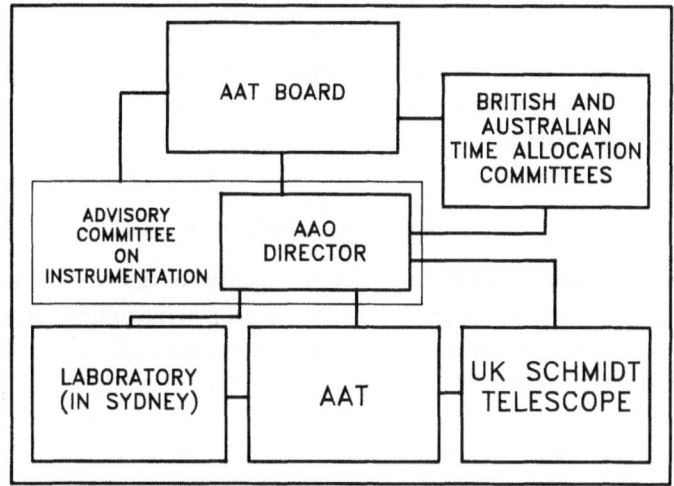

Fig. 2.   Organization of Anglo-Australian Observatory.

at 50 for several years, with about 20 based at Coonabarabran and 30 at the Laboratory (effectively the scientific and administrative base) in Sydney. (The staff total was expanded by a few when the UK Schmidt joined the AAO.)

8. Operational Support

A key element in the success of the AAT has been the high level of support given observers. Since anyone may apply for time and a run of more than 3 nights for the one group is unusual, the users in any year are very numerous. With the variety and sophistication of the instrumentation, virtually all of which is available to all applicants, efficient operation depends critically on supporting the visitors before their run (while they plan their observations and set up the instrument), during the observing, and after their nights (to reduce data).

Each observing team is notified, when their time is allocated, which AAO staff astronomer is to support them. This support astronomer is the contact for advice before the observer comes to the AAT. Then, unless the observer is fully confident he or she can handle the observations without such help, the support astronomer travels to the telescope to be with the observer during setting up and observing.

The four night assistants at the AAT are rostered on 7 nights at a time. Between these weeks each works at daytime duties, not as a fill-in, but as a full member of the technical staff. Four electronic technicians work a roster of evening shifts, a week at a time, starting at 2 pm and ending at about 10:30 pm, so that they overlap with the day staff and are present for the first few hours of observing, when faults are most likely. This evening technician is 'on call' after he ends his shift and over the weekend. Confidence that most faults can be cured on the night with minimal time loss has characterised the operation. Over the last five years the proportion of night hours lost due to AAT equipment failure has averaged 2.3%.

Minimising the distinction between day and night staff has helped a great deal in avoiding discord and it naturally makes the night assistants more knowledgeable about the technicalities of instruments and the day staff more appreciative of the problems facing astronomers at night.

9. Instrument Development

Since the telescope was commissioned, about a quarter of the AAO budget has been devoted to the purchase or development of instrumentation (including hardware for data reduction). Much of the instrument design and construction has been done by outside institutions under contracts from the AAO, but in a few key areas the bulk of the work has been done within the AAO. Examples are in infrared photometry, CCD applications, and multi-fibre feeds. Although most such in-house design and development is done at the Laboratory in Sydney, staff based at the telescope have been involved in some design and construction and a lot of the commissioning and subsequent development work.

The establishment of a strong instrument development capability within the Observatory has produced a valuable pool of technical expertise which helps in proposing, assessing, and planning new instruments and in getting the best out of all the instruments in operation. Doing some design and development at the telescope helps to maintain the interest of the very competent staff who are, in any case, needed on site to properly care for the instrumentation.

10. The Site and the People

Siding Spring mountain, at only about 1200 m altitude, lacks the very good infrared performance of much higher mountains and it has a smaller proportion of excellent seeing than a few of the best sites. But the fact that good staff are happy to live indefinitely within half an hour's journey from the telescope and work standard hours for normal rates of pay is a very strong advantage of the site. The median period of service of the present staff at Coonabarabran is 11 years.

11. The End Product

The end product of operating a telescope is scientific data (it is up to our customer – the astronomer – to make good use of those data). So it is important that the aim, to produce the best possible data within the constraints of time and available resources, be kept in mind always.

SOME NEW ASTRONOMICAL FACILITIES IN CHINA*

WANG SHOUGUAN

Beijing Astronomical Observatory, Beijing, China

(Received 26 December, 1988)

Abstract. For the 1990's, plans for some astronomical facilities and related researchers are being carried out in China. In this report we describe in some details the plans of radio astronomical facilities, 150/220 cm Schmidt telescope, and experiments on porcelain mirror material.

1. Radio Astronomical Facilities

1.1. THE MIYUN METRE-WAVE APERTURE SYNTHESIS TELESCOPE

This instrument was constructed in 1985. Its general feature is described by Wang *et al.* (1986). It consists of an E–W array of 28 elements, each of 9 m aperture. One-hundred ninety-two spacings are available. Figure 1 shows the configuration of the array. At the present stage, it works at a frequency of 232 MHz, giving a resolution of $3\rlap{.}'8 \times 3\rlap{.}'8 \cosec \delta$ with a field of view of $8° \times 8°$. Figure 2 is the record of SNR G78.2 + 2.1, obtained with $u - v$ coverage of 96 spacings rotated (Zheng, 1988). As the SNR lies only 4° from the strong Cyg A, we estimate that the dynamic range of the synthesis system is well above 1000.

Sky survey of selected area is scheduled as the first sequence of mapping for all sky survey of $\delta \geq +30°$. A typical map of selected area obtained with 2×12 hr Earth rotation synthesis records more than 400 sources in a $8° \times 8°$ field (amount to about 6 sources per square degree). Many of them are new. Precisions of position and flux density are at

$$1 \text{ Jy, position: } \pm 3'', \text{ flux density } \pm 5\% ,$$

Second-epoch observations of the selected areas will be carried out after an interval of time in order to detect variable sources. A second frequency – 327 MHz – will be added in the same array to work simultaneously with the first so as to increase the credence of variability detection and to help in the study of the spectrum during variation.

1.2. THE CHINESE VLBI NETWORK (CVN)

The VLBI Station was built in Shanghai and has been operational since 1987. Its 25 m antenna is equiped with receivers working at wavelengths of 18, 13, 6, 3.6, and 2.8 cm. MK-2 and MK-3 terminals are available. In Table I, joint operations it has participated in so far are listed.

* Paper presented at the Symposium on the JNLT and Related Engineering Developments, Tokyo, November 29–December 2, 1988.

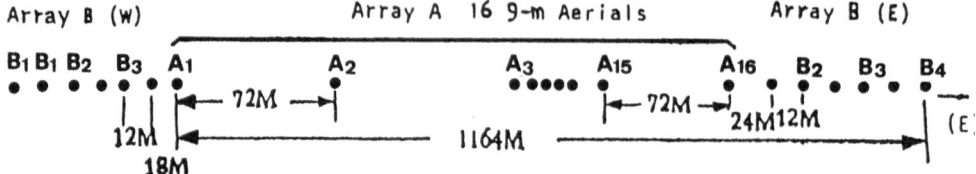

Fig. 1.   Configuration of Miyun metre-wave aperture synthesis array.

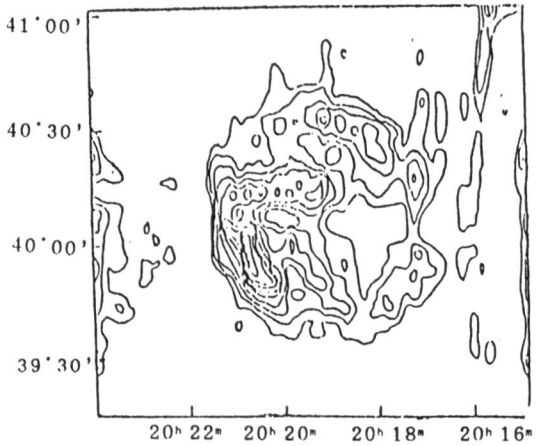

Fig. 2.   SNR G78.2 + 2.1 232 MHz MAP (Cyg A is about 4 from the centre of the field).

TABLE I

Date	Stations	Wavelength	Terminal	Purpose
9 June, 1987	Shanghai, Kashima, Gilereck, Kanai	3.6 cm	MK III	base geo. coordinates
23 Oct., 1987	Shanghai, Kashima	3.6 cm	MK III	ditto
30 Nov., 1987	Shanghai, Kashima, Tidbinbilla	3.6 cm	MK III	ditto
9 Apr., 1988	Shanghai, Kashima, Gilereck, Kanai	13/3.6 cm	MK III	ditto
14 June, 1988	Shanghai, EVN (4 stations)	6 cm	MK II	Nucleus NRAO 150
21 July, 1988	Shanghai, Wettzell, Kashima	13/3.6 cm	MK III	baseline geo. coordinates

Facilities for wavelengths 92, 50, and 1.3 cm are being developed.

For operation in the 1990's the second 25 m antenna is now in construction, together with small telescopes available at different places, a Chinese VLBI network will be formed (with Shanghai–Urumqi–Kunming as its basic configuration).

Figure 3 shows the locations of the stations and Figure 4 illustrates the $u - v$ coverage when the Chinese VLBI network is combined with the European network.

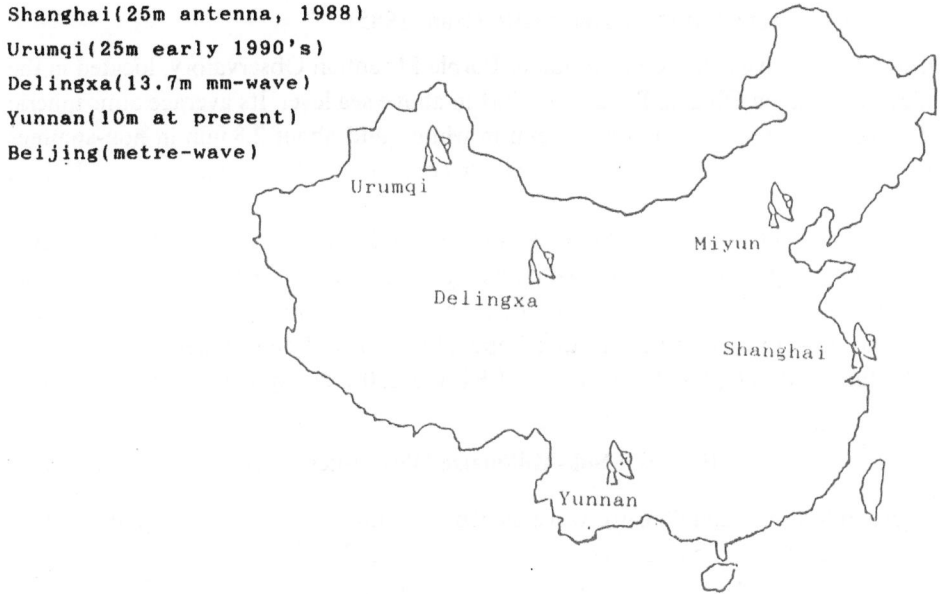

Shanghai(25m antenna, 1988)
Urumqi(25m early 1990's)
Delingxa(13.7m mm-wave)
Yunnan(10m at present)
Beijing(metre-wave)

Fig. 3.   Location of Chinese VLBI stations.

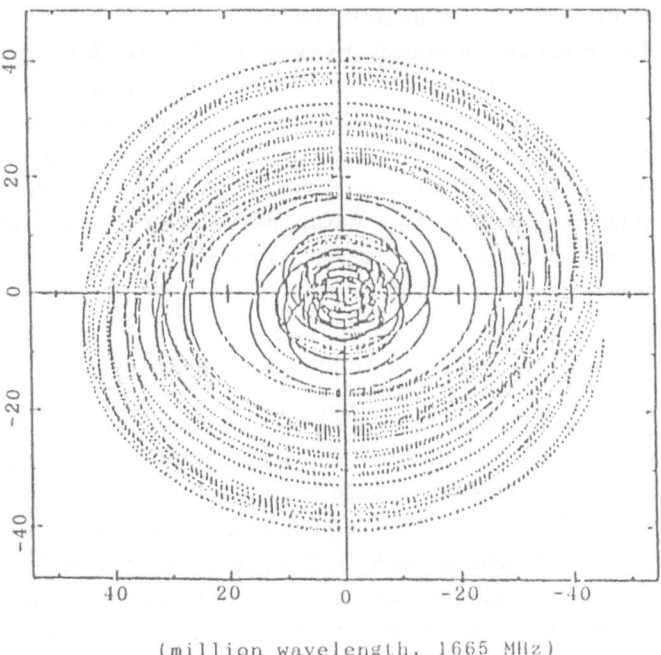

(million wavelength, 1665 MHz)

Fig. 4.   (UV) Coverage of CVN and EVN.

1.3 MM-WAVE TELESCOPE AT DELINGXA (Han, 1985)

Delingxa mm-wave astronomy station of Purple Mountain Observatory, located in the Zaidamu Basin on Qinghai Plateau, is 3204 m above sea level. Its average atmospheric water vapor content at zenith is 1.4 mm in winter, and about 2.8 mm in non-summer seasons. Number of days with water vapour content less than 1 mm is about 30 in a year.

The 13.7 m antenna with ESSCO A240 type randome, as well as 22 GHz receiver and computer systems were installed on site (37°22'.4 N, 97°33'.6 E) in 1986. Further adjustment of the 22 GHz system and the calibration of the telescope will be completed this year. 85–115 GHz front end quasi-optical system will be installed in 1989.

For further development, mm-wave VLBI and 250 GHz system are contemplated.

2. Potential Sub-Millimetre Wave Sites in Tibet

For ground-based submillimetre wave astronomy, atmospheric water vapour content becomes the chief parameter to consider. The scale height of atmospheric water vapour is about 2 km. In Tibetan Plateau, the average altitude of the 'ground' is as high as 4000 m. Preliminary investigation has indicated that several places in this region may be suitable candidates for sub-millimetre sites. Among them Dingri (28°38' N, 87°05' E, 4300 m) seems to be most promising (Wang, 1988).

Crude estimation of atmospheric water-vapour content was made with the help of radiosonde and ground absolute humidity data. For Dingri, such estimation together with other meteorological data are listed in Table II.

Dingri may also be suitable for optical observations. However, it has drawbacks: In winter and spring, the best seasons for astronomical observation, the temperature variation at night time is high, it is also the time when the gale days occur most. However, gales in this region always blow in the afternoon and subside at night.

Dingri is about 200 km for Rikaze, the second largest city in Tibet. In recent years, Dingri is frequented by foreign tourists, mostly mountaineers. As the sunshine-hour there is profuse, and direct radiation of the Sun is strong, solar energy may be economically utilized to generate electricity for the installations.

3. The 150/220 cm Schmidt Telescope Project

For astronomical research, telescopes with large collecting area and telescopes with large sampling capacity are complementary. For many decades, great amount of support has been given to the development of large or giant reflecting telescopes (often cost 10–100 million US dollars). In contrast, attention paid to the second category (cost much less) is by far insufficient. It would be of value to astrophysics in the 1990's to devise a large Schmidt-type telescope to cope with the rapid development. The project of 150/220 cm Schmidt telescope was thus proposed. This is an instrument of classical type, and is the result of choice among several suggestions. Material for the main mirror is a spare blank of the Chinese 2 m reflector, and the facilities in Nanjing Astronomical

TABLE II

	Jan.	Feb.	Mar.	Apr.	May	June	July	Aug.	Sept.	Oct.	Nov.	Dec.	yr
Numbers of													
clear nights	25.7	24.0	24.7	20.0	16.3	12.0	0.7	2.0	10.0	21.0	25.0	24.0	209.7
half clear nights	3.3	2.7	5.0	5.7	9.7	13.0	9.3	6.7	7.0	5.5	1.3	5.0	74.2
H_2O content													
$W \leq 0.5$ mm (days)	6.3	6.3	2.3	0	0	0	0	0	0	0	0	1.0	15.9
$W \leq 1.0$ mm (days)	25.3	21.3	13.3	1.0	0	0	0	0	0	1.0	4.0	19.5	85.4
$W \leq 2.0$ mm (days)	31.0	27.3	29.6	18.0	8.0	0	0	0	0	4.0	27.3	30.5	175.7
Sunshine hours	284	265	296	305	338	286	242	221	250	300	296	290	3373
Gale day (days)	13.2	16.4	13.8	12.2	7.2	1.0	1.2	0.2	0.6	1.9	4.0	9.6	81.2
Wind vel. (m s^{-1})	3.2	3.9	3.9	3.8	3.5	2.6	2.2	1.9	1.9	2.3	2.2	2.2	2.8
Temp. variation													
$T_{00} - T_{min}$ (°C)	6.2	7.5	7.6	6.7	5.5	4.9	3.0	3.2	3.8	4.7	5.4	6.1	5.4

Instrument Factory are adequate for polishing and testing mirrors up to this size. See Report of Scientific Objective Group of CST, 1988 (in Chinese), and Report of Engineering and Technical Group of CST, 1988 (in Chinese).

Preliminary-phase work is under way. Main features of the telescope are:

Aperture of main mirror: 220 cm.

Aperture of correcting plate: 150 cm.

Focal ratio: $f/2.5$.

Receiving system: (1) total field photographic plate,
                  (2) total field optical fibre system.

Objective prism.

Spectrograph and CCD system for multi-object detection with the help of optical fibre system (about 10 sets of CCD for simultaneous detection of 500 objects).

Pointing accuracy: 6″.

Tracking accuracy: 0″.1.

Field of view: 5°.

Working wavelength range: 3200–10 000 Å.

Among the technical problems, fabrication of the 150 cm correcting plate needs some new technological considerations. It will be an achromatic plate of K9 and QF2 glasses. A factory in Chengdu is willing to cooperate and adapting its facilities for manufacturing thin plates of this size. The question of adhesive for bonding the glasses is also under investigation.

The introduction of optical fibres to the Schmidt system undoubtedly presents great advantage. The experience of Steward Observatory, ESO and UK Schmidt will be followed. Prototype research is being planned and the system will be fitted in the 60/90 cm Schmidt at Xinglong for experiment.

To avoid large expenditure and long time required for building a new site, the site for the Schmidt Telescope is chosen among the existing sites. The condition of Xinglong Station has been studied. The main parameters are found to be comparable to that of the site of UK Schmidt at Siding Spring.

It is expected to complete the installation in about five years.

4. Experiments on Porcelain Mirror

Dr Yang Shijie of Purple Mountain Observatory, whose work we quote here (Yang, 1987), has for many years studied the technique of making optical mirrors with porcelain. As it is well known, the raw material for porcelain is cheap and abundant, and the material can be shaped into required form before firing. Yang uses general industrial 'hard' porcelain for his experiments. This type of porcelain is in mass production, so the production technique is readily available. Its properties are compared in Table III with other material for mirror fabrication.

We see in the table that the 'hard' porcelain has a thermal expansion coefficient (α) comparable to that of Pyrex. The important parameters, besides α, for porcelain mirror blanks are material homogeneity and low air bubble content. An ideal approach is to

TABLE III

Material	Thermal expansion (1°/C)	Modulus of elasticity (kg cm^{-3})	Thermal conductivity (cal/°C cm s)	Specific heat (cal/g °C)	Density (g cm^{-3})
Pyrex	32×10^{-7}	6.8×10^5	0.0027	0.160	2.23
Quartz	5×10^{-7}	7.0×10^5	0.0032	0.170	2.21
Glass ceramic	0.4×10^{-7}	9.2×10^5	0.0039	0.196	2.52
'Hard' porcelain	$30–50 \times 10^{-7}$	$6–8 \times 10^5$	0.0025	0.220	2.3–2.5
Cordierite	$20–30 \times 10^{-7}$	?	0.0037	?	2.60
Aluminum	230×10^{-7}	6.9×10^5	0.5300	0.215	2.70
Beryllium	124×10^{-7}	28×10^5	0.3800	0.516	1.82
Titanium	25×10^{-7}	11.6×10^5	0.042	0.126	4.54
Super	$1 \times 10 \times 10^{-7}$	13.8×10^5	0.026	0.120	8.13

use a vacuum kneading machine to keep the air bubble away while the clay is being mixed under vacuum.

Porcelain is a crystalline material. The size of the crystals are relatively large. In the process of becoming porcelain, there is a volumetric reduction, creating large spacings between the crystals. For this reason, one cannot achieve a high-quality reflecting surface. It is, hence, desirable to add a layer of polishable material on the top to achieve a good reflecting surface. Glazing over the porcelain has been tried, but was found to introduce a large number of bubbles. Other methods evolved are: (1) Glass fused over the porcelain, (2) Chrome-plating the surface after an initial layer of silver deposit, (3) Using chemical deposits of nickel and phosphorous alloy, and (4) Porcelain with vacuum deposit of silicon dioxide.

Yang has reported his experiments on all these methods. Some samples of glass fusing has been produced (Yang, 1987).

Elsewhere (mainly in Jingdezhen), porcelain of zero-thermal expansion coefficient is being processed.

References

Han Pu: 1985, *Acta Astron. Sinica* **26**, 188.
Moran, J. N. *et al.*: 1984, *Smithsonian Astrophys. Obs.* **000**, 000.
Wang, H. *et al.*: 1986, *Chin. Astron. Astrophys.* **10**, 3.
Wang, S. G. *et al.*: 1988, *Preliminary Investigations on Potential Astronomical Sites in Tibet* (in press).
Yang Shijie: 1987, *The Making of Porcelain Mirrors* (in press).
Zhang Xizhen: 1985, *Miyun 232 MHz Survey on Regions [α 00°41, δ 41'00], [α 00°06, δ 34'40].*
Zheng Yija: 1988, *Kexue Tongbao* **11**, 844.

KOGURE – Your site searching in Tibet seems concentrated to the condition of mm-wave observation. How do you think about the optical condition, particularly seeing condition? Do you have some plan to make a site search for optical observation in this area?

WANG – Despite of the complexity of the meteorology over 'the roof of world' which makes the assessment of astronomical seeing condition difficult, the chance of finding a site excellent for optical astronomy there is not be under-rated when one thinks of the great height, the dryness, and the variety of topography of the Plateau. Meteorological and topographical data of the region around Ding-ri are being studied for the preliminary assessment to help in the decision on the setting up site testing devices in the region.

HÜGENELL – Professor, what about the wind forces up to 4000 m altitude in Tibet? Could you perhaps say something about this problem?

WANG – Gales (wind velocity ≥ 17 m s^{-1}) occur on the Tibetan Plateau much more frequently than in other places in China. In some place, gale-days can reach 100–150 a year (about 80 gale days yr^{-1} in Ding-ri). About 75% of the gales happen in winter and spring-time. However, monthly average of wind speed there is not very high, e.g., less than 4 m s^{-1} in the most windy months at the most windy place – Ga-er. Gales always blow in the afternoon and early evening and subside at night. They are often gusty in nature. Momentary strength of 34 m s^{-1} has been recorded in history at Ding-ri.

SIXTH SESSION

INSTRUMENTATION

OPTICAL INSTRUMENTS FOR JNLT*

SADANORI OKAMURA

Kiso Observatory, The University of Tokyo, Japan

(Received 18 January, 1989)

Abstract. The plan for optical instruments for the Japanese National Large Telescope is described. Performance of the first-generation instruments is computed on the basis of tentative designs, and the capability of the telescope is demonstrated.

1. Introduction

Observing instruments for any telescope must evolve. The lifetime of the up-to-date instruments will be 5–10 years because of the rapid progress in technologies in various fields including detectors, electronics, mechanics, optics, and computers. This is generally valid for observatory instruments, i.e., user instruments. Some sophisticated iunstruments may survive longer. It is, thus, very important to establish well-defined renewal cycles of observing instruments for the efficient use of a telescope, although much attention should be paid to consistency for long-term scientific programs. Experiences gained in the design study and initial tests of the previous-generation instruments should be incorporated into the design study of the next-generation instruments, which should start immediately after the commissioning of the previous-generation instruments.

It will take seven years to construct the Japanese National Large Telescope (JNLT). The final year of its construction will be spent on the system test. Design study of the first-generation instruments for JNLT should be completed 2–3 years before the completion of JNLT. Much of our efforts have been spent so far on the design study of JNLT itself. We are now gradually turning our efforts to the design study of the first-generation instruments. Several Working Groups (WGs) on JNLT instrumentation were formed by the Group of Optical and Infrared Astronomers (GOPIRA) in June 1988. They include WGs on wide field study, low-dispersion spectroscopy, high-dispersion spectroscopy, speckle and interferometry, infrared instruments, guide and acquisition system, and data analysis. These WGs are to produce the reports early 1989 in which preliminary design of the first-generation instruments are included. The design will be used as a 'strawman' to identify possible conflicts with the telescope design and items that need further design studies and/or technical developments. There exists no concrete design for any instrument approved by the WGs at the moment. I describe the optical instrument plan for JNLT as of November 1988 based upon the discussions of respective WGs communicated to me.

* Paper presented at the Symposium on the JNLT and Related Engineering Developments, Tokyo, November 29–December 2, 1988.

Our philosophy on the first-generation instruments that emerged from the discussions in the WGs are the following.

(1) They should be ready for use in the latest stage of JNLT construction for the system test, and some of them should be considered as an integral part of JNLT.

(2) They should consist of a reasonable number of simple, well-understood, standard instruments. Incorporation of too many technical innovations into the first-generation instruments might lead to the significant delay of JNLT itself.

(3) However, they should not be too obsolete. They should have a good match to frontier technology available at the time of the design study.

(4) They should be associated with dedicated and efficient software for data acquisition and reduction.

In this paper, I will demonstrate that even with such simple standard instruments JNLT will provide exciting scientific opportunities that no other existing telescope can do.

2. JNLT Foci and Instruments

Table I lists the foci considered for JNLT together with the fundamental parameters. Field rotation will be compensated by instrument rotators at prime focus and Cassegrain

TABLE I

JNLT foci

	F-ratio	Focal length (m)	Image scale		Field of view	
			"/mm	μm/"	arc min	mm
Prime	2.3 [a]	17.25	11.9	84	30	151
Cassegrain						
visible/IR	12.5	93.75	2.19	455	6	164
IR	35	262.5	0.78	1274	1	77
Nasmyth 1	12.5	93.75	2.19	455	4	110
Nasmyth 2	12.5	93.75	2.19	455	4	110
Coudé						

[a] With a three-lens wide field corrector.

focus, and by image derotators at Nasmyth foci. The coudé focus will be used mainly for interferometry in future and specifications for this focus have not been fixed. A Shack–Hartmann wave-front analyzer will be provided at each focus. Incorporation of an atmospheric dispersion corrector (e.g., Wynne and Worsick, 1986, 1988a) and an image stabilizer (e.g., Maaswinkel *et al.*, 1987) into the acquisition/guide systems for these foci is considered. Introduction of an adaptive optical system (e.g., ESO, 1987) will be a major challenge after the initial commissioning of JNLT.

The prime focus with specially designed three-lens correctors will provide a field-of-view of 0.5 deg. This is a very promising focus with a scale of the Hale 5 m telescope, a speed faster than a typical Schmidt telescope and the resolution of 0.5–0.25". This is the most efficient place for faint object studies.   A still somewhat preliminary list of first-generation optical instruments considered for these foci includes the prime focus camera (PFC), multi-object fiber spectrograph (MOFS) at prime focus, faint object spectrograph and camera (FOSC) at Cassegrain focus, high-dispersion spectrograph at Nasmyth focus, and speckle camera to be placed at either Cassegrain or Nasmyth focus.

3. Computation of Signal-to-Noise Ratio

Before going to the description of the individual instruments, I will explain how I compute signal-to-noise (S/N) ratio in order to compare the performance of JNLT instruments with that of instruments for the Hubble Space Telescope (HST). The procedure adopted here is a standard one described, for example, by Tyson (1984).

The signal flux f is given by

$$f = 1.06 \times 10^3 \times 10^{-0.4m_v} \quad (\text{photons cm}^{-2}\,\text{s}^{-1}\,\text{Å}^{-1} \text{ at } \lambda = 5500 \text{ Å}), \quad (1)$$

where m_v is the apparent magnitude of an object in the V band. The corresponding signal count received by a detector is

$$S = fA\varepsilon\Delta\lambda t = n_* t \quad (e/\text{image area}), \quad (2)$$

where A is the telescope aperture in units of cm^2, ε the overall efficiency, $\Delta\lambda$ the band width in ångstroms, and t is the integration time in seconds. Similarly, the flux and count from the sky is computed as

$$f = 1.06 \times 10^3 \times 10^{-0.4m_s} \quad (\text{photons cm}^{-2}\,\text{s}^{-1}\,\text{Å}^{-1}) \quad (3)$$

and

$$N_s = fA\varepsilon\Delta\lambda a_p t = n_s t \quad (e/\text{pixel}), \quad (4)$$

where a_p is the pixel size in arc sec^2, and m_s is the sky brightness in the V band. I assume $m_s = 22.0$ mag arc sec^{-2} for JNLT and 22.5 mag arc sec^{-2} for HST (cf. Tyson, 1984).

I assume a CCD as the detector and the following internal noise sources are taken into account: readout noise N_R (e/pixel), digitization noise N_D (e/pixel), and dark current n_d ($e/\text{pixel s}$). The total internal noise of the detector becomes

$$(N_R^2 + N_D^2 + n_d t)^{1/2} \quad (e/\text{pixel}), \quad (5)$$

and the total noise is given by

$$\{N_R^2 + N_D^2 + (n_s + n_d)t\}^{1/2} \quad (e/\text{pixel}). \quad (6)$$

The signal-to-noise ratio per pixel $(S/N)_p$ is computed by

$$\left(\frac{S}{N}\right)_p = \frac{n_* t}{\{N_R^2 + N_D^2 + (n_s + n_d)t\}^{1/2} n_p}, \quad (7)$$

where n_p is the number of pixels included in the image area or a resolution element under consideration. The signal-to-noise ratio per image area, or per resolution element, is then given by

$$\frac{S}{N} = \left(\frac{S}{N}\right)_p n_p^{1/2} .$$ (8)

4. Description of the Instruments

4.1. PRIME FOCUS CAMERA (PFC)

PFC will be the key instrument for the investigation of a variety of fundamental cosmological problems such as formation and evolution of galaxies, origin of large-scale structure of the Universe, non-uniformity of the Hubble flow, the cosmic distance scale, nature of dark matter, and physics of active galactic nuclei.

Hardware of PFC that we have in mind is a simple CCD camera with a suit of many selectable filters and polarizers. The design of PFC will be similar to that of the Advanced Radial Camera (ARC) for HST (Griffiths *et al.*, 1986), as shown schematically in Figure 1. In spite of its extreme simpleness, PFC will meet requirements for almost all imaging programs up to ~ 1 μm except for narrow-band imaging ($\Delta\lambda < 30$ Å) and programs benefited by oversampling (> 6 pixels/resolution element).

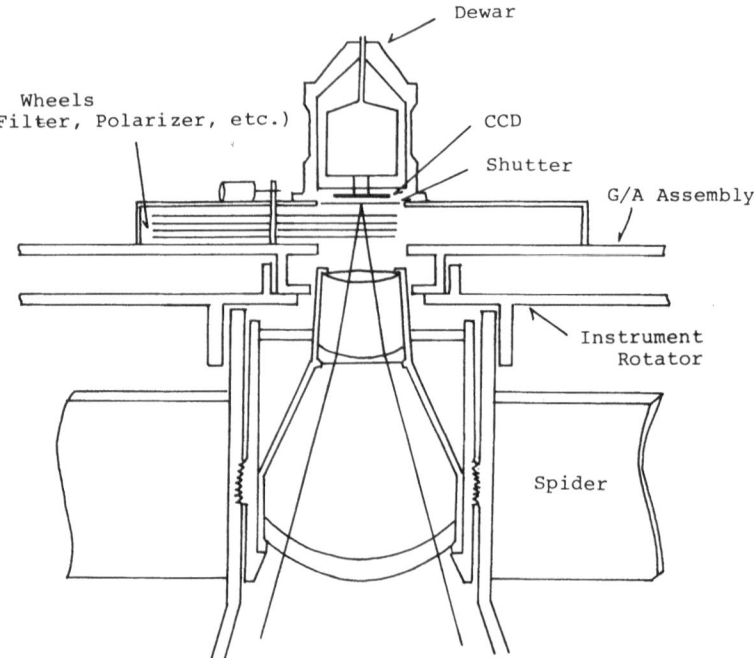

Fig. 1.   The schematic drawing of the Prime Focus Camera.

The final performance parameters of PFC depend upon detectors. Tektronix succeeded in manufacturing the large-format CCD with very high quantum efficiency and very low readout noise (Blouke et al., 1985). This is really an exciting technical innovation. However, progress in detector technology is so rapid that one Japanese company, Hamamatsu Photonics Co. Ltd., is interested in manufacturing the 'one-wafer CCD' made from a whole silicon wafer of 20 cm diameter. If successfully made, the CCD will have an effective area of 12×12 cm^2 with the pixel format of $\sim 8000 \times 8000$, pixel size being 15 μm square. It would be an ideal detector for PFC and it may not be a dream at the time of JNLT operation.

TABLE II

Characteristic parameters used in S/N computation [a]

	HST (WFC)	HST (ARC)	JNLT (PFC)
Telescope aperture	2.4 m	2.4 m	7.5 m
F-ratio	$F/13$	$F/24$	$F/2.3$
Focal length	30.9 m	57.6 m	17.25 m
Image scale	6″67 mm^{-1}	3″70 mm^{-1}	11″9 mm^{-1}
Field of view	2′6 × 2′6	3′3 × 3′3	6′0 × 6′0
Pixel resolution	0″10	0″096	0″18
Dynamic range	~ 2000	250 000	60 000
Overall efficiency [b]	0.25	0.38	0.40
CCD	TI	Tektronix	Virtual CCD
Pixel size	15 μm	27 μm	15 μm
Pixel format	800 × 800 × 4	2048 × 2048	2048 × 2048
Size	12 mm × 12 mm × 4	55 mm × 55 mm	30 mm × 30 mm
Readout noise	15e	3e	3e
Digitization noise	7.5e	7.5e	7.5e
Dark current	$< 0.01e$ s^{-1}	0.001e s^{-1}	0.001e s^{-1}

[a] References: Griffiths (1985), Griffiths et al. (1986), Tyson (1984).
[b] (Atmosphere) × (telescope) × (camera) × (detector) at 5500 Å.

Performance of PFC is compared with that of HST cameras, i.e., Wide Field Camera (WFC: Tyson, 1984; Griffiths, 1985) and ARC. Characteristic parameters used for the comparison are summarized in Table II. A virtual CCD is assumed for PFC that have the same noise characteristics as the Tektronix CCD but have pixel size twice as small as it. The overall efficiency ε of PFC is assumed to be 0.40, the breakdown being 0.8 for atmosphere, 0.85 for telescope, 0.85 for camera, and 0.7 for detector.

Figure 2 shows the limiting magnitude of aperture photometry of point sources for $S/N = 3$. The aperture size of 0.2″ is assumed for HST (Tyson, 1984), and two cases, 0.5 and 1.0″, are simulated for PFC. It is seen in Figure 2 that PFC always wins over WFC and that PFC outperforms ARC for objects down to $m_v \sim 27$ mag. Figure 3 shows the similar comparison in slightly different way, where S/N is plotted as a function of magnitude of a point source. Figure 4 is the same plot as Figure 3 but for extended sources. Here the aperture size of 2″ is assumed for all the cameras. This value is the size of a typical galaxy at $z \sim 1.5$ (Tyson, 1984). As expected, superiority of PFC is

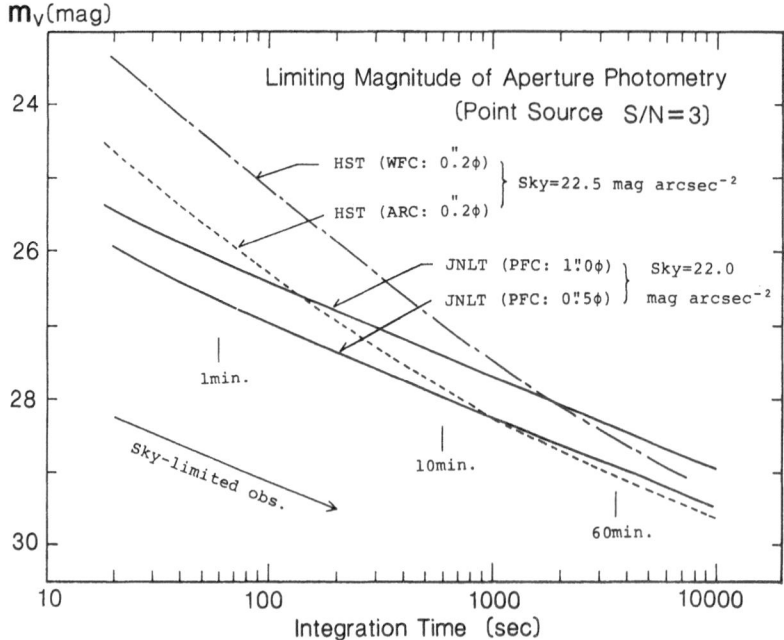

Fig. 2. Limiting magnitude of aperture photometry of point sources. The Prime Focus Camera (FPC) of JNLT is compared with the Wide Field Camera (WFC) and the Advanced Radial Camera (ARC) of the Hubble Space Telescope (HST).

evident in Figure 4 for extended sources where difference of the image size between JNLT and HST has little to do with the limiting magnitude.

The conclusion here is that when the resolution less than $\sim 0.1''$ is not important, (1) PFC will be an extremely powerful imaging instrument that outperforms WFC and ARC by a wide margin in photometry of extended sources, and (2) even for point sources, PFC always wins over WFC and outperforms ARC for $m_v \lesssim 27$ mag.

Data processing will be a crucial point for the exploitation of wide field-imaging instruments such as PFC. Preprocessing of the data, including bias/dark subtraction and flat-fielding, should be carried out on the real time basis during the observation using some hard-wired logic or dedicated array processors.

4.2. MULTI-OBJECT FIBER SPECTROGRAPH (MOFS)

We hope to get the highest efficiency and to exploit the wide field of view (0.5 deg) of the prime focus with this instrument. Figure 5 shows the optical layout of the prototype of MOFS which we are testing now. The unique feature of this prototype spectrograph is the simpleness. The spectrograph has only one component, i.e., a flat field grating that works both as a disperser and a camera. No collimator is present. A fiber manipulator controls the positioning of 48 fibers. The (r, θ) positioning system (e.g., Hill *et al.*, 1982) is being tested. The fiber manipulator will position the fibers during the telescope's slewing to the target. In order to match the spectrograph to JNLT, use of a faster flat field grating is crucial. A design study for such gratings is now under way.

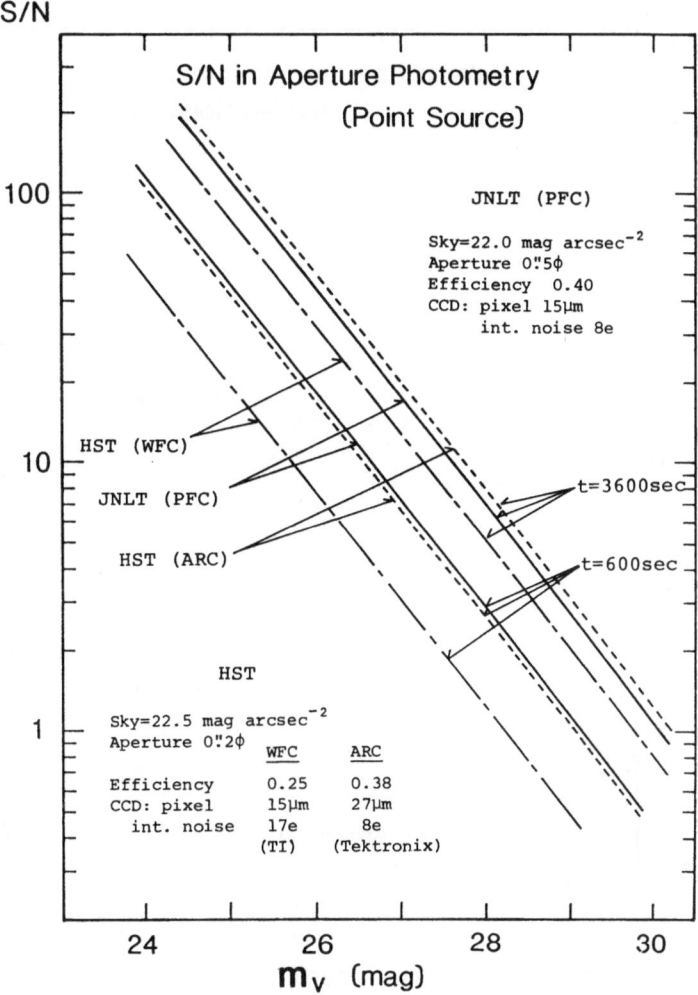

Fig. 3.   Signal-to-noise ratio in aperture photometry of point sources. Abbreviations are the same as in Figure 2.

In order to compute the performance of MOFS, I assume a flat field grating of $F/2$ with 300 g mm^{-1} and a diameter of 104 mm. With a demagnification factor of unity, it produces a linear dispersion of 160 Å mm^{-1}. The fiber will have a core diameter of 50 μm, which corresponds to 0.6″ at the prime focus. With these parameters, MOFS will give a spectral resolution of 8 Å ($R = \lambda/\Delta\lambda = 500-1000$). The overall efficiency is estimated to be 0.23 as the multiplication of 0.8 (atmosphere), 0.85 (telescope), 0.8 (fiber), 0.6 (spectrograph), and 0.6 (detector). The Tektronix 2048 × 2048 CCD with 27 μm pixels is assumed as the detector. The results are shown in Figure 6 together with those for PFC and FOSC. MOFS will be sky-limited within 30 min.

The well-known disadvantage of fiber spectrographs is the poor sky subtraction due to the transmission difference between fibers (e.g., Dennefeld and Fort, 1986). Because

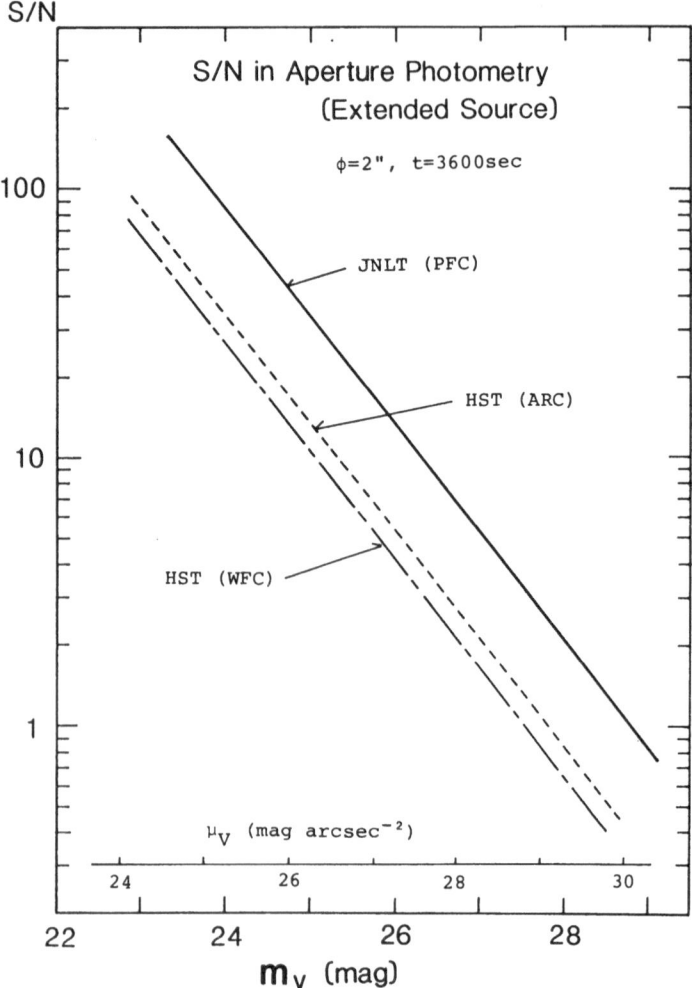

Fig. 4.   Signal-to-noise ratio in aperture photometry of extended sources. Abbreviations are the same as
in Figure 2.

of this, MOFS may not be suitable to quantitative spectroscopy of extremely faint
objects unless substantial technical improvements are achieved. However, MOFS will
be an ideal instrument for programs that involve a large number of objects such as
redshift surveys of galaxies and spectral classification of many faint objects.

4.3. FAINT OBJECT SPECTROGRAPH AND CAMERA (FOSC)

Hardware for this instrument considered by the WG is the scaled-up version of ESO
Faint Object Spectrograph and Camera (EFOSC: Enard and Delabre, 1982). It is
essentially a focal reducer with both spectroscopic and imaging capabilities. Filters,
grisms, and polarizers are supported on the wheels and they can be inserted into the
collimated beam. There is a choice to be decided between multi-slits (e.g., Geary *et al.*,

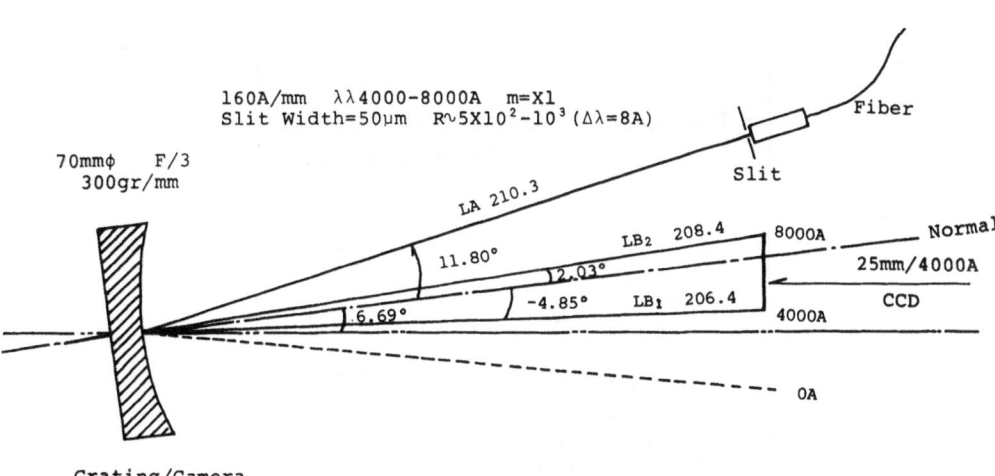

Grating/Camera

Fig. 5.   Optical layout of a prototype multi-object fiber spectrograph.

1986; Wynne and Worsick, 1988b) and multi-apertures (e.g., Fort et al., 1986; Dupin et al., 1987). Since we will be using FOSC at the Cassegrain focus, use of polarizers is rather straightforward. Polarimetric spectroscopy using FOSC may become very important in future.

First, I run through the usual process of parameter matching. The relation between the angular image size θ and the linear size s of the image produced by the telescope at the entrance slit of a spectrograph is given by

$$s = D_{tel}\theta F_{cam} , \qquad (9)$$

where D_{tel} is the telescope aperture and F_{cam} is the focal ratio of the camera of the spectrograph. For $\theta = 0.5''$ and $D_{tel} = 7.5$ m, $F_{cam} = 1.5$ will give $s = 27$ μm. The Tektronix CCD is again assumed as the detector. The spectral resolution is computed by

$$\Delta\lambda = \frac{s}{F_{cam}D_{cam}A} = \frac{D_{tel}\theta}{D_{cam}A} , \qquad (10)$$

where A is the angular dispersion of the spectrograph and D_{cam} is the diameter of the camera that is nearly the same as the beam size.

In order to estimate the performance of FOSC, I tentatively choose $D_{cam} = 100$ mm and $F_{cam} = 1.5$ that match the 27 μm pixel size of the Tektronix CCD. The efficiency is assumed to be $\varepsilon = 0.13$. Table III gives the characteristic parameters of FOSC for different gratings considered. The performance of FOSC is shown in Figure 6 together with those of MOFS and PFC, where the limit of the Faint Object Spectrograph (FOS) of HST is also shown for comparison. As seen from Figure 6, low-dispersion spectroscopy with JNLT can go fainter than HST by 2–4 mag.

Figure 7 shows the apparent magnitude of distant galaxies as a function of z (Spinrad, 1986). The practical limits of photometry and low-dispersion spectroscopy with JNLT

S/N

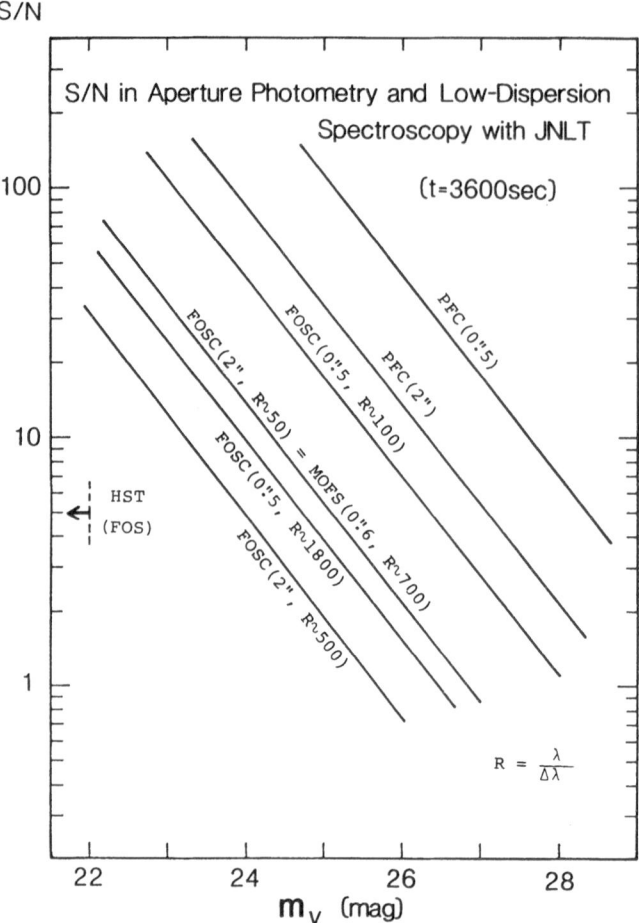

Fig. 6. Summary of signal-to-noise ratio in aperture photometry with the Prime Focus Camera (PFC) and in low-dispersion spectroscopy with Multi-Object Fiber Spectrograph (MOFS) and Faint Object Spectrograph and Camera (FOSC). The limiting magnitude for the Faint Object Spectrograph (FOS) of the Hubble Space Telescope (HST) is also shown for comparison.

TABLE III

Characteristic parameters of FOSC[a]

Grating (g mm^{-1})	A ("/Å)	Disp. (Å mm^{-1})	$\theta = 0''.5$		$\theta = 2''$	
			$\Delta\lambda$ (Å)	t^b (s)	$\Delta\lambda$ (Å)	t^b (s)
35	0.7219	1905	52	270	208	60
75	1.547	889	24	600	97	130
150	3.094	444	12	1200	48	260
300	6.188	222	6.1	2400	24	530
600	12.376	111	3.0	5100	12	1030

[a] $D_{tel} = 7.5$, $D_{cam} = 100$ mm, $F_{cam} = 1.5$, Tektronix CCD.
[b] Integration time for (sky counts/pixel) = 5 × (CCD noise)^2.

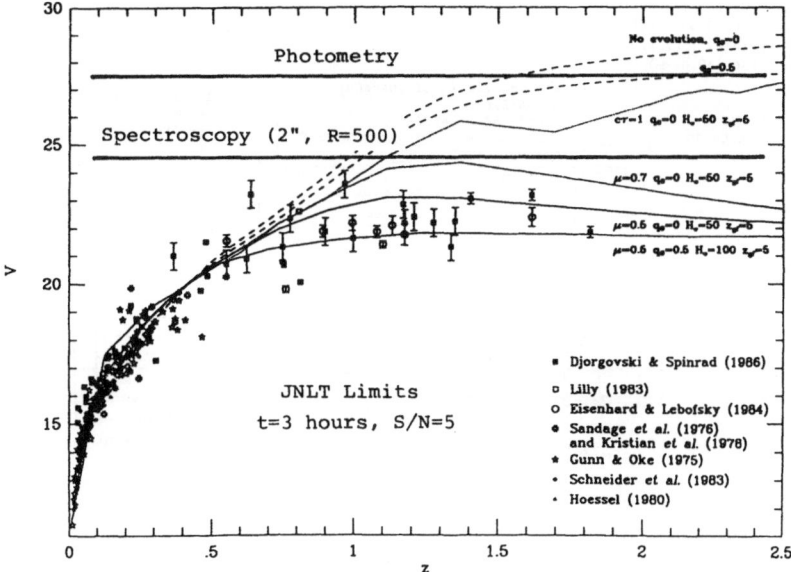

Fig. 7. Practical limits of aperture photometry and low-dispersion spectroscopy with JNLT superposed on the diagram by Spinrad (1986), which shows the V magnitude of distant galaxies as a function of z together with some model predictions.

are indicated. Here, an integration time of 3 hours and $S/N = 5$ are assumed. This figure demonstrates the power of JNLT that can penetrate deep into the Universe.

4.4. HIGH-DISPERSION SPECTROGRAPH

This instrument will take full advantage of the 'photon-limited gain' of JNLT. The list of scientific programs to be carried out with this instrument includes studies of chemical composition and atmospheric structure of stars to understand the chemical evolution of the Galaxy and the Universe, stellar winds and circumstellar environments for the better understanding of interactions between stars and interstellar matter, stellar velocity field, stellar seismology, stellar magnetic field, kinematics of active galactic nuclei, and Lα forests and absorption lines of QSOs. All these studies require high spectral resolution of $R = 10^3$–10^5. The WG emphasizes the importance of high S/N rather than super resolution of $R > 10^5$. The importance of high spatial resolution attainable with JNLT is also stressed in terms of spectroscopic programs.

A strawman's design was made by Prof. Yamashita, which is a modification of a design originally made by REOSC. It is a double-beam spectrograph with mosaic echelles, the blue part of which is shown in Figure 8. As Equation (9) indicates, a fast camera is essentiall to a high dispersion spectrograph for a large telescope in order to minimize the light loss. The design of fast cameras is described by Nariai and Yamashita (1987) and by Nariai (1989). The key point of this design is the use of a field mirror. The advantage of the use of a field mirror is that (1) the optical system can be made to be of auto-collimation type, (2) angular dispersion of the echelle can be magnified or

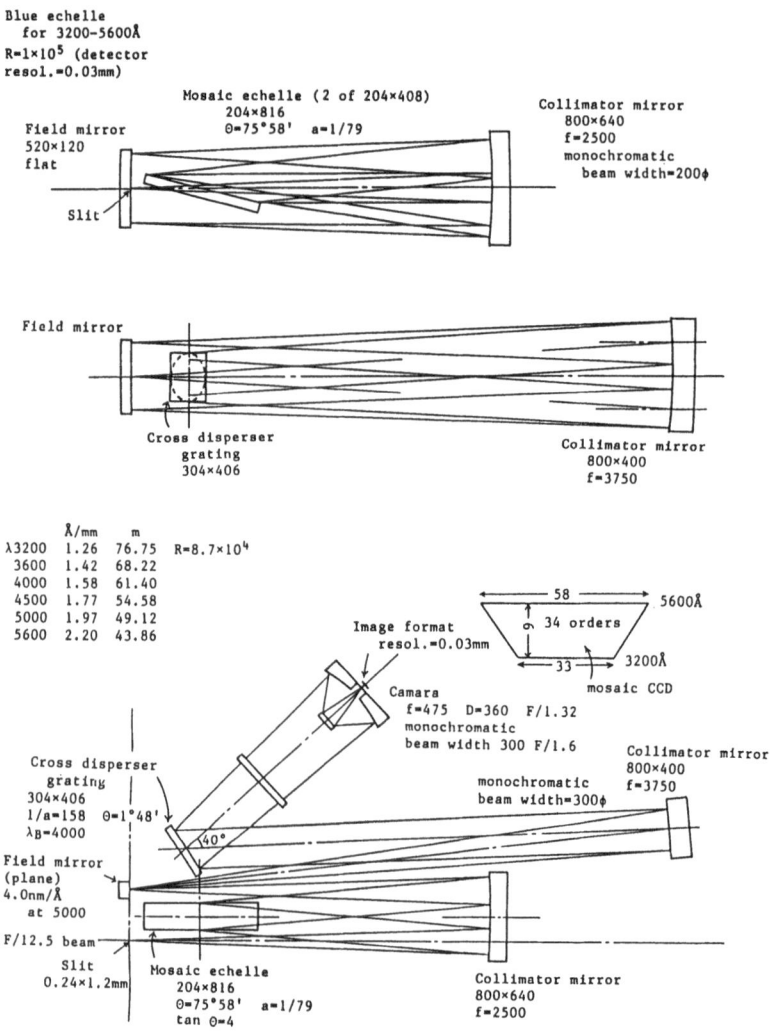

Fig. 8.   The optical layout of the blue part of the double beam high-dispersion spectrograph.

de-magnified so that the matching of the spectral image with the CCD format becomes easier, and (3) beam size can be made smaller. The disadvantage is, however, the increase of reflection surfaces. How to minimize the light loss due to multiple reflections by the introduction of special coatings is the point for further study. The detailed performance parameters have not been computed.

4.5. SPECKLE CAMERA

The resolution up to the diffraction limit of 7.5 m JNLT (0.02″ at the V band) can be achieved by a speckle camera. The limiting magnitude in speckle observations with JNLT is estimated to be about 18 mag. Observations at this resolution of active galactic nuclei and star-forming regions in the Galaxy would bring completely new pictures on these objects. Observations of many binary stars would eventually lead us to the understanding of initial mass function.

The following four modes of operation are considered: ordinary speckle interferometry, spectroscopic speckle interferometry, differential speckle interferometry, and pupil plane speckle interferometry. Observations have been successfully made in ordinary speckle interferometry mode using a prototype (Isobe, 1987). Laboratory experiments of spectroscopic speckle interferometry is reported by Baba (1989). A prototype of the differential speckle interferometer is under construction. A design study is under way of a prototype for the pupil plane speckle interferometer. Some experiments on the data processing, including those to increase the speed of FFT (Isobe *et al.*, 1988), are being carried out.

The four modes have been tested using separate prototype cameras, but they will be integrated into a single speckle camera for JNLT. Experiences currently being gained are the key factor to the construction of the speckle camera for JNLT.

5. Key Projects

The Working Group on wide field study is going to submit a recommendation that some fraction of JNLT observing time should be put aside for a small number of key projects for which JNLT outperforms other existing telescopes.

One of such key projects would be a deep survey of the Universe. This is a coordinated photometric and spectroscopic survey involving tens of thousands of galaxies that will be carried out by a consortium consisting of observers, theoreticians, software people, and instrument developers. With the power of JNLT that I shown in this paper, it is possible to delineate the three-dimensional distribution of galaxies and velocity field in $z \lesssim 0.04$ with an accuracy of $\Delta v \lesssim 50$ km s^{-1} using a sample of galaxies complete down to $M \sim -16$ mag. With a smaller sky coverage and a slightly lower velocity resolution ($\Delta v \sim 50$–300 km s^{-1}) it is possible to carry out a CfA-type magnitude-limited redshift survey (Davis *et al.*, 1982) with a limiting magnitude of $m \sim 20.5$, i.e., 5 mag below that of CfA2 survey (Geller *et al.*, 1987). This survey will reach up to $z \sim 0.5$. With a velocity resolution of $\Delta v \sim 1000$ km s^{-1} we could extend the survey even beyond $z > 1$.

Figure 9 shows the domain of JNLT survey in comparison with existing redshift surveys. This kind of large survey belongs to the territory of ground-based large telescopes. JNLT will be one of very few telescopes that can derive the basic parameters of the Universe based upon such a coordinated large survey as described above.

6. Concluding Remark

Since the design study of the first-generation observing instruments for JNLT has just begun, we would greatly appreciate any suggestion or comment on our instrument plan. International collaborations on instrumentation for JNLT are also welcome.

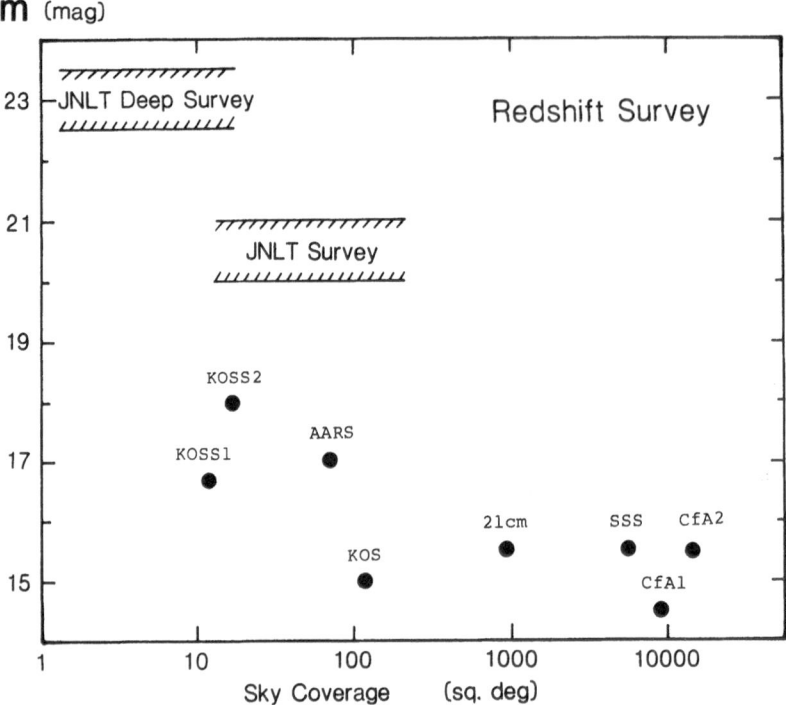

Fig. 9.   Deep redshift surveys that would be possible with JNLT. The ordinate is the limiting magnitude and the abscissa is the sky coverage. Existing representative surveys are shown. KOSS1: Kirschner *et al.* (1982), KOSS2: Kirschner *et al.* (1987), AARS: Peterson *et al.* (1986), KOS: Kirschner *et al.* (1978), 21 cm: Giovanelli and Haynes *et al.* (1985, 1986), SSS: da Costa *et al.* (1988), CfA1: Davis *et al.* (1982), CfA2: Geller *et al.* (1987).

Acknowledgements

I wish to express my sincere thanks to all the people who communicated to me valuable information necessary to prepare this manuscript, and K. Tarusawa for help in drawing figures.

References

Baba, N.: 1989, *Astrophys. Space Sci.* **160**, 373 (this issue).

Blouke, M. M., Heidtmann, D. L., Corrie, B., Lust, M. L., and Janesick, J. R.: 1985, *Proc. SPIE* **570**, 82.

da Costa, L. N. *et al.*: 1988, *Astrophys. J.* **327**, 544.

Davis, M., Huchra, J. P., Latham, D., and Tonry, J. L.: 1982, *Astrophys. J.* **253**, 423.

Dennefeld, M. and Fort, B.: 1986, in S. D'Odorico and J.-P. Swings (eds.), *ESO's Very Large Telescope*, ESO, Garching, p. 151.

Dupin, J. P., Fort, B., Mellier, Y., Picat, J. P., Soucail, G., Dekker, H., and D'Odorico, S.: 1987, *ESO Messenger*, No. 47, p. 55.

Enard, D. and Delabre, B.: 1982, *Proc. SPIE* **445**, 522.

ESO: 1987, *Proposal for the Construction of the 16-m Very Large Telescope*, ESO, Garching, pp. 275–289.

Fort, B., Millier, Y., Picat, J. P., Rio, Y., and Lelievre, G.: 1986, *Proc. SPIE* **627**, 321.

Geary, J. C., Huchra, J. P., and Latham, D. W.: 1986, *Proc. SPIE* **627**, 509.

Geller, M., Huchra, J. P., and de Lapparent, V.: 1987, in A. Hewitt *et al.* (eds.), 'Observational Cosmology', *IAU Symp.* **124**, 301.

Giovanelli, R. and Haynes, M. P.: 1985, *Astron. J.* **90**, 2445.

Giovanelli, R., Haynes, M. P., Myers, S. T., and Roth, J.: 1986, *Astron. J.* **92**, 250.

Griffiths, R. E.: 1985, *Instrument Handbook for the Wide Field and Planetary Camera*, Space Telescope Science Institute, Baltimore.

Griffiths, R. E. *et al.*: 1986, *Proc. SPIE* **627**, 591.

Hill, J. M., Angel, J. R. P., Scott, J. S., Lindley, D., and Hintzen, P.: 1982, *Proc. SPIE* **331**, 279.

Isobe, S.: 1987, *Rev. Mexicana Astron. Astrof.* **14**, 722.

Isobe, S., Otsubo, J., Takemori, T., and Fujita, K.: 1988, in F. Merkle (ed.), *Proc. NOAO–ESO Conference on High-Resolution Imaging by Interferometry*, ESO, Garching, p. 401.

Kishner, R. P., Oemler, A., Jr., and Schechter, P. L.: 1978, *Astron. J.* **83**, 1549.

Kishner, R. P., Oemler, A., Jr., Schechter, P. L., and Shectman, S. A.: 1983, *Astron. J.* **88**, 1285.

Kishner, R. P., Oemler, A., Jr., Schechter, P. L., and Shectman, S. A.: 1987, *Astrophys. J.* **314**, 493.

Maaswinkel, F., D'Odorico, S., Huster, G., and Bortoletto, F.: 1987, *ESO Messenger*, No. 48, p. 51.

Nariai, K.: 1989, *Astrophys. Space Sci.* **160**, 159 (this issue).

Nariai, K. and Yamashita, Y.: 1987, *Publ. Astron. Soc. Pacific* **39**, 505.

Peterson, B. A., Ellis, R. S., Efstathiou, G., Shanks, T., Bean, A. J., Fong, R., and Zen-Long, Z.: 1986, *Monthly Notices Roy. Astron. Soc.* **221**, 233.

Spinrad, H.: 1986, *Publ. Astron. Soc. Pacific* **98**, 269.

Tyson, J. A.: 1984, *Publ. Astron. Soc. Pacific* **96**, 566.

Wynne, C. G. and Worswick, S. P.: 1986, *Monthly Notices Roy. Astron. Soc.* **220**, 657.

Wynne, C. G. and Worswick, S. P.: 1988a, *Monthly Notices Roy. Astron. Soc.* **230**, 457.

Wynne, C. G. and Worswick, S. P.: 1988b, *Observatory* **108**, 161.

SEKIGUCHI – The number of fibers (~ 50) seems to be too few compared to number of the objects JNLT can observe.

OKAMURA – I guess Professor Kodaira would have a comment on this matter.

KODAIRA – This limitation comes from the prototype design using a single holographic concave grating. When an additional optic is incorporated to widen the field of view, more fibers can be adopted by the spectrograph.

LAING – I am worried about the efficiency of fibers at the fast focal ratio of the JNLT prime focus. Do you know what the losses induced in getting light into the fiber might be?

OKAMURA – The numerical aperture of the fiber which we are currently using is large enough to accept an $F/2.3$ beam. So the loss at the input end of the fiber is insignificant.

ELLIS – I predict much of the fiber optics spectroscopy on 8 m telescopes will be done at higher dispersions than your prime focus MOFS instrument could accommodate. What possibilities are there for efficient multiple object spectroscopy at resolutions of 1 Å or better?

OKAMURA – One obvious possibility is to narrow the slit width at the cost of light loss (for bright objects). Other possibility involves the whole design of the spectrograph and I cannot make a comment on this immediately. I do appreciate the importance of intermediate-dispersion ($\Delta\lambda \sim 1$ Å) spectroscopy.

INFRARED INSTRUMENTS FOR JNLT*

TOSHINORI MAIHARA

Kyoto University, Kyoto, Japan

(Received 18 January, 1989)

Abstract. The principal features of the JNLT as an infrared telescope are presented along with its ultimate performance of detectivity in typical methods of imagery and spectroscopy. Some infrared instruments: infrared camera, grating spectrometer, and Fabry–Pérot-based imager, currently proposed as the first generation instruments are also discussed in relation to the scientific objectives of the JNLT.

1. Introduction

It is widely accepted that large ground-based telescopes should have as fully as possible infrared observation capability. The planned telescope JNLT has also been expected to work on a vast area of infrared astronomical studies with various potential observing techniques in the infrared spectral range from near-infrared to submillimeter wavelengths. JNLT can be simply said to have two major features: a large light-collecting area, about four times the area presently available 3 to 4 m class telescopes, and a high imaging quality of the mirror system. In the visible region, the telescope has a feature of wide field mainly at the prime focus, and in the infrared, it is characterized by a relatively fast F-ratio of infrared foci ($F/12.5$) to guarantee a better imaging capability than the conventional slow F-ratio system. This feature is practically quite important to ensure efficient instrument adaptability, as well.

Concerning the instrumentation, a number of instrument concepts have been put forth in the past general discussions about scientific goals to be pursued by JNLT together with various observational methods, in which only two or three of infrared instruments (all are in the near-infrared region) have so far been proposed. The instruments in the initial phase when JNLT is going to be operational must be simple, reliable and also capable of testing the fundamental and ultimate performance of the telescope system. They should also be useful and unique in contributing to the production of the first astronomical results of JNLT.

In this report, I would like to present a current list of the first-generation instruments with specific design studies under way as well as inherent scientific issues to be investigated with them.

* Paper presented at the Symposium on the JNLT and Related Engineering Developments, Tokyo, November 29–December 2, 1988.

Astrophysics and Space Science **160**: 313–323, 1989.

2. Infrared Optimization of JNLT

Typical parameters or quantities of JNLT characterizing the infrared telescope specifications are first described as shown in Table I.

TABLE I

Infrared related specifications of JNLT

F-ratio	F/12.5	Good image quality
	(~100 cm dia.)	0.2″ within ~5′ dia.
	F/35	Wobbling secondary
	(~50 cm dia.)	For thermal IR
Emissivity coating	0.1	
	(0.08 expected at Cassegrain)	
	Primary:	Al
	Secondary and tertiary:	$Ag + Al_2O_3 + SiO_x$
Aquisition/guide	Cassegrain:	Instrument rotator
	Nasmyth:	Image derotator
Wobbling secondary	Light weight mirror	50 cm dia.
	Typical quantity	$\theta \sim 60''$
		$f \sim 10$ Hz
		$\varepsilon \sim 0.85$

(1) *F*-ratio: Two types of infrared secondary mirrors are planned to be prepared. The focal ratio of the first one for both Cassegrain and Nasmyth foci has been chosen as $F/12.5$ to maintain good image quality of about 0.2″ or better within the central 5′, but it does not have a wobbling mechanism, and this is assumed to be suitable in most near-infrared observations and even in the mid-infrared region, if we can use relatively large arrays which may provide sky cancellation within the array. The second one is the wobbling secondary with focal ratio of $F/35$, same as the infrared secondaries on the presently available infrared telescopes on Mauna Kea.

(2) Emissivity: The emissivity of the telescope is expected to be smaller than 0.1, desirably about 0.08 at the Cassegrain focus, by a careful design of suspension as well as mirror surface coating. The primary mirror is coated by aluminum, while the secondary and tertiary mirrors dedicated to infrared are presently considered to be coated by silver with the overcoat of Al_2O_3 together with SiO_x.

(3) Acquisition and guide system: (a) At the Cassegrain focus an instrument rotator will be attached to ensure accurate correction of the rotation of observing field. The instrument rotator accommodates a couple of individual observational instruments together with an acquisition and guide system which consists of a dichroic mirror and an optical offset guider with autotracking capability. (b) At the Nasmyth focus, we intend to provide an image derotator to obtain stationary images with respect to

Nasmyth instruments during observation. An acquisition/guide system similar to that at the Cassegrain will be provided.

(4) Wobbling mechanism: The $F/35$ secondary mirror is equipped with a wobbling mechanism to perform efficient sky cancellation. If we assume a light-weight mirror (a factor of two or three lighter than a conventional thick mirror), the required power to perform 10 Hz chopping with a typical 60″ throw having an 85% efficiency (the transient time being 15%) is about 15 W. We could use even a more light-weighted mirror such as the frit-bonded mirror to reduce the power requirement.

3. Prospect of Infrared Capability

In this section we have a quick view of the infrared capability of JNLT. One is the limiting magnitude in observing the faintest celestial objects by means of photometric and spectroscopic techniques, and the other is the high spatial resolution including imaging and spectroscopy as well. They should also be compared to the prospective ranges which the presently available telescopes as well as future ground-based and space-borne facilities will cover.

(1) Limiting magnitudes: Presented in Table II are limiting magnitudes of some

TABLE II

Limiting magnitudes

Band	Photometric (mag)	Spectroscopic (mag)	
		1000	10 000
H	23	20	18
K	22.5	20	18
L	19	16	14.5

Notes on assumptions:
$D = 7.5$ m.
Efficiency = 0.3 (photometric),
            0.2 (spectroscopic).
FOV = 1 × 1 arcsec^2.
Background flux = 3 × 10^4 (H-band),
                 1.5 × 10^4 (K-band),
                 5 × 10^6 (L-band).
Background unit: photons s^{-2} m^{-2} arcsec^{-2} μm^{-1}.
Read noise = 100e^- r.m.s.
SN-ratio = 3σ by 1 hr integration.

typical observations, assuming typical parameters under certain observational conditions with potential advanced instruments together with the state-of-the-art detectors of highest performance. The actual astronomical objects with such faint brightness will be discussed later in relation to the major scientific goals of JNLT (Figure 1).

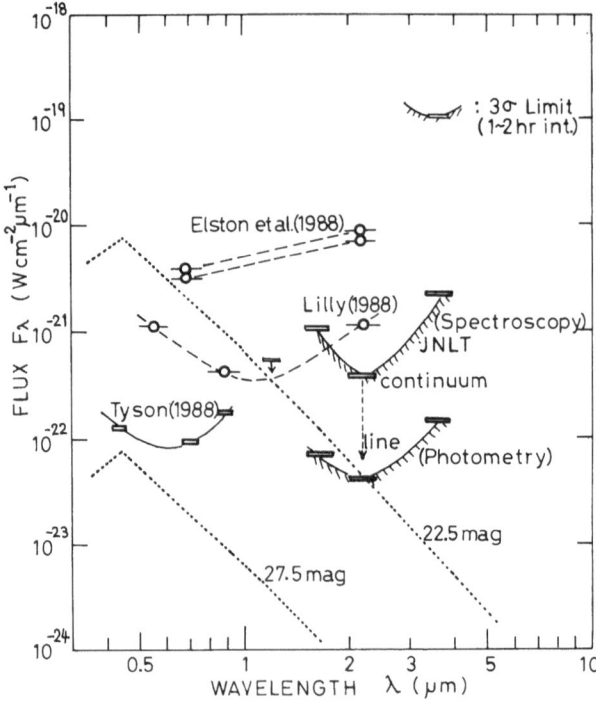

Fig. 1.   Detection limit of infrared photometric and spectrometric observations. Recently discovered galaxies at the farthest distances are plotted by open circles.

(2) High spatial resolution: Another feature of JNLT for infrared observations is the high spatial resolution in infrared by virtue of its excellent imaging capability achieved by the active mirror control system. In Figure 2, I roughly draw a curve for the anticipated image blur of the JNLT telescope system including the atmospheric seeing effect on Mauna Kea, which is also compared to the diffraction limited image size expected by Hubble Space Telescope. As seen from the figure, the spatial resolution of JNLT surpasses in the wavelength range longer than 3 μm even in the case of conventional imagery suffering from the seeing effect. The observation using a technique such as the adaptive optics-based method could hopefully make an appreciable factor of improvement in image size.

4. Representative Instruments and Scientific Objectives

A number of preliminary infrared instruments has so far been proposed in the past general discussions on the scientific goals expected to be pursued by JNLT. Such instruments are assumed to be manufactured and maintained as standard observatory instruments for common users, if the situation is allowed. However, it would be very difficult to construct and set up all these instruments in a short time-scale, and is not practical. In fact, optimal instrument concepts would still evolve along with the current rapid progress in optics and detector technologies.

In the present stage of the project, we have tentatively selected a couple of infrared instruments to be designated as the first generation facilities, which are: (1) Infrared camera, (2) Infrared grating spectrometer, and (3) Fabry–Pérot-based line imager. Although the status of design study of the individual instrument differs from one to another, some details of the instrument specifications are to be presented below including the respective scientific justifications.

(1) Infrared camera: The concept of infrared camera could be divided into two categories: i.e., wide-field camera for deep survey and high spatial resolution camera in conjunction with speckle interferometry or adaptive optics. Both types of camera are absolutely necessary as versatile instruments applicable to a wide variety of observations.

A wide-field camera is to be designated to have the highest sensitivity in detecting faintest infrared objects like the most distant galaxies. Though we do not have a specific design of infrared camera for JNLT, some desirable features of the instrument to realize major scientific objectives are to be mentioned. As shown in the previous section, the limiting magnitude of the camera is around 23 mag (3σ) in the H- and K-band for a typical one-hour observing time. To achieve the maximum sensitivity, a reasonable field size per pixel would be about 0.5" and the frame size could be about 1', if the array size is assumed as 128×128, for instance.

One of the major scientific objectives of the infrared imager with relatively wide field is the deep survey of protogalaxy candidates. A deep optical survey has already been

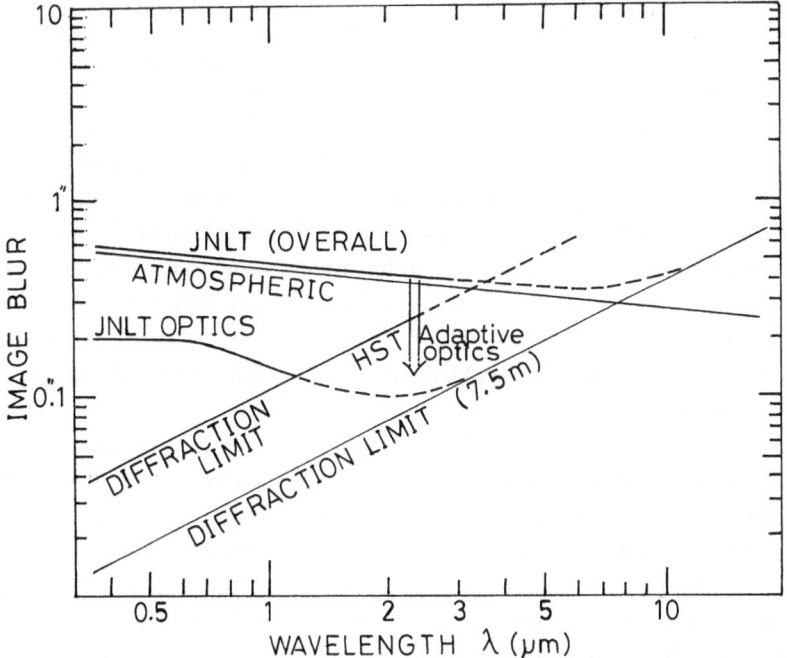

Fig. 2.   Image size expected by JNLT compared to that by HST.

revealing abundant existence of very faint blue galaxies supposedly a sort of juvenile
galaxies or even protogalaxies (Tyson, 1988). Infrared surveys have also discovered a
couple of very interesting objects (Elston *et al.*, 1988; Lilly, 1988), sometimes claimed
as protogalaxies (Figure 1).

Meanwhile, there is a possibility that galaxies at great distances ($z \approx 3-5$) may suffer
appreciably reddening effect due to the absorption by dust either in disks of intercepting
glaxies, or more hypothetically, in the intergalactic space. Calculations have so far been
made by Ostriker and Heisler (1984) and by Lilly and Cowie (1986) deriving the
reddening effect. Under reasonable assumptions, the probability that a line-of-sight to
an object with red shfit of $z = 3-5$ is intercepted by foreground galactic disks could
increase to the order of unity. If such effect holds, the infrared survey would become
the most crucial method for exploration of the extremely deep space relating to the galaxy
formation and its earliest evolution. There is also a possibility that the abundance of
quasar or quasar-like (dust rich) objects may dramatically increase.

High-resolution imaging is exploited in investigation of inner structure of star-forming
regions, galactic nuclei, inner part of compact stellar systems, etc. It is needed to develop
two kinds of instruments: one is an infrared speckle interferometer and the other is a
camera in conjunction with the adaptive optics to realize quick compensation of the
atmospheric seeing effect. In Figure 2 we compare an image size of JNLT to that of
Hubble Space Telescope, showing that JNLT could have a higher spatial resolution in
the near-infrared than HST.

Middle-infrared cameras complementary to the above near-infrared ones are also
very important. They should be developed in the relatively early phase of telescope
operation.

(2) Grating spectrometer: Near-infrared grating spectrometer is an instrument very
powerful in covering wide wavelength range with highest sensitivity. It is suitable for
spectrophotometric measurements of faintest celestial objects like distant young galaxies
or protogalaxy candidates, because such objects are supposed to be similar to emission
line galaxies whose line widths are around 300 km s^{-1} corresponding to a spectral
resolution of about 10^3. Ionized O-lines: [O II]3727 and [O III]5007, and a couple of
H-recombination lines, for instance, are anticipated to come to the near-infrared range,
having large equivalent widths under reasonable conditions. Spectroscopic information
about line intensities as well as red-shifted line positions are extremely important for the
study of the earliest evolution of galaxies and, hence, for the examination of various
galaxy formation scenarios.

The minimum detectable fluxes of the objects using an optimized grating spectrometer
are 20 mag in the H- and K-bands as shown in Table II and Figure 1. However, if the
line intensities have large equivalent widths like emission line galaxies, the observable
objects in spectroscopy could be as faint as 23 mag in the H/K-bands.

The detectivity is primarily restricted by the strong OH airglow emission in the
H- K-bands. However, it is noteworthy to see that each spectral element on array
detector would have different detection limit due to the difference of incident
background photon flux corresponding to the individual line position of the OH

TABLE III

Features of the spectrometer

Resolving power	300–3000, typically 1000
Field of view	1.5″ (slith width) × 45″ (slight length) typical
Detector	InSb 128 × 128 ~ 256 × 256
	HgCdTe 256 × 256
Pixel size	50 × 50 μm² ~ 100 × 100 μm²
Grating	new type holographic grating unequi-spaced and curved grooves
Infrared imaging channel	simultaneous image
Visual monitor channel	simultaneous optical image
Cryogenics	closed cycle cooler
Baffling	Lyot stop
Optics	all reflective optics
System throughput	~20%

emission. This means that the detectivity of an appreciable fraction of pixels could be better than the estimated average by a factor of about 3. It is regarded as an advantage of slit spectrometer.

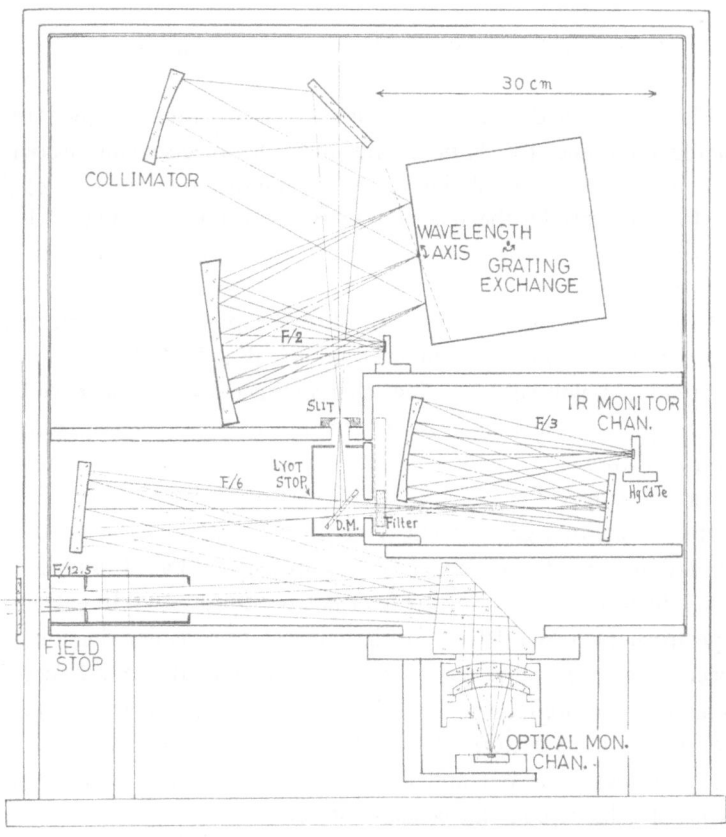

Fig. 3.   Optical layout of the proposed grating spectrometer.

GRATING

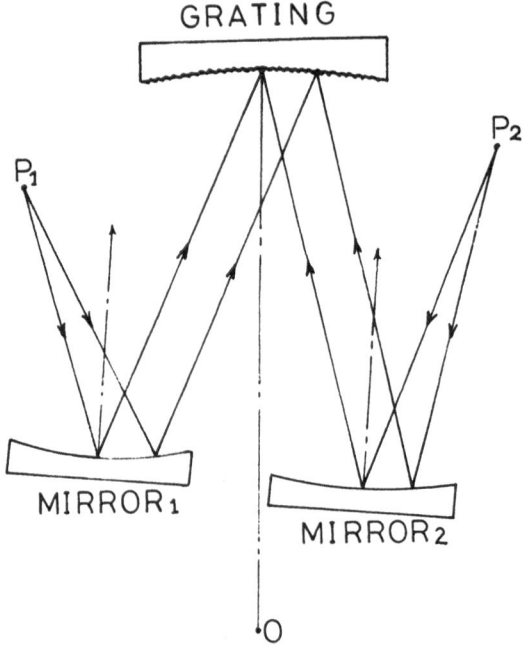

Fig. 4.   Concept of generating the new type of holographic grating.

Also in Figure 1 the 3σ detection limits are shown for spectroscopy at the resolution of 10^3 by integration time of 1–2 hours. The intriguing objects so far discovered by deep infrared survey observations relating to the formation and evolution of galaxies are presented. It is feasible by the use of the spectrometer as proposed here to make extensive spectroscopic studies on the similar proto-galactic objects with much fainter apparent magnitudes.

Figure 3 shows a specific optical layout of the proposed grating spectrometer. The specifications and the features are summarized in Table III. In Figure 3 we consider to use a new type of holographic grating, characterized by unequi-spaced and curved grooves on a slightly concave substrate. The method to create such a grating is schematically shown in Figure 4. We have adopted this new holographic grating in our spectrometer design because of its excellent performance capable of reducing the coma and spherical aberrations to a great extent (Harada *et al.*, 1988).

(3) Fabry–Pérot-based line imager: A Fabry–Pérot-based imager utilizing an acousto-optic filter as an order sorting filter has been proposed. In galactic and extragalactic spectroscopic observations for the purpose of revealing dynamical properties of the objects with appreciable spatial extent, the combination of a Fabry–Pérot interferometer and an infrared camera is obviously the best choice. There are already numbers of observations detecting fairly strong infrared lines such as H_2-lines (*S*- and *Q*-branches), H-lines (Brackett and Paschen series), ionized Fe-lines, etc., towards various types of galactic and extragalactic objects.

As an example, Figure 5 displays the detected fluxes of the H_2 line vs beam size for

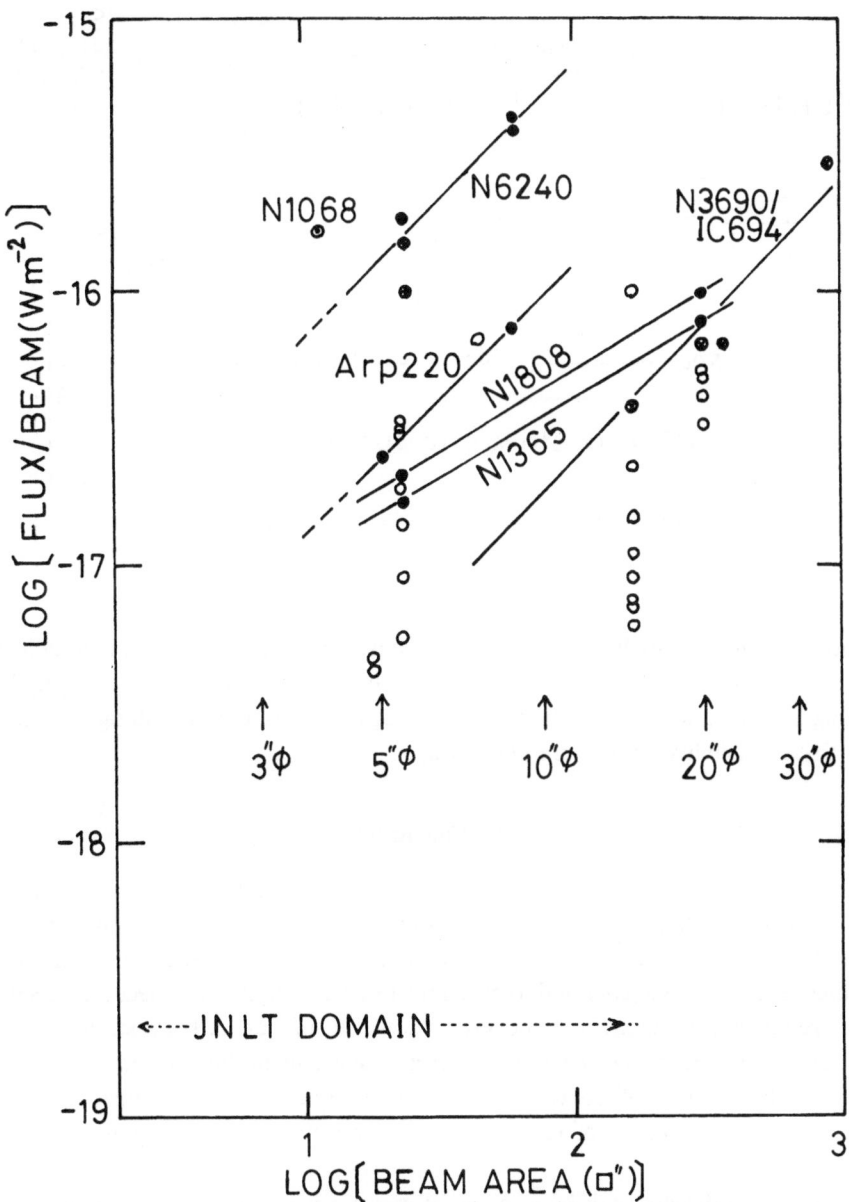

Fig. 5.   Typical observations of the $H_2v = 1$–$0S(1)$ line in external galaxies.

external galaxies. The proposed instruments are supposed to extend the spectroscopic observations by adding spatially- as well as velocity-resolved informations on the active star formation regions or the intriduing galactic nuclear regions, for instance. Even at the line flux level of about 10^{-19} W m^{-2}, a Fabry–Pérot imager on the JNLT is capable of detecting the lines with a SN-ratio of about 10 or more by a typical 300 s integration.

The conceptual optical configuration of the proposed instrument is shown in Figure 6,

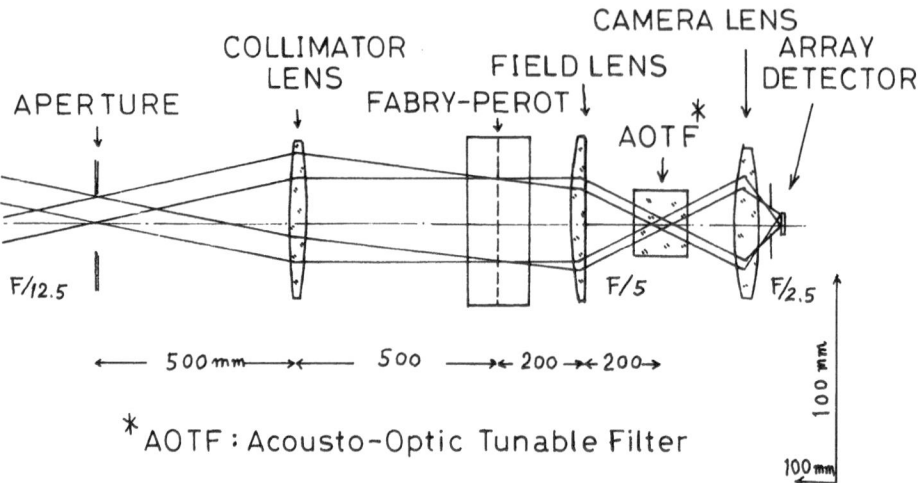

Fig. 6. Conceptual optical configuration of a Fabry–Pérot-based imager. An acousto-optical filter is
utilized as an order sorting prefilter.

a unique feature of which is the use of an acousto-optical filter having high efficiency
as well as tunability. The filter can be inserted in either converging or diverging beam
keeping the resolution of a few hundreds. Though a concrete optical design has not yet
been fixed, feasibility studies are under way.

5. Conclusion

In this talk I have presented major features of JNLT as an infrared telescope, and some
infrared instruments currently proposed as the first generation instruments in relation
to the scientific objectives of interest. The area of astrophysics mentioned in this talk
is rather restricted to galaxies and to near-infrared wavelengths. It is needless to say that
there are much more areas to be covered. However, in view of the restricted resource
of manpower and technological basis in Japan, especially in the infrared arena, it is not
easy to realize such a full coverage of vast areas rapidly. In these considerations, we
have recognized necessity of international collaborations in developing such instruments.

   The status of design work of a couple of instruments I mentioned is different from
one to another. The instrument concepts and, of course, their specific designs have
inevitably been evolving continually, but the three instruments: the infrared camera, the
grating spectrometer, and the Fabry–Pérot imager are the presently selected candidates
of the first generation instruments. We hope that at least these three kinds of instruments
will be constructed by around the time of the beginning of the JNLT operation.

References

Elston, R., Rieke, G. H., and Rieke, M. J.: 1988, *Astrophys. J.* **331**, L77.

Harada, Y., Koike, M., and Noda, H.: 1988, *Shimadzu Review* **44**, 83 (in Japanese).

Lilly, S. J.: 1988, *Astrophys. J.* **333**, 161.

Lilly, S. J. and Cowie, L. L.: 1986, in C. G. Wynn-Williams and E. E. Becklin (eds.), *Infrared Astronomy with Arrays,* Univ. Hawaii Press, Honolulu, Hawaii, p. 473.

Ostriker, J. P. and Heisler, J.: 1984, *Astrophys. J.* **278**, 1.

Tyson, J. A.: 1988, *Astron. J.* **96**, 1.

LAING – How much field is available for guide stars with the image derotations at the Nasmyth foci of the JNLT?

MAIHARA – It somewhat depends on the detailed optical design of the JNLT infrared mirror system, especially the design of back-focus length. Currently we hope to secure a 6′ field of view to guarantee efficient offset guiding capability.

INFRARED DETECTOR FOR MID-INFRARED ASTRONOMY*

TETSUO NISHIMURA

University of Arizona, Steward Observatory, Tucson, U.S.A.

(Received 1 February, 1989)

Abstract. Recent developments of infrared detectors and arrays for mid-infrared astronomical observations are discussed with an emphasis on technical issues in designing and fabricating photometers and cameras. The discussion includes a small-scale silicon bolometer array being tested at the Steward Observatory.

1. Introduction

Almost all future large telescope projects list the capability of producing diffraction-limited images at mid-infrared wavelengths (10–30 μm) among their essential objectives. The new telescopes should achieve higher angular resolving power with larger collecting area than most of the planned space telescopes in this spectral range. The sensitivity of ground-based large telescopes should match those in space with the use of cryogenically cooled spectrometers. Yet it seems mid-infrared instrumentation loses its priority to others in the actual planning of new telescopes. There also exists some indication that the percentage of research utilizing mid-infrared instrumentation on existing telescopes has declined in recent years. Why is mid-infrared instrumentation not being considered more seriously?

The advantages of expensive large telescopes are best realized by utilizing instruments with array detectors. Not only have the arrays obvious superiority in the number of sensitive elements, in some cases the performance of individual detector elements was found to exceed that of the conventional single detector probably due to improved process control. When applied to imaging, its inherent accuracy of two-dimensional mapping is trivial merit over traditional, clumsy synthetic method from data obtained by successive sky choppings. As the format of arrays becomes larger and as on-chip integration performance is improved, further progress in image processing technique can be expected in the future. Unfortunately the array detector technology in mid-infrared is still in its infancy both from device and the system points of view. In addition, most devices are not easily accessible to the general astronomical community. This is the case for the types that the U.S. export control is applicable, such as IBC and BIB arrays described in Section 3. These are contributing factors to the slow development of mid-infrared instrumentation utilizing arrays, together with recent draining of available resources toward the much faster developments of near-infrared array imaging. We will discuss the prospect of the technology in mid-infrared in regard to the application to future large telescopes.

* Paper presented at the Symposium on the JNLT and Related Engineering Developments, Tokyo, November 29–December 2, 1988.

2. Design Consideration for Mid-Infrared Array System

To illustrate the potential of a prospective 7.5 m class telescope at 10 μm as an example, Table I shows some of the relevant parameters for two hypothetical array cameras with low spectral resolution (spectral resolving power $R = 10$) and with moderate resolution ($R = 2000$). The sample performance is characterized as limited by both diffraction (pixel size λ/D) and thermal photon noise from telescope, assuming here emissivity of 10% with telescope temperature at 273 K. The detector is a generic photoconductor with a quantum efficiency of 30%, its validity being discussed in Section 3. The system transmission is taken to be 50%. In comparison, the performance of a near-infrared system is usually affected by seeing disk size and either by detector noise or OH airglow photon noise. The determination of the beam size for an array camera is straightforward compared to traditional single detector photometer. The magnification of the optics is usually set so that pixel dimension is either corresponding to spatial cutoff frequency of a telescope (D/λ which gives 0.27" pixel field-of-view for 7.5 m telescope and comparable to 10 μm seeing disk size (Hall, 1989)) or one half of it. For single a detector system, on the contrary, some allowances have to be made to the aperture diameter for tracking error and mediocre seeing condition, resulting in higher background photon level than theoretically possible.

TABLE I

Spectral resolving power	10	2000
Expected background Q (photons s^{-1})	1×10^9	6×10^6
Detector NEP (W Hz$^{-1/2}$)	2×10^{-15}	1×10^{-16}
System NEFD (mJy Hz$^{-1/2}$)	3	40

A good N band (spectral resolving power of 2) bolometer photometer used on the Steward Observatory 61-inch telescope optimized for infrared observation with 3" field-of-view has peak-to-peak limiting magnitude of about 4.5 or r.m.s. NEFD of 110 mJ Hz$^{-1/2}$, and after one hour integration about 9th magnitude or 2 mJy. The figures in Table I show remarkable improvement. If we could assume that sky-chopping is not necessary any longer as we speculate later, this brings a 5σ detection limit of 12th magnitude after 1000 s of integration for the narrow N band ($R = 10$) imaging. How realistic is this large factor? Aside the increase of telescope collecting area, the largest improvement comes from reduction of angular throughput for background photons and promise of maintaining the telescope emissivity to less than 10%. The former depends strongly on the careful design of the systems. Thus a prospective support laboratory for a large telescope such as JNLT has to study telescope thermal characteristics; effectiveness of Lyot stop, routing of secondary heat dissipation, optical surface emissivity monitor for infrared which is suspected to degrade relatively in short time resulting in decreased sensitivity of systems.

Let us now briefly discuss technical consideration to realize performance described above. It is much more demanding to design, fabricate, and align optics for an array camera compared to a single element photometer because of tight imaging specifications. Optical and mechanical properties of materials for windows, filters, and lenses at cryogenic temperature is only poorly documented. If changing of magnification is required, which is a reasonable requirement for observatory instruments, the complexity of cryogenic optics is greatly increased. To cover more than one spectral window reflective optics can be used, but with more difficulty than the lens system. On the other hand, the fact that any flaw in the imaging optics manifests clearly by itself makes its evaluation easy.

Usually existing infrared cameras have magnification less than unity. This is the result of large F/number of conventional infrared telescopes built in the single detector era. Smaller pixel size compared to traditional single detector causes unnecessary over-sampling and narrow FOV coverage. It is then necessary to decrease the angular resolution by demagnification. This requires fast optics which brings a problem in image quality. Whether the high F/number ($F\# > 20$) is still optimum approach requires careful investigation. The situation might be improved if large format arrays become available so that a portion of the array field-of-view can always find some off-source sky. In near-infrared a large array makes it possible to abandon the conventional sky-chopping technique by using in-frame sky subtraction technique. For this reason a long one-dimensional array format is also attractive. Still in the mid-infrared background sky noise is much more severe and, therefore, a more sophisticated method such as interpixel correlation might have to be developed. Also an undersized secondary without the surrounding baffle tube will remain as an essential feature of the infrared configuration.

3. Photon Device

Based upon the previous discussion we shall now look into photon devices, in which individual photons generate charge carriers in either photovoltaic or photoconductive mode. Readout technique of an array is as critical as the detector itself because of the high photon arrival rate shown in Table I. After the discussion on the readout rate there is a brief introduction to Impurity Band Conduction device which is currently receiving attention.

Infrared focal-plane array architecture has more varieties than optical device (mostly monolithic CCD) in order to interface various detector materials needed for wider spectral range at different operating conditions. The main reason why CCD is not used in mid-infrared is degradation of charge transfer efficiency at cryogenic temperature. Instead, hybrid structures are frequently used in which silicon multiplexers are bonded on the backside of the detector. The multiplexers are called Direct Read-Out (DRO). The front side of the detector has a transparent electrode that allows incident photons to generate a pair of charges in the detector substrate at a certain quantum efficiency. The advantage of the hybrid architecture is the flexibility of design in the readout circuit that can adjust to the latest silicon integrated circuit technology, independent of detector material development. DRO can usually access individual pixels randomly and non-destructively through MOS–FETs, which are used as source followers. The random access capability would become important in the future when elaborate data acquisition and processing such as interpixel correlation method is seriously considered. Because most of the current detector materials are also silicon based, thermal mismatch at low temperature is not a serious problem in this structure, although it is still one of the reasons why the formats of infrared arrays are much smaller than those of visible CCD array.

The charges generated by photons are accumulated in the sense node capacitance (usually about 0.1 pF at MOSFET gates) of individual pixels. The capacitance limits the total number of charges that can be stored to the order of 10^5 electrons before the onset of significant debiasing effect causing nonlinear response to the input flux. The well capacity might be large enough for arrays in cooled spectrometers or space application. However, the flux levels indicated in Table I suggest that for narrow N band ($R = 10$) imaging each pixel has to be read out every milli-second or more frequently on the ground-based telescopes. This readout rate is one of the difficult technical challenges to the current camera system as described in Section 5. Improvement in speed is expected by turning into other material such as GaAs but would not be available in the immediate future.

Recently devices named Impurity Band Conduction (IBC) array or Blocked Impurity Band (BIB) array were introduced as the superior alternative to simple photoconductive or photovoltaic detector used in the mid-infrared arrays. These arrays have a thin layer of undoped pure silicon and another thicker layer of heavily doped infrared sensitive layer between the two electrons. The front surface electrode is transparent to the incident photons while the backside electrodes defining each pixel are bump bonded to the inputs of the DRO multiplexer. The doping concentration of the active layer is high so that an impurity level conduction band is formed through which ionized donors drift toward the electrode under the bias field. The electrons in the regular conduction band are swept away through the undoped blocking layer toward the other electrode. The depletion of ionized donors in infrared sensitive layer and the absence of donor conduction band in the blocking layer next to the contact prevents recombination of charge pairs and reduces the dark current noise. The high doping also helps to improve the quantum efficiency of the sensitive area greatly over conventional detectors. In addition, this type of detector has better linearity and radiation hardness.

4. Bolometer Array

Thermal detectors have been exclusively helium-cooled bolometers in infrared astronomy. Although for a long time since the invention of the liquid helium-cooled bolometer, germanium has been a dominant detector material, silicon is now replacing it quickly. Recently it became feasible to fabricate arrays by using silicon bolometer because of its high quality. There is no documented attempt of serious comparison of the performance of the two materials, although one might expect that Si should benefit from the advancement of silicon based integrated circuit technology in areas such as purification and doping, uniformity, therefore higher yield, and also from availability of associated technology such as ion implantation for accurate doping, transparent electrode, degenerated photon absorption layer, etc. One visible effect of these changes is greatly reduced $1/f$ noise down to the 1 Hz domain by improved electrode technique.

In applications where sensitivity is limited by background noise, which is the case for wide to medium spectral resolution photometry/imaging at 10 µm as discussed earlier, the bolometer has two advantages over photon detectors: (1) large quantum efficiency: because of lack of optimum impurity for the 10 µm window the quantum efficiency of PC/PV is much lower compared to bolometers which is higher than 80%, (2) smaller noise: because of recombination process photoconductors have additional square-root-two factor in the NEP. Furthermore, the bolometer array is essentially immune from cross-talk because of spatial separation among elements. Photon detector arrays, on the contrary, suffer from cross-talk due to capacitive coupling between the closely spaced output FETs on the array.

Fabrication of a bolometer array is completely different from the photon detector array and only small-scale arrays are available at present. A discrete monolithic bolometer from a single silicon wafer has been successfully fabricated by applying silicon integrated circuit technology (Downey et al., 1984) with a good prospect of expanding the technique to fabricate one-dimensional arrays. Two-dimensional monolithic bolometer arrays, however, would require more advanced silicon micromachining technology and would take considerable undertaking. An alternative approach is to extrapolate the construction technique of conventional discrete bolometers. A 1×8 helium-3 cooled bolometer array was thus fabricated for the Kuiper Airborne Observatory far-infrared super-resolution photometer (Low et al., 1987). The photometer has NEFDs of 16 Jy Hz$^{-1/2}$ per pixel ($23'' \times 11''$) at 100 µm and the overall sensitivity for a point source of a factor 1.5 better than the pixel NEFD (Joy, 1987). The intrinsic NEP of this type of bolometer at $T = 0.35$ K is mainly determined by phonon noise contribution $2T\sqrt{kG}$ and is 4×10^{-16} W Hz$^{-1/2}$ for thermal conductance $G = 25$ nW K^{-1}. The -3 dB response frequency of the KAO array is approximately 80 Hz which implies heat capacitance of elements 50 pJ K^{-1}. By using similar design we have now at the Steward Observatory a 2×10 helium-3 cooled bolometer array in the cryostat designed for 10 and 20 µm super-resolution observations. The system expects center pixel NEFD 500 mJy Hz$^{-1/2}$ for a point source at 11 µm.

The data acquisition scheme for bolometer array has to be also modified from that of the single bolometer system. The current bolometer data acquisition scheme we are testing at the Steward Observatory consists of an all DC-coupled multiplexer and amplifier system. Filtering is controlled by a data acquisition computer for flexibility. Twenty JFET J230s connected to each element with voltage gain of 3.5 feed into two RCA CD4067 multiplexers, contained in a module of $1'' \times 1.5'' \times 0.75''$ cooled at 77 K. Two additional JFETs are mounted in the same module for reference. Outside the dewar, the two channels are routed through an offset subtraction circuit using a small-scale RAM with readout synchronized to the MUX switching. After overall gain of 1000 signals are fed o 16-bit ADC in an 80386 based computer sampling at the rate of about 6 kHz.

The bolometer can also be used to 350 μm or 450 μm windows, where no photon detectors available for lack of suitable impurity levels. A large telescope equipped with a large area composite bolometer array will have important uses, for example, for the study of low brightness extended continuum sources, even with the existence of several submillimeter telescope in the near future. Estimated 350 μm NEFD with pixel size of 5″ is limited by the bolometer phonon noise and about $10 \, \mathrm{mJy \, Hz^{-1/2}}$ for the background level of 1×10^{-11} W.

For lower background applications in the submillimeter or with cooled spectrometers it is possible to improve the bolometer NEP beyond the value quoted above. However, it requires development of telescope compatible adiabatic demagnetization refrigerator or $\mathrm{He^3/He^4}$ dilution refrigerator. Below 100 mK extrapolation of 0.3 K bolometer NEP predicts better than $3 \times 10^{-17} \, \mathrm{W \, Hz^{-1/2}}$.

Finally we note that bolometer arrays have additional merits to these scientific ones, namely its accessibility in the astronomical community and relatively low cost of developing devices optimized for astronomical use.

5. Mid-Infrared Camera

Table II summarizes two mid-infrared cameras currently operational (Arens *et al.*, 1987; Hoffmann *et al.*, 1987). Other active mid-infrared camera groups in the U.S.A.

TABLE II

	SAO/UA/GSFC	UC-Berkeley
Detector	Aerojet/AMCID/Si:Bi	Hughes/IBC/Si:As
Format	32×32	10×64
Pixel size (μm²)	200×200	120×120
Well capacity (electrons)	2×10^5	2×10^5
Read noise (electrons)	< 600	50
Responsivity (A/W)	0.5	0.9
NEFD at 10 μm		
for a point source	1 Jy in 60 s on IRTF	$1.3 \, \mathrm{Jy \, Hz^{-1/2}}$

which the author is aware of are D. Gezari's at GSFC using Hughes 58×62 Si:Ga/IBC, C. Townes and E. Bloemhof's (now at CfA) at UC–Berkeley using 32×32 Rockwell HgCdTe, F. Low's at Steward Observatory using 2×10 0.3 K bolometers and C. Telesco's at MSFC using 5×4 bolometers. Also groups at AFGL (P. LeVan and P. Tandy), Cornell University (J. Houck) and University of Texas at Austin (J. Lacy) are using arrays in the spectrometers for 10μ region. In addition NRL and WIRO (K. Shivanandan and H. Thronson) uses Rockwell 16×50 Si:As/BIB and Hughes 20×64 Si:As/IBC for their camera which is just starting telescope tests and a joint team from SAO, NRL, and the Steward Observatory (G. Fazio, K. Shivanandan, and W. Hoffmann) has started the construction of a camera using Hughes 20×64 Si:As/IBC.

The sensitivity of mid-infrared array camera to a point source in the literature misses a factor of at least two (Arens *et al.*, 1987) and sometimes more (Hoffmann *et al.*, 1987) compared to best bolometer photometers, even if the difference of quantum efficiency is considered. For example, Hoffmann *et al.* states that the sensitivity of their AMCID camera on IRTF is about 1 Jy after 1 min with 5% spectral bandwidth. This performance is explained mainly by the need to drive the inherently slow array at high frame rate to avoid saturation. The readout noise of DRO has already been decreased down to the order of 100 electrons per read and is well below the excess noise we are concerned here. The excess noise has to be eliminated before long integration could become effective for the detection of weak sources by using these cameras. Recently the SAO/UA/GSFC camera was tested on the Steward Observatory 90 inch to investigate the performance at lower background by interfacing to a helium-cooled Fabry–Pérot interferometer in the optics chain. It was found that the frequency response was slowed down even further. However, on a positive side, the crosstalk observed in higher background seems to be non-existent in this configuration (Hoffmann *et al.*, 1988).

6. Conclusions

Let us conclude by giving more general thoughts to the support and improvement of mid-infrared performance of a future large telescope on a premier location. A well-equipped laboratory is essential near the telescope site for maintaining and up-grading the instruments. In mid-infrared this is particularly true because the performance of the instruments is strongly coupled to the thermal environment of the individual telescope and atmospheric condition of the site. The experience of the laboratory with detectors applied to astronomical observations should help keeping pace with the progress of industrial development of detector technology. The ability to interact with detector manufacturers based upon the observatory environment tests and the astronomical data/image processing would be mutually beneficial and essential for the future progress.

In addition to the activities directly related to the astronomical observation there are other applications of infrared array detector technology for telescope operation or performance monitoring, such as (1) two-dimensional cloud/precipitable water monitor; (2) seeing and/or telescope surface monitor using a point source image onto an array; (3) infrared guiding system for observation of the galactic plane sources and observation on bright nights or even daytime observation.

Acknowledgements

The author is grateful to the organizing committee of the symposium for the invitation and to their hospitality. Discussions on various subjects with Dr F. J. Low and Dr K. Shivanandan were helpful in preparing the manuscript.

References

Arens, J. F., Jernigan, J. G., Ball, R., Peck, M. C., Gaalema, S., and Lacy, J.: in C. G. Wynn-Williams and E. E. Becklin (eds.), *Infrared Astronomy with Arrays*, p. 256.

Arens, J. F., Jernigan, J. G., Ball, R., Peck, M. C., Gaalema, S., and Lacy, J.: in C. G. Wynn-Williams and E. E. Becklin (eds.), *Infrared Astronomy with Arrays*, p. 256.

Downey, P. M., Jeffries, A. D., Meyer, S. S., Weiss, R., Bachner, F. J., Donnelly, J. P., Lindley, W. T., Mountain, R. W., and Silversmith, D. J.: 1984, *Appl. Optics* **23**, 910.

Hall, D. N. B.: 1989, *Astrophys. Space Sci.* **160**, 243 (this issue).

Hoffmann, W. F., Fazio, G. G., Tresch-Feinberg, R., Deutsch, L. K., Gezari, D. Y., Lamb, G. M., Shu, P., and McCreight, C. R.: 1987, in C. G. Williams and E. E. Becklin (eds.), *Infrared Astronomy with Arrays*, University of Hawaii, p. 241.

Hoffmann, W. F., Nishimura, T., Hora, J. L., Fazio, G. G., Deutsch, L. K., and Shivanandan, K.: 1988, *Report on Observing Run with AMCID Camera and Fabry–Pérot Interferometer,* Steward Observatory, Internal Report.

Joy, M.: 1987, (priv. comm.).

Low, F. J., Nishimura, T., Davidson, A. W., and Alwardi, M.: 1987, in C. G. Williams and E. E. Becklin (eds.), *Infrared Astronomy with Arrays*, University of Hawaii, p. 91.

KODAIRA – Could you specify more the causes of the high background emissivity of the UA telescopes at 10 micrometers?

NISHIMURA – I don't think we have a good understanding of what causes higher than expected emissivity. It would be a combination of mirror surface condition, baffle designs, poor matching of user instrument... The mirrors are recoated more frequently and this seems to have improved a bit.

INFRARED DETECTOR ARRAYS AND SOME APPLICATIONS
TO SPECTROSCOPY*

A. T. TOKUNAGA

University of Hawaii, Honolulu, Hawaii, U.S.A.

(Received 16 December, 1988)

Abstract. The read noise, dark current, and pixel sizes of state-of-the-art infrared arrays for astronomy are presented. Considerations for instrument development utilizing infrared arrays are discussed, with emphasis on the background emission and expected sensitivity. A simple method of estimating the background emission on the JNLT and some applications to spectroscopy are presented.

1. Introduction

A turning point in infrared (IR) astronomy occurred 18 months ago. At a detector conference surveying the state-of-the-art in IR detector technology for astronomy, some of the first results of astronomical instruments utilizing very sensitive IR arrays were presented (Wynn-Williams and Becklin, 1987). The advances represented by these developments are enormous: the very best of the detectors reported at the conference, if used with the JNLT, will provide an increase in sensitivity of up to 1000 over current instruments in use at existing telescopes. It is, therefore, the challenge of astronomers of our generation to determine how this technical advance can be applied toward the discovery of new cosmic phenomena.

The purpose of this paper is to discuss the current state of detector array technology for astronomy, the limiting flux levels one can expect to achieve on the JNLT, and some applications to spectroscopy. Previous reviews of IR technology have been given by Low and Rieke (1974), of IR detectors by Gillett *et al.* (1977), and of arrays by Gillett (1987).

2. General Considerations

2.1. THE OH AIRGLOW AND THERMAL BACKGROUND

Of paramount importance in the design of IR instrumentation is the need to understand the background levels to be encountered. At $1-2.2$ μm, the major contributer to the background emission is the OH airglow, while at longer wavelengths it is the thermal emission from the telescope and sky.

From measurements at the IRTF, UKIRT, UH 2.3-m telescope, and the Wyoming Infrared Observatory, the simple average of the OH emission was found to be

---

* Paper presented at the Symposium on the JNLT and Related Engineering Developments, Tokyo, November 29–December 2, 1988.

Astrophysics and Space Science **160**: 333–343, 1989.

13.7 mag arc sec^{-2} at 1.65 μm (E. Becklin, priv. comm.). This is equal to 9000 photons s^{-1} m^{-2} arc sec^{-2}, with a variation of 60%. This figure is adopted for this paper.

Since the OH background is variable, the time-scale for variability will determine the maximum integration time possible before reading out the array. In general, this is dependent on the latitude of the observatory and nightly variations during the time of the observations (this is discussed in detail by McCaughrean, 1988). Rieke *et al.* (1987) have reported on some practical limitations on reducing the sky emission.

In the following, a simple method of estimating the background photocurrent is presented. The number of signal electrons s^{-1} generated by the detector is

$$S_e = A\eta \left(\frac{F_\lambda \Delta\lambda}{h\nu} \right) = A\eta \left(\frac{\lambda F_\lambda}{h\nu} \right) \left(\frac{\Delta\lambda}{\lambda} \right) = M_\lambda \left(\frac{1}{R} \right) \quad \text{electrons s}^{-1}, \tag{1}$$

where $R = \lambda/\Delta\lambda$ is the resolving power; M_λ, the number of detected electrons per second for $R = 1$; A, the telescope area (m^2); η, the optical efficiency, including the detector quantum efficiency; $h\nu$, energy per photon; and F_λ, the flux density (W m^{-2} μm^{-1}). This expression is useful because the number of detected signal electrons can be readily calculated for any resolving power and also compared to the read noise and dark current of the array.

In a similar manner, the thermal background in electrons s^{-1} is given by

$$M_{TB} = A\Omega\eta\varepsilon \left(\frac{\lambda B_\lambda(T)}{h\nu} \right) \quad \text{electrons s}^{-1}, \tag{2}$$

where Ω is the solid angle viewed by the instrument, ε is the telescope emissivity, and T is the temperature of the telescope. For the JNLT, we assume $A = 44.18$ m^2, $\varepsilon = 0.1$, $\eta = 0.5$, and $T = 273$ K. The background emission is also determined by the pixel size. For convenience, 0.5″ pixels are chosen so that the field of view for each pixel is 0.25 arc sec^2.

The OH background is given by

$$M_{OH} = A\eta(\lambda N_\lambda) \quad \text{electrons s}^{-1}, \tag{3}$$

for which at $\lambda = 1.65$ μm:

$$N_\lambda = 9000 \,(0.25 \text{ arc sec}^2 \text{ pixel}^{-1})/\Delta\lambda =$$
$$= 7500 \quad \text{ph s}^{-1}\text{ m}^{-2}\text{ μm}^{-1}. \tag{4}$$

The passband of the OH emission observations at 1.6 μm was $\Delta\lambda = 0.3$ μm, and this was used in Equation (4).

The resulting background levels for the OH airglow and thermal emission from the telescope is shown in Figure 1. From this figure the background for any resolving power, instrumental efficiency, and pixel size can be estimated for the JNLT. For example, if the resolving power is 100, the M_λ scale should be divided by 100. If the optical efficiency

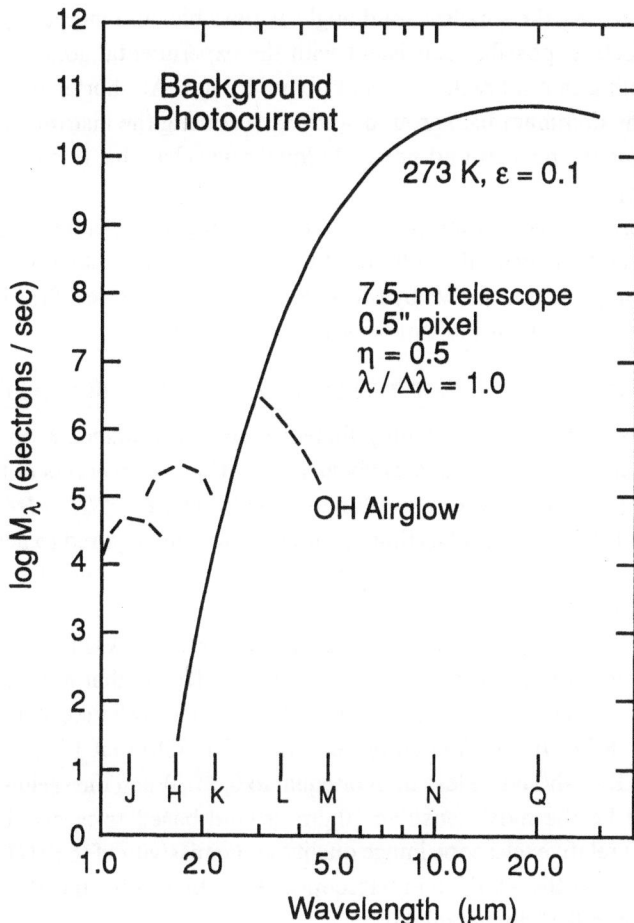

Fig. 1. The background photocurrent for the JNLT. For this calculation, it is assumed that the pixel size is 0.5" (square); the optical efficiency including detector quantum efficiency is 0.5; the resolving power ($\lambda/\Delta\lambda$ = 1.0; the telescope temperature is 273 K; and the telescope emissivity is 0.1. The solid line shows the background photocurrent from the telescope emission and the dashed line shows the background from the OH airglow emission. The infrared photometric bands are shown at the bottom of the plot: J (1.25 μm), H (1.65 μm), K (2.20 μm), L (3.45 μm), M (4.80 μm), N (10.0 μm), and Q (20.0 μm).

(η) is 0.1, then the M_λ scale should be further divided by 5. The background current in electrons s^{-1} can then be estimated from the figure and compared to the signal current or detector dark current.

From Figure 1, the estimated OH background for a broad-band IR camera operating at 1.65 μm (with R = 5), would be about 5×10^4 electrons s^{-1} pixel^{-1} and at 10 μm the background current would be approximately 5×10^9 electrons s^{-1} pixel^{-1}. Because of the high background emission at IR wavelengths, the instrument design approach must always seek to reduce the background emission as much as possible. This is accomplished by cooling the optics (typically to liquid nitrogen temperatures), making the pixels view the smallest solid angle as possible, and increasing the spectral resolution as much as possible consistent with the experimental goals.

It is worth noting that a break occurs at about 2.2 μm. At shorter wavelengths, the OH airglow is the dominant background source, so cooling the instrument optics does not help to reduce the background noise. At longer wavelengths, it is essential to cool the instrument optics.

Another important point is that the OH emission is highly dependent on the spectral resolution since at high spectral resolution, the OH emission breaks up into individual lines. A good example is that shown by Moorwood (1987). The level of the OH emission shown in Figure 1 is only an indication of the average level.

2.2. The advantage of a large telescope at thermal IR wavelengths

At thermal IR wavelengths ($\lambda \geq 2.5$ μm), there is a large advantage for operating at the diffraction limit of the telescope, especially at 10 μm. For observations at the telescope diffraction limit and under background-limited conditions, the $S/N \sim D^2 \sqrt{t}$ (see, for example, Gehrz, 1987). This implies that the integration time required to achieve a given S/N decreases as D^{-4}, and this is a major advantage for any large telescope that can work at the diffraction limit.

For the JNLT, the 10 μm diffraction-limited field of view (full width at half maximum) is 0.34″. Under conditions when the seeing at Mauna Kea is that good, observations of a point source will have 6.25 times more signal-to-noise or will take 39 times less time than on the 3 m IRTF. It is estimated by Azouit *et al.* (1988) that 10% of the time the seeing at Mauna Kea should be less than or equal to 0.3″. When the seeing is that good, the JNLT could be the most sensitive 10 μm ground-based telescope in the world, depending on the relative telescope image quality and emissivity of the JNLT compared to the Keck 10 m and the VLT 7.5 m telescopes. To achieve this, the JNLT should be designed to provide 0.3″ image quality.

3. Detector Survey

A small survey was undertaken by the author to assess the current state-of-the-art in detector arrays for astronomy. The results of this survey is shown in Table I in which some of the most sensitive arrays in the hands of astronomers are shown. Details on other arrays can be found in *Infrared Astronomy with Arrays* (Wynn-Williams and Becklin, 1987). Also Nishimura (this conference) discussed a variety of 10 μm arrays that are available. In addition to the 10 μm array shown in Table I, a 10 × 64 array has been used for astronomical observations (Arens, priv. comm.; Lacy, 1987).

Certain parameters, such as fill factor, the wavelength-dependence of the quantum efficiency, and readout rates, should also be considered in comparing the relative merits of these arrays for any given application. However, Table I provides a guide as to what can be obtained with current detector arrays on the JNLT, and this is discussed in the next two sections. The primary emphasis will be on maximizing sensitivity and how to achieve it.

TABLE I

Selected IR arrays with highest sensitivity

Spectral range (µm)	1–2.5	1–5	5–14
Detector material	HgCdTe	InSb	Si:Ga
Type	PV	PV	PC
Pixel size	60 µm	75 µm	75 µm
Multiplexer type	DRO	DRO	DRO
Format	128 × 128	58 × 62	58 × 62
Manufacturer	Rockwell	SBRC	SBRC
Quantum efficiency	70%	45%	30%
@ λ (µm) =	2.3	3.8	10.0
Dark current (electrons s^{-1})	9	<6	<630
@ temperature (K) =	77	10	10
Read noise (electrons)	50	280	<200
Full well (electrons)	3 × 10^5	10^6	7 × 10^5
Reference	1	2	3

Notes:
(a) PV = photovoltaic; PC = photoconductor.
(b) DRO = direct readout.
(c) SBRC = Santa Barbara Research Corp.
(d) References:
   (1) M. Rieke, Univ. of Arizona (priv. comm.).
   (2) J. Pipher, Univ. of Rochester (priv. comm.).
   (3) D. Gezari et al. (1988).

Large format IR arrays are becoming available to IR astronomy, and those known to the author are summarized in Table II. Although the PtSi arrays have lower quantum efficiencies than the arrays shown in Table I, the low read noise of PtSi arrays make them competitive with the best InSb DRO arrays for large areal mapping projects or for experiments requiring short integration times such as speckle imaging. The technology used for the Mitsubishi PtSi array offers the possibility of even larger arrays, such as 1024 × 1024.

4. Broad-Band Imaging

In this section, an estimate of the limiting magnitude for broad-band imaging is presented. Assuming that all of the light from a point source falls on one pixel, then the signal-to-noise can be estimated from

$$S/N = \frac{S_e t}{[r_n^2 + (S_e + d + b)t]^{1/2}}, \tag{6}$$

where S_e is the signal current from Equation (1); r_n, read noise (electrons); d, dark current (electrons s^{-1}); and b, background-generated current (electrons s^{-1}). This equation is appropriate for the photovoltaic case. See Gillett (1987) for the general case including the photoconductive case. Appropriate corrections should be applied if the

TABLE II

Large format IR arrays

Spectral range (μm)	1–2.5	1–2.5
Detector material	PtSi	PtSi
Type	SBD	SBD
Pixel size	20 μm	20 × 26 μm
Multiplexer type	DRO	CSD
Format	256 × 256	512 × 512
Manufacturer	Hughes	Mitsubishi
Quantum efficiency	15%	13%
@ λ (μm) =	1.2	1.2
Dark current (electrons s^{-1})	10	100[a]
@ temperature (K) =	50	65[a]
Read noise (electrons)	40	40[a]
Full well (electrons)	10^6	7×10^5
Reference	1	2, 3

[a] Predicted.

Notes:
(a) SBD = Schottky Barrier Diode; PV = photovoltaic.
(b) DRO = direct readout; CSD = charge sweep device.
(c) The spectral response extends to 5 μm, but with very low quantum efficiency.
(d) References:
   (1) A. Fowler, National Astronomy Obs. U.S.A.
   (2) S. Sato and M. Ueno, National Astronomical Lbs. of Japan.
   (3) M. Kimata et al. (1988).

point source does not fall entirely on one pixel. For example, if the point source falls on 4 pixels (as at the intersection of 4 pixels), then the S/N should be reduced by $\sqrt{4} = 2$.

If we substitute Equation (1) into Equation (6) and solving for F_λ we obtain

$$F_\lambda = \frac{(S/N)R[r_n^2 + (d + b)t]^{1/2}}{1.11 \times 10^{20} \lambda^2 t} , \tag{7}$$

where λ is the wavelength in μm. The efficiency factor η was taken to be 0.5. The signal current S_e has been assumed to be small compared to $(d + b)$. From this equation, the limiting flux density and magnitude as a function of integration time can be calculated. For the 1–5 μm spectral region, the limiting magnitude was calculated assuming $S/N = 1.0$ and $R = 5.0$. The maximum integration time on chip was determined by the full well capacity divided by $(d + b)$ in the background-limited case. To obtain the integration time for 1800 s, it was assumed that multiple reads would be done for a total integration time of 1800 s. The limiting magnitude obtained is shown in Table III. A maximum on-chip integration of 600 s was assumed. This may be optimistic, as the sky variations such as from the OH emission, may change on a time-scale more rapidly than this, in which case the on-chip integration would have to be shorter.

TABLE III

Broad-band imaging limiting magnitudes

Wavelength (λ)	1.25	1.65	2.20	3.45	4.80	10.0
Detector material	HgCdTe	HgDdTe	HgCdTe	InSb	InSb	Si:As
R	5.0	5.0	5.0	5.0	5.0	100.0
η	0.35	0.35	0.35	0.25	0.25	0.15
r_n (electrons)	50.	50.	50.	280.	280.	100.
b (electrons s^{-1})	10^4	3.9×10^4	2.2×10^4	4.0×10^6	10^8	9.5×10^7
Full well (electrons)	3×10^5	3×10^5	3×10^5	10^6	10^6	10^5
Limiting magnitude in 1800 s	26.4	25.1	24.8	20.7	17.1	13.0

Notes:
(a) $S/N = 1.0$ in all cases.
(b) η is the total instrumental efficiency, including the detector quantum efficiency.

These limiting magnitudes can be compared to the IRTF, for which the limiting magnitudes (1σ, 1 hr), are: $J - 21.8$; $H - 20.9$; $K - 20.6$; $L - 15.9$; $M - 12.4$; $N - 9.7$. Thus the JNLT can reach 4.6 mag fainter at J and 3.3 mag fainter at N compared to conventional single-detector IR photometers.

As mentioned previously, the background photocurrent at 1.65 µm ($R = 5$) is expected to be about 5×10^4 electrons s^{-1}. For a 25.1 mag point source at 1.65 µm, the signal current is only 6 electrons s^{-1}. Thus the sky emission fluctuations have to be below one part in 10^4 in order for sources of this faintness to be detected. In addition, the flat-fielding must be better than one part in 10^4. This is in fact been proven to be possible in long integrations on the faint sources by Elston *et al.* (1988) and Cowie *et al.* (1988). Thus extremely high sensitivity in broad-band imaging is possible, but only because the flat-fielding required is achievable and the sky fluctuation can be virtually eliminated.

5. Spectroscopy

In a similar fashion, the limiting magnitudes for spectroscopic observations can be estimated. Maihara (this conference) shows what can be achieved with moderate resolution dispersive spectrometers ($R = 1000$) and with imaging Fabry–Pérot spectrometers. In this section, I will concentrate on what can be achieved with low- and high-resolution dispersive spectrometers. Such instruments can provide the highest sensitivity to point sources since (1) the background emission can be dispersed (and, therefore, reduced to a minimum), and (2) there is the advantage of observing all spectral elements simultaneously. The latter advantage is limited, of course, by the detector array size.

The limiting magnitude for *dispersive* spectroscopic instruments on the JNLT are shown in Table IV for $R = 100$ and $R = 20\,000$. For these calculations, the detector parameters from Table I was used with the HgCdTe array for 1–2.2 µm, InSb for 3.45–4.8 µm, and Si:Ga for 10.0 µm. A maximum integration time of 600 s was

TABLE IV

Spectroscopic limiting magnitudes

λ (µm)	$R = 100$			$R = 20\,000$	
	η	Limiting magnitude		η	Limiting magnitude
1.25	0.20	24.5		0.10	19.5
1.65	0.20	23.1		0.10	18.9
2.20	0.20	22.9		0.10	18.5
3.45	0.11	18.7		0.05	15.1
4.80	0.11	16.3		0.05	13.0
10.0	0.15	13.0		0.05	9.6

Notes:
(a) Limiting mag for 1σ 1 hr in 1800 s.
(b) η is the total instrumental efficiency, including the detector quantum efficiency.
(c) The maximum on-chip integration before reading out is assumed to be 600 s.

assumed, but this may be higher or lower in practice, depending on the time-scale for sky or background emission variations.

In the following, two classes of spectrometers are discussed to indicate the new types of instruments infrared arrays are making possible. Because of space limitations, the discussion is very brief.

35.1. THE CRYOGENIC PRISM SPECTROMETER

For low-resolution work, a prism spectrometer approach has the advantage of offering a wide spectral range. One application of this approach is being implemented by S. Sato and H. Takami at the National Astronomical Observatory of Japan. They are using a quartz prism to cover the 1–2.5 µm spectral range with a linear 16-element detector array. The resolving power is low, approximately 10. Its primary purpose is to observe the polarization at 16 resolution elements simultaneously, thus it is an extremely efficient spectropolarimeter. First tests of this instrument will occur by the end of 1988.

It is also possible to consider covering the 1–2.5 µm or 3.0–5.4 µm spectral range at a resolving power of 50–100 using a prism spectrometer. Such an instrument could utilize a linear 128 or 256 element array or a two-dimensional array of this size. The main problem would be to obtain a prism material with sufficiently high index of refraction in order to make the instrument as compact as possible. One possible material is strontium titanate ($SrTiO_3$); however it cannot be obtained in a large enough size for a reasonable price at this time. If this instrument could be built, it would be a very efficient survey instrument.

5.2. THE CRYOGENIC ECHELLE SPECTROGRAPH

This class of instrument offers the possibility of large improvements in sensitivity over current instruments and, therefore, several observatories are presently constructing such instruments (at the UKIRT and IRTF, both at Mauna Kea). The best way of appreciating this improvement is to compare the expected performance of a cryogenic echelle spectrograph on the JNLT to present-day spectrometers that provide comparable spectral resolution.

At a wavelength of 4.8 μm, the UKIRT infrared Fabry–Pérot spectrometer has a 1σ 1 hr magnitude limit of 7.3 at a resolving power of 20 000. Comparing to Table IV, this is 5.7 mag less in sensitivity compared to a cryogenic echelle spectrograph on the JNLT. The improvement at shorter wavelengths is even more dramatic. At a wavelength of 2.2 μm, the KPNO FTS at the 4 m telescope has a 1σ 1 hr magnitude limit of 10.4. From Table IV, this is 8.1 mag lower sensitivity. Of course, these comparisons are not fair in that the UKIRT and KPNO instruments were designed for use with larger entrance apertures and the latest detector arrays are not currently in use with these instruments. However, this comparison shows the dramatic increase in sensitivity we can expect in the future. At the wavelengths where the cryogenic echelle spectrograph is not background-limited (1–3 μm), further advances in lower read noise will lead to even greater increases in sensitivity than that projected in Table IV.

In order to realize the full potential of these sensitivity increases, the JNLT and its spectrometers must be designed to operate at the seeing limit, 0.5" or less. In addition, the very best infrared detector arrays must be incorporated into the instrumentation.

5.3. OTHER IDEAS

G. Rieke (priv. comm.) is presently constructing a high-performance 1–16 μm spectrometer using 2×32 Ge diode array for use with a variety of telescopes, including the Multiple Mirror Telescope. The spectral resolution is 100–3600, depending on the grating that is chosen. This spectrometer takes full advantage of the very high quantum efficiency and low read noise of the Ge diode array (Rieke *et al.*, 1987). Another advantage is that the spectrometer does not have to be cooled and, therefore, the cost of construction is relatively low.

The use of fiber optics will also be eventually incorporated into infrared spectrographs. While it may be straightforward in concept to utilize aperture plates or movable fibers for multi-object spectroscopy in the infrared, there is a difficult problem of cooling the entire mechanism for optimum performance at thermal wavelengths ($\lambda > 2.5$ μm). The additional cost of incorporating such a facility would be worthwhile given the cost of telescope time for a 7.5 m telescope. Another possible use of fiber optics would be to place the fibers in a linear format along the dispersion of the spectrograph and then using it to place the entire spectrum in a two-dimensional format on the infrared array. Such design is being developed at the Goddard Space Flight Center (Glenar *et al.*, 1988).

6. Summary

During the past 5 years infrared astronomy has advanced from the era of single-detector technology to that of extremely sensitive two-dimensional arrays. This is truly a new era for infrared astronomy. The technical advances in infrared array technology continues unabated. Furthermore, during the next 5 years we can expect to see the construction of 8–10 m class telescopes – literally a new generation of telescopes for ground-based astronomy. These technical advances, combined with the high angular resolution provided by the Mauna Kea site, will lead to improvements in sensitivity of as much as more than 1000 over current instrumentation. Such an increase in sensitivity should lead to the discovery of new astrophysical phenomena.

In order to make the most advantage of the large collection area of the JNLT, an image quality of 0.3″ is required. At infrared wavelengths, the primary advantage is that the background emission for the sky and telescope can be kept to a minimum and thereby maximize sensitivity. The largest gains in sensitivity will come from high-resolution spectroscopy using dispersive spectrometers. Finally, the availability of two-dimensional infrared arrays is stimulating the development of new types of infrared instruments, and many of them are very similar to instruments developed for optical wavelengths.

References

Azouit, M., Cowie, L., Erasmus, A., Lugten, J., Roddier, C., Roddier, F., Songaila, A., and Vernin, J.: 1988, 'A Description of Results from the November 1987 Mauna Kea Site Campaign', Preprint.
Cowie, L. L., Lilly, S. J., Gardner, J., and McLean, I. S.: 1988, *Astrophys. J.* **332**, L29.
Elston, R., Rieke, G. H., and Rieke, M. J.: 1988, *Astrophys. J.* **331**, L77.
Gehrz, R. D.: 1987, in C. G. Wynn-Williams and E. E. Becklin (eds.), 'Matching Infrared Array Instruments to Future Large Telescopes', *Infrared Astronomy with Arrays*, Univ. of Hawaii, p. 499.
Gezari, D. Y., Folz, W. C., Woods, L. A., and Woolridge, J. B.: 1988, *Proc. SPIE* **973** (in press).
Gillett, F. C.: 1987, in C. G. Wynn-Williams and E. E. Becklin (eds.), 'Infrared Arrays for Ground-Based Astronomy', *Infrared Astronomy with Arrays*, Univ. of Hawaii, p. 3.
Gillett, F. C., Dereniak, E. L., and Joyce, R. R.: 1977, *Opt. Eng.* **16**, 544.
Glenar, D., Mumma, M. J., Jennings, D. E., and Weaver, H. A.: 1988, *Bull. Am. Astron. Soc.* **20**, 841.
Lacy, J. H., Arens, J. F., Peck, M. C., and Gaalema, S. D.: 1987, in C. G. Wynn-Williams and E. E. Becklin (eds.), 'A Mid-Infrared Cryogenic Echelle Spectrometer', *Infrared Astronomy with Arrays*, Univ. of Hawaii, p. 402.
Kimata, M., Denda, M., Yutani, N., Iwade, S., and Tsubouchi, N.: 1988, *IEEE J. Solid-State Circuits* **SC-22**, 1124.
Low, F. J. and Rieke, G. H.: 1974, in N. Carleton (ed.), 'The Instruments and Techniques of Infrared Photometry', *Methods of Exp. Physics*, Vol. 12, Academic Press, New York, p. 415.
McCaughrean, M. J.: 1988, 'The Astronomical Application of Infrared Array Detectors', Ph.D. Thesis, Univ. of Edinburgh, p. 107.
Moorwood, A. F. M.: 1987, in C. G. Wynn-Williams and E. E. Becklin (eds.), 'IRSPEC: Design, Performance, and First Scientific Results', *Infrared Astronomy with Arrays*, Univ. of Hawaii, p. 379.
Rieke, G. H., Elston, R. J., Lebofsky, M. J., and Walker, C. E.: 1987, in C. G. Wynn-Williams and E. E. Becklin (eds.), 'Germanium Diodes as High Performance Near Infrared Detectors', *Infrared Astronomy with Arrays*, Univ. of Hawaii, p. 69.
Rieke, M. J., Rieke, G. H., and Montgomery, E. F.: 1987, in C. G. Wynn-Williams and E. E. Becklin (eds.), 'Rockwell HgCdTe Arrays as Imagers', *Infrared Astronomy with Arrays*, Univ. of Hawaii, p. 213.
Wynn-William, C. G. and Becklin, E. E.: 1987, *Infrared Astronomy with Arrays. Proc. of the Workshop on Ground-Based Astronomical Observations with Infrared Array Detectors*, Univ. of Hawaii.

MAIHARA – To achieve ~ 1000 times of improvement in sensitivity at very high resolution, it would be necessary to use a considerable long time of individual integration. How long of integration time is assumed in your calculations?

TOKUNAGA – I assumed a 600 s maximum integration before reading out the array. In practice, this will be determined by the time-scale for significant sky variations. I am, therefore, estimating that this will be about 600 s. If you can integrate longer before reading out the array, then the sensitivity will be higher.

LEE – In your computations for detection limit did you include the flat fielding time?

TOKUNAGA – In the case of the 12 hr integration experiment, the flat-fielding was achieved by displacing the array slightly between exposures and applying a median filter through all of the images to get the flat-field.

ADAPTIVE OPTICS FOR LARGE TELESCOPES*

JACQUES M. BECKERS and FRITZ MERKLE

European Southern Observatory, Garching, F.R.G.

(Received 26 December, 1988)

Abstract. The performance of large telescopes is determined both by their angular resolution and by their collection area. It is, therefore, important to achieve as high an angular resolution as possible by site selection, by avoiding image deterioration by the telescope and its environment, and by real time image restoration by adaptive optics. We summarize the principles of adaptive optics, their predicted performance and the current programs underway to implement adaptive optics for astronomical purposes.

1. Introduction

The increased collecting area of very large telescopes results in an increase in sensitivity over smaller telescopes. Their increased (diffraction limited) angular resolution results, however, in a unique new capability of the telescopes which cannot be obtained in other ways with the smaller telescopes. This angular resolution can be achieved by eliminating seeing effects either by image recovery techniques using post-detection analysis of speckle observations (see, e.g., Alloin and Mariotti, 1988; Chelli, 1989; Baba, 1989) or by image reconstruction before detection using adaptive optics. The latter method is much to be preferred when possible. Its astronomical benefits lie, in addition to direct high-resolution imaging of astronomical objects from planets to distant galaxies, in improved spectral resolution because of the narrower spectrograph slits that can be used, in improved detection of point sources against the sky background, and in the vastly improved conditions for interferometric imaging with telescope arrays.

A number of papers have already been written reviewing the principles and properties of astronomical adaptive optics (e.g., Beckers, 1988a; Beckers and Goad, 1988; Merkle, 1988a, b, c; Hardy, 1982). We refer to those papers for detailed description. This paper summarizes the principles and expected performance of adaptive optics and describes the programs for their astronomical implementation currently underway.

2. Principles of Adaptive Optics

The principle of adaptive optics (see Figure 1) is very similar to that of the active optics described elsewhere in this symposium (Iye, 1989). The main difference is the speed of the system which is high for the adaptive optics (up to ≈ 250 Hz), since it corrects for rapidly varying wavefront disturbances due to atmospheric seeing, and low for active optics (slower than ≈ 1 Hz) since it corrects for the slow changes in the wavefront due to the gravitational and thermal changes of the telescope optics. Both types of systems

---

* Paper presented at the Symposium on the JNLT and Related Engineering Developments, Tokyo, November 29–December 2, 1988.

Astrophysics and Space Science **160**: 345–351, 1989.

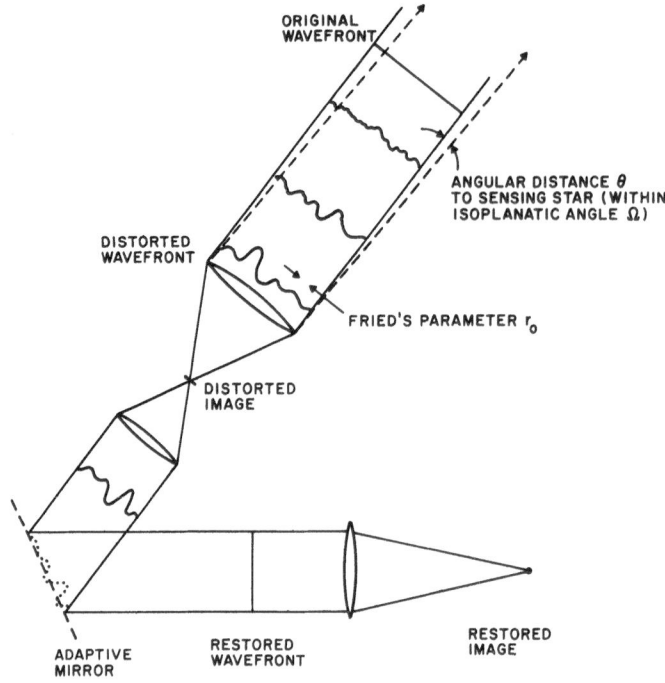

Fig. 1.   Principle of adaptive optics.

use similar wavefront sensors (mostly Hartmann–Shack sensors used at visible wave-lengths) to sense the wavefront tilts on stars or on small (< few arc sec) extended objects within the usable field of view of the telescope. In the case of adaptive optics this field of view is limited to the so-called isoplanatic patch (= the area on the sky within which the atmospheric wavefront disturbances are approximately the same). In adaptive optics the wavefront correction is made with a small, agile optical component (generally a deformable mirror) placed at an image of the pupil or of an average atmospheric seeing layer to achieve the speed required.

Both the wavefront sensor and the adaptive mirror need to have sufficient spatial and temporal resolution to resolve the significant wavefront spatial and temporal variations. The former are of the magnitude of the Fried's parameter. This parameter increases with wavelength from 10 cm in the visible for 1 arc sec seeing to 60 cm in the K band (2.2 µm) and 380 cm at 10 µm. The durations of the temporal variations also increase propor-tional to the Fried's parameter so that the number of photons available for the wavefront sensing increases as the cube of this parameter. One can use, therefore, fainter and fainter stars for wavefront sensing when going to longer and longer wavelengths. In addition to the larger number of stars which become available, the diameter of the isoplanatic patch also increases leading to the rapid increase in sky coverage for astronomical adaptive optics shown in Figure 2.

In addition to the wavefront sensor and the adaptive mirror, an adaptive optics system includes a digital controlled servo system, which couples the wavefront error

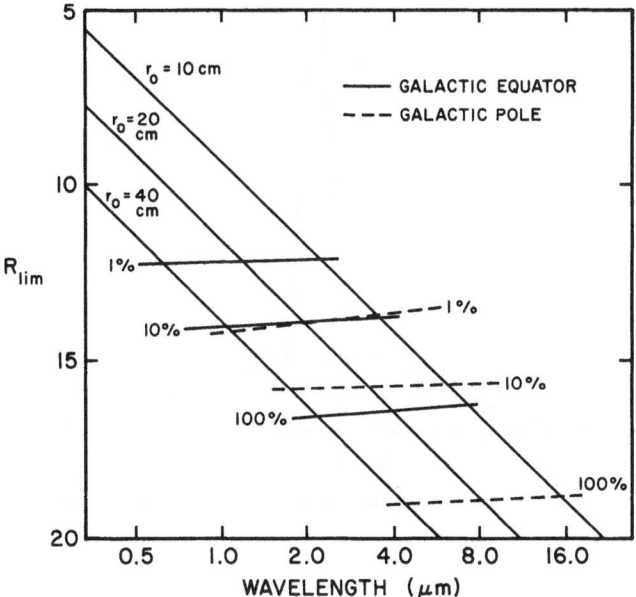

Fig. 2. Sky coverage by adaptive optics for different values of the Fried parameter.

signals to the mirror actuators using sophisticated control algorithms, and an optical system which couples the adaptive optics to the telescope and the astronomical instrument.

3. Expected Performance of Adaptive Optics

A perfect adaptive optics system will fully restore the wavefront to give an image for a point source equal to the Airy disk. A number of imperfections will cause this to be not the case. Among these are: (a) limited spatial and temporal resolution of the wavefront sensor and adaptive mirror, (b) time lag between the wavefront measurement and the wavefront correction, (c) deviation from perfect isoplanatism, (d) failure to correct for amplitude variations (scintillation), (e) chromatic effects related to the wavefront sensing at visible wavelengths while observing in the IR, (f) noise in the wavefront sensing, and (g) imperfections in the wavefront control algorithms.

Roddier and Roddier (1986), Gaffard and Boyer (1987), and Smithson and Peri (1987) have determined the point-spread function resulting for most of these imperfections. It is schematically shown in Figure 3. It closely approximates a combination of the original seeing disk with the Airy disk of a fully corrected telescope. The fraction of the energy (Σ) in the Airy spike is a good measure of the performance of the adaptive optics system. It is approximately equal to the so-called Strehl ratio S. For a perfect system $\Sigma = 1.0$, for a system with a spatial resolution near the Fried parameter $\Sigma \approx 0.8$ when a wavefront sensing object is used at the center of the isoplanatic patch, and $\Sigma \approx 0.4$ when the object is at the edge of this patch.

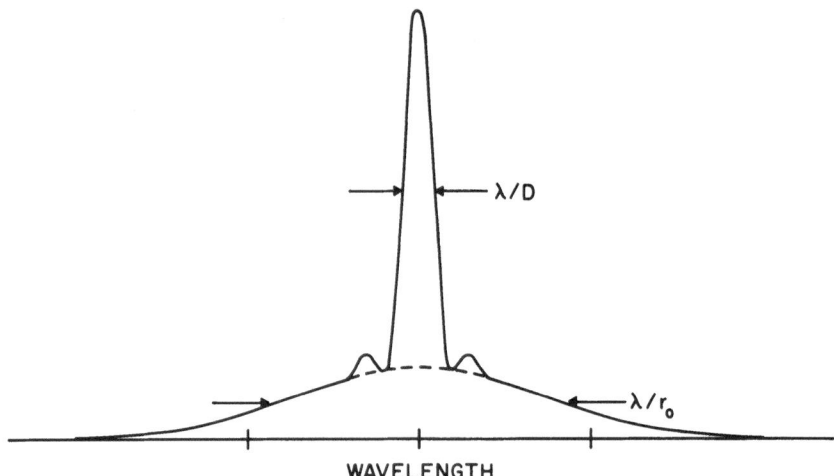

WAVELENGTH

Fig. 3.   Point spread function for imperfect adaptive optics.

Since the seeing changes with time one expects the value of Σ to also change. Absolute photometry using the Airy spike will, therefore, be difficult. Relative photometry over a small area at the center of the isoplanatic patch should be possible. The main use of adaptive optics will, therefore, be in studies of morphology, high-resolution spectroscopy, relative photometry, and interferometry (using the coherent radiation in the Airy spike).

4. Current Programs in Astronomical Adaptive Optics

Table I summarizes the programs which are now in progress which attempt to implement adaptive optics on astronomical telescopes. In other types of optical systems (military related) adaptive optics systems appear to have been successfully used for some time giving one confidence that this technology will become soon a component of astronomical telescopes.

The first three of these programs are being interfaced with their telescopes now. We fully expect adaptive optics to become a common feature of large telescopes in the 1990s.

5. Use of Partial Adaptive Optics

An adaptive optics system designed to work at the 2.2 μm K-band at an 8 m diameter telescope with 0.75″ seeing at visible wavelengths (Fried parameter = 13.5 cm at 0.5 μm and 80 cm at 2.2 μm) will contain about 100 adaptive elements. When using an on-axis object for wavefront sensing it results in $\Sigma \approx 80\%$ at 2.2 μm. At shorter wavelengths Σ will decrease to about 50% at 1.2 μm, 20% at 0.9 μm, and 9% at 0.7 μm because of the decrease in Fried's parameter. However, the width of the Airy spike (Figure 3) decreases and the width of the seeing disk slightly increases with decreasing wavelength

TABLE I

Current programs in adaptive optics

Location	Principal investigator	Number of elements	Proposed use (telescope diam.)	Ref.
ESO/Meudon/ ONERA/CGE	Kern/Merkle/ Fontanella/Gaffard	19	IR stellar at La Silla (3.6 m)	a
NOAO–KPNO	Goad	37/55	IR stellar (3.8 m) and IR solar (1.5 m) at Kitt Peak	b
Lockheed–PARL	Smithson	19/37	Visible solar	c
NOAO–SPO	Dunn	?	Visible solar Sac. Peak (0.76 m)	
Center for Astrophysics	Nissenson	?	Visible speckle interferometry	
Chengdu PR China	Jiang Wenhan	?	Visible solar (LEST, 2.4 m)	e

References:
(a) Merkle (1988a, b).
(b) Beckers et al. (1986).
(c) Smithson et al. (1987).
(d) Dunn (1987).
(e) Jiang Wenhan et al. (1987).

resulting only in a small decrease of its contrast against the background seeing halo (from ≈ 400 at 2.2 μm to ≈ 390, 220, and 160 at 1.2, 0.9, and 0.7 μm, respectively). For a number of astronomical programs such a partial adaptive optics system would, therefore, be quite powerful. The advantage of such a partial system lies of course in the small, realizable number of actuators. A full adaptive optics system for example at 0.7 μm would require 1600 or more actuators. In addition, brighter stars would be needed for the wavefront sensing so that its sky coverage would be much less.

A partial adaptive optics system would be especially beneficial for interferometric imaging with telescope arrays. At the ESO VLT 8 m telescopes, for example, it results at 0.7 μm in the light gathering power corresponding to that of a 2.4 m telescope (equal to the HST) but with the coherent radiation inside an Airy disk corresponding to that of an 8 m telescope with a width of about 0.02″ (vs $\approx 0.1″$ for the HST).

6. Extending the Sky Coverage

Foy and Labeyrie (1985) proposed the use of artificial stars for wavefront sensing. These stars would be created by illuminating the neutral sodium layer at ≈ 100 km elevation with a laser tuned to one of the sodium D lines. The incoherent back-scattered radiation would create an extended small (1 to 2 arc sec) object in the center of the field of view which could be used for wavefront sensing at any position in the sky. Thompson and

Gardner (1987) experimented with this technique and showed that a star with $V = 14$ was feasible. Kibblewhite (1988) estimate that current laser technology would be capable of creating at best a $V = 10$ star. From Figure 2 it can be seen that this would increase the sky coverage substantially (e.g., from $< 0.1\%$ at 0.75 µm to 100%). The trick in getting bright artificial stars is to make the laser duty cycle (pulse width × number of pulses s^{-1}) as large as possible since only increasing the energy does not increase the artificial star brightness significantly due to a saturation of the upper atomic level population by stimulated emission in the sodium layer.

7. Extending the Isoplanatic Patch

Beckers (1988b) suggested a way to increase the size of the isoplanatic patch using a technique called 'Multiconjugate Adaptive Optics'. It achieves this by imaging different layers of the atmosphere onto different adaptive mirrors. The array of adaptive mirrors, therefore, corrects the atmospheric wavefront distortions in detail, at each height, resulting in the increase of the area on the sky that is corrected. In order to do this it is necessary to measure the wavefront distortion as a function of height in the atmosphere. That is accomplished by a technique called 'Atmospheric Tomography' which measures the wavefront distortion using an array of articial stars. No numerical simulation or experimentation of this technique have as yet been done. Preliminary analysis indicate an improvement of the isoplanatic patch diameter of about two times the number of layers used, the amount of improvement being very dependent on the atmospheric optics structure and on the heights being imaged.

8. Conclusion

Adaptive optics coupled to large 8 m class telescope promises to lead to major changes in ground-based optical astronomy. The incorporation of the present first steps in adaptive optics on astronomical telescopes will already bring major advances in our capabilities to do astronomy. From thereon one might foresee an ongoing program of improvements using more actuators, artificial stars, multi-conjugate systems, and other refinements still unthought of. Ground-based telescopes will as a result compete favourably with space-based systems but at a fraction of the cost (however, only at wavelengths transmitted by the atmosphere). In the next few years the completion of the first adaptive optics systems will be at least as exciting to astronomical research as will be the completion of large telescopes. The combination of both will move astronomy forward with a major enhancement of capabilities.

References

Alloin, D. and Mariotti, J. M.: 1988, *Cargese Summer School on Diffraction Limited Imaging with Large Telescopes* (in press).
Baba, N.: 1989, *Astrophys. Space Sci.* **160**, 373 (this issue).
Beckers, J. M.: 1988a, in R. G. Kron and A. Renzini (eds.), *Fifth Erice Workshop on Towards Understanding Galaxies at Large Redshifts*, Kluwer Academic Publ., Dordrecht, Holland, p. 319. .

Beckers, J. M.: 1988b, in M. H. Ulrich (ed.), *ESO Symposium on Large Telescopes and Their Instrumentation* (in press).

Beckers, J. M. and Goad, L.: 1988, in L. B. Robinson (ed.), *Ninth Santa Cruz Summer Workshop on Instrumentation for Ground-Based Optical Astronomy*, Springer-Verlag, Berlin, p. 315.

Beckers, J. M., Roddier, F. J., Eisenhardt, P. R., Goad, L. E., and Shu, K. L.: 1986, *Proc. SPIE* **628**, 290.

Chelli, A.: 1989, *Astrophys. Space Sci.* **160**, 369 (this issue).

Dunn, R. B.: 1987, in F. Merkle, O. Engvold, and R. Falomo (eds.), *Adaptive Optics in Solar Observations*, LEST Technical Report No. 28, p. 243.

Foy, R. and Labeyrie, A.: 1985, *Astron. Astrophys.* **152**, L29.

Gaffard, J. P. and Boyer, Corinne: 1987, *Appl. Optics* **26**, 3772.

Hardy, J. H.: 1982, *Proc. SPIE* **332**, 252.

Iye, M.: 1989, *Astrophys. Space Sci.* **160**, 149 (this issue).

Jiang Wenhan, Lui Yueai, Shi Fang, and Tang Guomao: 1987, in F. Merkle, O. Engvold, and R. Falomo (eds.), *Adaptive Optics in Solar Observations*, LEST Technical Report No. 28, p. 137.

Kibblewhite, E.: 1988, (priv. comm.).

Merkle, F.: 1988a, in L. B. Robinson (ed.), *Ninth Santa Cruz Summer Workshop on Instrumentation for Ground-Based Optical Astronomy*, Springer-Verlag, Berlin, p. 366.

Merkle, F.: 1988b, in M. H. Ulrich (ed.), *ESO Symposium on Large Telescopes and their Instrumentation* (in press).

Merkle, F.: 1988c, *J. Opt. Soc. Am.* **A5**, 904.

Roddier, F. and Roddier, C.: 1986, *Proc. SPIE* **628**, 298.

Smithson, R. C. and Peri, M. L.: 1987, in F. Merkle, O. Engvold, and R. Falomo (eds.), *Adaptive Optics in Solar Observations*, LEST Technical Report No. 28, p. 193.

Thompson, L. A. and Gardner, C. S.: 1987, *Nature* **328**, 229.

KODAIRA – I suppose that the information for wavefront correction must come from a bright enough object within the same isoplanatic angles as the target object itself is. How large do you expect to be such a field of view?

BECKERS – That depends on the wavelength used for the observation, on the seeing and on the height of the seeing layer at visible wavelength. The isoplanatic angle equals about 5 arc sec. In the *K* band it as large as 30 arc sec.

LABEYRIE – If adaptive optics are achieved in *coherent* arrays, using a reference star in the isoplanatic field, the limiting magnitude improves, owing to the narrower lobe observed against the sky background. Instead, fringe observations with 'seeing' have a lower magnitude limit than incoherent imaging.

BECKERS – I agree.

INTERFEROMETRY

CAN THE OPTICAL VERY LARGE ARRAY BE COUPLED WITH THE JNLT?*

ISABELLE BOSC, DENIS MOURARD, and ANTOINE LABEYRIE

CERGA/OCA, Saint Vallier de Thiey, France

(Received 16 February, 1989)

Abstract. Coupling the JNLT with the Keck telescope is of considerable interest. Further enhancement may be possible with auxiliary small telescopes, as planned for ESO's VLT. Current plans for installing the optical very large array at Mauna Kea provide opportunities for extra OVLA telescopes near the JNLT.

A coudé field slicer is proposed for interferometric observing of a reference star together with the main object. Additions to the JNLT coudé spectrograph are also suggested for use as a speckle camera with multiple spectral channels.

1. Introduction

As long-baseline optical interferometers begin to operate, the insights gained into more intimate details of stellar or galactic structures will help understanding these and other exotic objects. Photometrists will become involved in photometry with milli arc-sec spatial resolution. Spectroscopists will be happy to utilize the extra spatial information in resolved star images or maps of active galactic nuclei showing details of the broad-line emissions regions. Experts in galactic morphology will assemble wide-field images from narrow-field ones obtained by aperture synthesis techniques.

Most astronomers will realize that the new array instruments are also suitable for programs of the traditional kind, requiring only the atmosphere-limited resolution of about 0.5″. Valuable conventional spectroscopic programs will probably be pursued on arrays when the interferometric mode is hampered by fast winds. Indeed, equal areas of collecting mirrors can be utilized with the same efficiency if monolithic or segmented, on a single or on many mounts.

Future large telescopes should, therefore, be designed as components of long-baseline arrays optimized for high-resolution imaging. However, at ESO, the weight of tradition made it difficult to really optimize the VLT for its new uses: the large telescopes should perhaps be mobile, an option studied by O. Citterio. D. Enard mentioned that telescopes Nos. 2, 3, and 4 could possibly be made mobile after the construction of No. 1.

2. Coupling the JNLT with an Optical Very Large Array

The JNLT project presented at this meeting features some interesting possibilities of connections with surrounding large instruments, as discussed in the presentation of

---

* Paper presented at the Symposium on the JNLT and Related Engineering Developments, Tokyo, November 29–December 2, 1988.

Astrophysics and Space Science **160**: 355–360, 1989.
© 1989 *Kluwer Academic Publishers.*

S. Isobe. This is likely to boost its performance in an age of increasing use of array systems.

A few kilometers away from the JNLT site is a sub-summit plateau where the Optical Very Large Array (OVLA) may be installed in 1995. As described in Labeyrie *et al.* (1988), OVLA is dedicated to efficient high-resolution imaging, with versatile baseline configurations. The initial OVLA will have perhaps 27 small apertures of 1.5 to 2 m, but larger components are likely to be selected for second-generation OVLAs.

If suitable sites are found, instruments such as the VLT or the JNLT can possibly be coupled to an OVLA: one or several large sub-apertures among the array of smaller elements can contribute to the high-resolution imaging performance, and reciprocally the array can enhance the performance of a large telescope. This is the philosophy behind VLT's auxiliary telescopes, which ESO is implementing in collaboration with European groups interested in high-resolution observing.

Our group has proposed that the auxiliary formation of small telescopes associated with the VLT will be in fact a second OVLA system, possibly having fewer elements than the main OVLA which is expected to be built on Mauna Kea's sub-summit plateau in 1995. Although initially small, this auxiliary OVLA could later grow as needed.

Similar possibilities may be worth considering for the JNLT. If the site allows, an OVLA-type array may be installed near the 7.5 m telescope. Instead of building a dedicated array near the JNLT dome, it would obviously be tempting to couple the JNLT, the Keck, and other local telescopes with the main OVLA system also installed on Mauna Kea. The distance between the flat sub-summit site considered for OVLA and the large telescopes would, however, require long delay-line elements. Extremely high angular resolutions would be achieved, of the order of 20 micro arc sec, but there is no experience yet in exploiting such long optical baselines. At CERGA, 70 m baselines are currently exploited. The longest baseline currently considered is 600 m, although D. Dravins proposed an interferometric link of approximately 80 km between Las Campanas and La Silla in Chile, using Brown and Twiss's intensity interferometry method.

Therefore, practical choices are the following: (1) find a location on Mauna Kea which can accommodate the JNLT together with a platform sufficiently large (500 m) for the main OVLA system; or (2) create a smaller platform (50–100 m) near the JNLT and equip it with a 'modest' OVLA having fewer telescopes.

In the second option, the main OVLA system would be built a few kilometers away, as initially considered, and some possibility of future connection through fiber optics may be envisaged, although space interferometers may then provide better solutions.

3. The Optical Very Large Array: Parts of a Prototype Telescope

The OVLA project described in Labeyrie *et al.* (1988) has received some initial funding through the Association of Laboratories for Optical High-Resolution Astronomy (ALOHA), from a donation by Instruments SA.

This allowed the construction of elements for a prototype telescope. A fiberglass/

epoxy sphere produced in a mold by the Stralpes corporation has been delivered to Haute Provence Observatory, where a prototype drive system is to be built and experimented in the coming months. The sphere weights 275 kg and measures 2.8 m in diameter. No final plans are yet made for procuring the 1.52 m mirror at $f/1.75$ which will fit in it. Replicated or polished mirrors are considered.

The mobility of the telescope will possibly involve 6 legs, as shown in Figure 1. The motion will imitate that of an insect. It will either use bare ground, or an array of ball fixtures affixed to the ground. A different mechanism of telescope translation, using crossed slides is also studied by D. Plathner at IRAM.

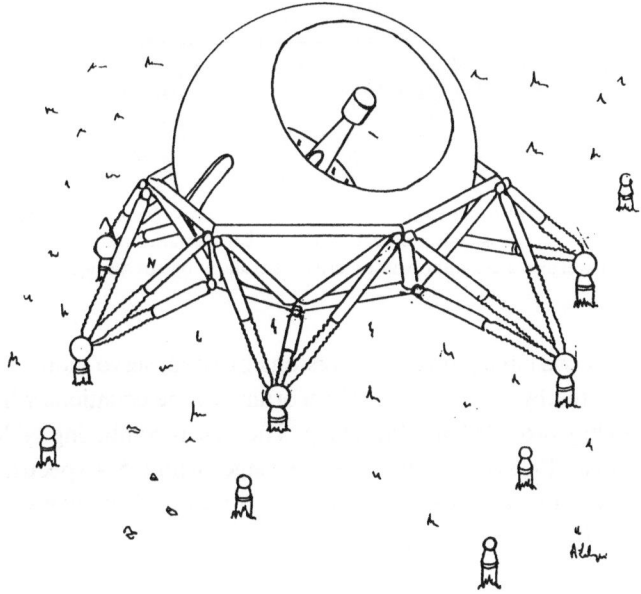

Fig. 1.   The walking telescope has 6 legs. This design option minimizes the platform structure investment. It is shown here with an array of balls as supports, but bare unprepared ground can also be utilized. During the motion, positions are controlled by 3 laser beams with accuracies of one or two microns.

The 'walking telescope' concept is also relevant to the moon-based interferometer project discussed at a forthcoming workshop in Albuquerque. It would be costly to install a flat platform or piers on the Moon. The laser metrology system of OVLA is expected to allow a very accurate linear motion on rough terrain for telescopes equipped with legs. The completed prototype telescope is expected to join the CERGA interferometer for operation with 3 apertures.

The laser metrology concept developped for OVLA (Labeyrie *et al.*, 1988) also undergoes preliminary testing at the CERGA GI2T interferometer. If successful, it should allow an active stabilisation of the interferometer geometry at the scale of a few microns, thus allowing the observation of faint objects in the range $m_v = 12$ to 18.

In recent months, the large interferometer GI2T at CERGA has started its observing program. Because the initial drive system, utilizing hydraulic pistons, had caused

Fig. 2.   Part of a prototype OVLA telescope, delivered to Haute Provence Observatory where the drive is to be built and tested. The fiberglass and epoxy sphere has a 2.8 m diameter and will contain a 1.52 m mirror. It weights 275 kg and will be driven angularly by special wheels.

vibration problems, we have replaced the pistons by small servo motors. The spherical drive software written by one of us (DM) according to the equations which he derived (Mourard, 1988) has provided good tracking. The new recombining table also built by one of us (IB) (Bosc, 1988) gave multi-speckle fringes with 1.5 Å spectral resolution on γ Cassiopeiae and β Persei (Algol). 500 000 exposure of 20 ms made with the CP40 camera (Blazit, 1987), are currently being analyzed.

4. Coudé Slicing Optics for Reference Stars

It appears feasible to phase several telescopes at infra-red wavelengths when a reference star is available. Such adaptive optical imaging can improve dramatically the limiting magnitudes if it becomes achievable with telescope arrays.

As discussed by P. Léna, infra-red reference stars are expected to be frequently available in fields of about 10 min which match somewhat the isoplanatic field in the infra-red beyond 2.2μ. Observing both stars simultaneously tends to be difficult if normal coudé systems serve to recombine light from several telescopes. The coudé optics cannot easily deal with the large fields required. Figure 3 shows a possible solution, in the form of a field-slicer optical system.

In the primary focal field, a field mirror FM has a central hole through which light from the observed star reaches the Gregorian mirror G, which collimates the beam towards the coudé focus. Light from the peripheral field is reflected laterally from FM towards spherical mirror M, where the pupil is imaged. M is tilted under computer control to bring the reference star image in the hole of FM, where a mirror facet reflects

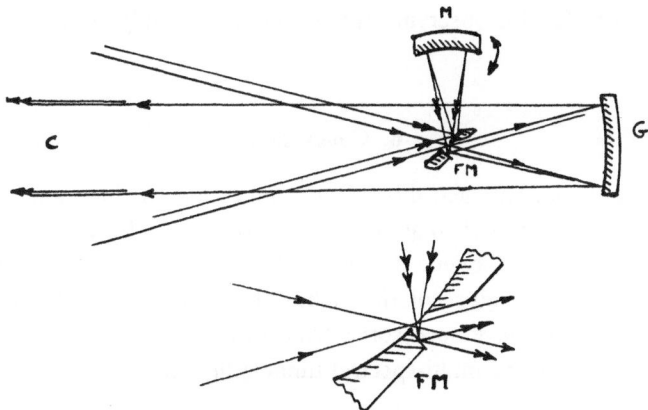

Fig. 3. Field slicer for observing simultaneously at coudé a reference star together with the object of interest. It can allow array phasing in the infra-red, for higher limiting magnitudes.

it towards G. Two stars in a 10′ field may thus be brought together, within a few arc sec, for compatibility with the severe field limitations of long collimated coudé trains.

5. A Speckle Mode for the JNLT Coudé Spectrograph

It is of appreciable interest to combine high spatial and spectral resolution. Differential speckle interferometry can probably reach beyond one milli arc sec resolution on certain objects, if utilized at a 7.5 m telescope (Beckers, 1982; Petrov, 1988).

The JNLT coudé spectrograph, according to the optical scheme presented at this meeting, can probably be equipped with options providing speckle interferometric capabilities with a spectral resolution of the order of 0.1 Å. The principle is shown in

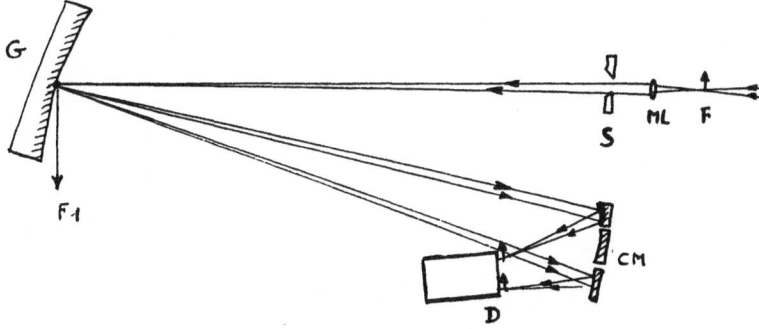

Fig. 4. Use of the JNLT coudé spectrograph for speckle interferometry with multiple narrow bands. A microlens ML in front of the entrance slit S focuses the pupil inside the slit aperture, assumed to be wide open, and the image on the grating G, with enough magnification to cover the grating. In the spectrum plane, a group of mirrors CM (or lenses) collects certain spectral bands and focuses the corresponding images side by side on one or several photon-counting cameras D. The wavelength and bandwidth of each image are adjusted by displacing the collecting mirror (or lens) along the spectrum and varying its width. The arrangement may thus be considered as a multi-band tunable filter. The minimum bandwidth is of the order of 0.1 Å.

Figure 4. Provisions for this observing mode should preferably be studied at the design stage.

6. Conclusion

Additional features can enhance the JNLT's future use for high angular resolution: connection with a sub-array of small movable telescopes such as a complete or partial OVLA, and a coudé train allowing the observation of reference stars. For best exploitation of the moderately high resolution achievable with a single 7.5 m aperture, the coudé spectrograph can have accessories transforming it into a speckle camera with high spectral resolution and multi-spectral tunable filtering.

References

Beckers, J.: 1982, *Opt. Acta* **29**, 361.
Blazit, A.: 1987, 'Comptage de photons bidimensionnel et applications astrophysiques', Thesis, Univ. de Nice.
Bosc, I.: 1988, in F. Merkle (ed.), *Conference Resolution Imaging by Interferometry,* Garching, F.R.G.
Labeyrie, A., Lemaitre, G., Thom, C., and Vakili, F.: 1988, in F. Merkle (ed.), *High Resolution Imaging by Interferometry,* Garching, F.R.G.
Mourard, D.: 1988, in F. Merkle (ed.), *High Resolution Imaging by Interferometry,* Garching, F.R.G.
Petrov, D.: 1988, in F. Merkle (ed.), *High Resolution Imaging by Interferometry,* Garching, F.R.G.

KODAIRA – When you consider such a large-scale array interferometer in the optical range, it is surely worth while to go to space. Please show us your slides of space interferometer.
LABEYRIE – If is of interest to go both in space and on the Earth for large interferometers.

OPTICAL INTERFEROMETER BETWEEN JNLT AND WMKT*

SYUZO ISOBE

National Astronomical Observatory, Tokyo, Japan

(Received 16 February, 1989)

Astronomical demands in making observations with high angular resolution are now very high in all wavelength ranges. At optical and infrared wavelengths, the VLT at ESO

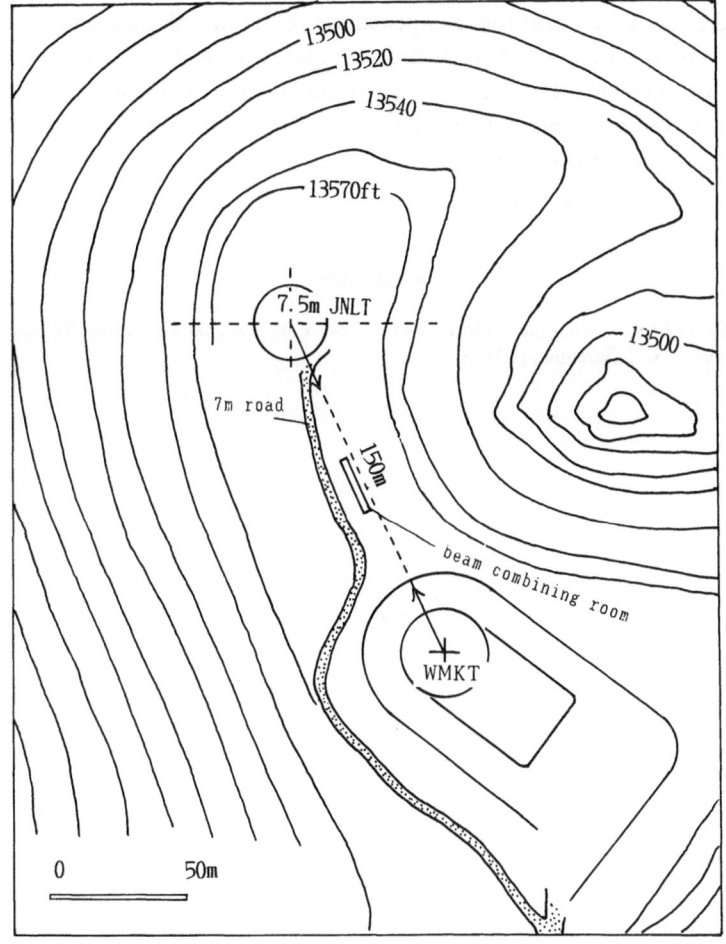

Fig. 1. Configuration of JNLT and WMKT.

* Paper presented at the Symposium on the JNLT and Related Engineering Developments, Tokyo, November 29–December 2, 1988.

will have an interferometric mode. The 10 m telescope (WMKT) and the planned 7.5 m Japanese National Large Telescope (JNLT) on Mauna Kea will have coudé foci. Therefore, we are making a proposal to build an interferometric facility between the JNLT and the WMKT (Isobe, 1988). The telescope will be set on a ridge having a shallow room to construct a tunnel for light beams from both the telescopes, as shown in Figure 1. Since a pathlength of light from both telescopes to a detector is over 150 m, one needs to keep variation of refractive index on lines-of-sight a minimum. We have a plan to put a vacuum tube with a diameter of 20 cm for light pass inside a tunnel with a height of 1.5 m for maintenance. To compensate the pathlength difference of the two beams by the diurnal motion of a celestial object, a fringe tracker will be needed. Since the length of baseline between both telescopes is about 130 m, the maximum pathlength difference is 10 m for a half hour observation. A building with a length of 10 m should be prepared under ground for this compensation of pathlength difference. Consideirng a situation of tight telescope time of both telescopes, we are also proposing to have two additional movable 2 m telescopes which give much large coverage of the $u - v$ plane. Although there remain some difficulties to construct a long baseline optical interfero- meter, we are solving them one by one with much help by engineering researchers at different institutions in Japan.

Reference

Isobe, S.: 1988, in M.-H. Ulrich (ed.), *Proceedings of ESO Conference on Very Large Telescopes and their Instrumentation*, ESO, Garching, p. 781.

THE INTERFEROMETRIC MODE OF THE EUROPEAN VERY LARGE TELESCOPE*

PIERRE LÉNA

Université Paris VII et Observatoire de Paris, France

and

FRITZ MERKLE

European Southern Observatory, Garching, F.R.G.

(Received 10 January, 1989)

Abstract. The European Very Large Telescope program has been approved in 1987. It aims to consists of an array of four 8 m telescopes, plus two additional 2 m class auxiliary telescopes, the latter being fully dedicated to optical (infrared and visible) interferometry, with possible combination of some and, in the long term, all large telescopes. We discuss the implementation of this program in the next ten years.

The concept of an interferometric mode, as one of the possible uses of the four 8 m telescope of the European Very Large Telescope, is a logical consequence of the choice of the array configuration, made for the general design of the program.

Although this concept appeared early in the plan (Léna, 1983), it was only proposed after a thorough analysis (ESO/VLT Working Group on Interferometry, 1986) and included in the final VLT Proposal (European Very Large Telescope Proposal, 1987). The current concept of the VLT interferometric mode may be found in Léna (1987) and Merkle (1988a) and is now the subject of detailed studies for its implementation.

Figure 1 shows the development of interferometric arrays around the world. Large telescopes form a well-identified class: although not primarily designed for interferometric purposes, they offer exceptional sensitivities due to their large collecting area.

A detailed analysis of interferometric beam combination (Roddier and Léna, 1984) shows that the sensitivity gain (limiting flux of limiting magnitude) can only be obtained when the individual pupils are fully phased. The Fried's parameter r_0 (λ) is strongly chromatic, from 10 cm in the visible (for $10''$ seeing), to over 7 m at 20 μm. Efficient operation of a large telescope interferometer requires, therefore, adaptive optics, the only method to phase individual pupils in the presence of atmospheric distorsions. Although this technique is not fully developed yet, phasing the VLT in the wavelength range 2–12 μm appears feasible with current technologies (Merkle, 1988b), and a strong development program is under way for this purpose (Kern *et al.*, 1988). At $\lambda \lesssim 2$ μm, phasing appears more difficult, the sensitivity is, therefore, not improved, but the use of large apertures provides a considerable gain in integration time.

---

* Paper presented at the Symposium on the JNLT and Related Engineering Developments, Tokyo, November 29–December 2, 1988.

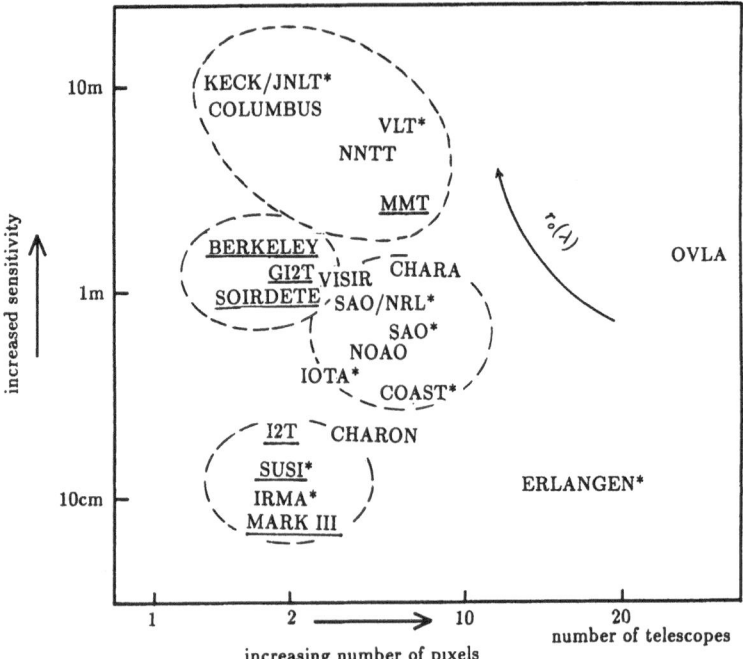

Fig. 1.   Existing (underlined), planned and funded (*) or planned optical interferometers in 1988. Abscissa: number N of telescopes in the array, ordinates: diameter D of individual telescopes. The underlined instruments have all produced visibility data, but are to date restricted to $N = 2$. Image quality improves with N, while sensitivity improves with D. Contours encircle families of instruments, either limited to $D = r_0$ (0.5 μm) at the bottom, or $D = r_0$ (infrared) at the center, or the large telescopes planned with some interferometric capabilities at the top. (From Léna, 1988, in Cargèse Summer School 'Diffraction-Limited-Resolution with Optical Telescopes'.)

The sensitivity achieved in the near-infrared (2–12 μm) with the VLT will be unprecedented. Flux levels of a few mJy should be detected in the short integration times imposed by the atmospheric turbulence, although later developments of fringe-tracking capabilities will allow to reach a higher sensitivity. The sensitivity gain has to be combined with adequate angular resolution, considering the size of the objects. The implementation of the VLT on the ground will be finalized only after the site selection (late 1990), but baselines for the linear array of 100–150 m are most likely. The resolution will span from 12–500 milli arc sec (single-dish operation from 0.5–20 μm) to 1–50 milli arc sec (interferometric operation of the full baseline). The combination of resolution and sensitivity appears particularly adequate for the study of star- and planet-forming regions within 2 kpc of the solar system, where objects of 0.1–10 AU in size, 200 to 300 K in brightness temperature, 10^{-2} to $1\ L_\odot$ emit most of their energy in the spectral range of the VLT interferometric operation.

   The particular configuration of a linear array leads to some restriction on imaging capabilities. The E–W alignment of the array, even combined with super-synthesis provided by the Earth's rotation, does not provide satisfactory $(u\ v)$ frequency coverage for low declination sources. This led to include in the VLT program two 2 m class additional telescopes, movable along and perpendicular to the main array line, and able to feed the common interferometric beam combination area. These auxiliary telescopes, fully dedicated to interferometry, provide a continuous operation of the interferometric mode with $N = 2$, and a flexible coupling with the 8 m telescopes with $N = 3$ up to $N = 6$. The final imaging capability, to be reached only after the completion of the VLT (beyond the year 2000) will then be excellent, as shown by computer simulation. The additional quality may be added by exploiting the short exposure times already mentioned, and moving the auxiliary telescopes at rapid speed, a mode called *hyper-synthesis* (Vivekanand *et al.*, 1988). The combination of uneven pupils in the infrared leads to an equivalent diameter being the geometric mean of the two, a situation identical to the one encountered at radio wavelength. It makes the 6 m + 2 m combination equivalent to a 4 m + 4 m telescope combination.

The flexibility provided by additional movable telescopes could be further exploited, either by adding extra telescopes on the existing stations and/or tracks, or, if site allows, by providing a fixed station at a kilometric distance from the main array, for higher resolution dedicated to unresolved objects. As a general remark, this concept of adding smaller interferometric telescopes to the large instruments under construction should be investigated further.

Preliminary studies are devoted to the movable telescope concept (Plathner, 1988) and fairly conclusive on its feasibility (Figure 2).

A long-range development program on interferometry is, therefore, undertaken since the approval of the VLT, in order to enable the implementation of the interferometric

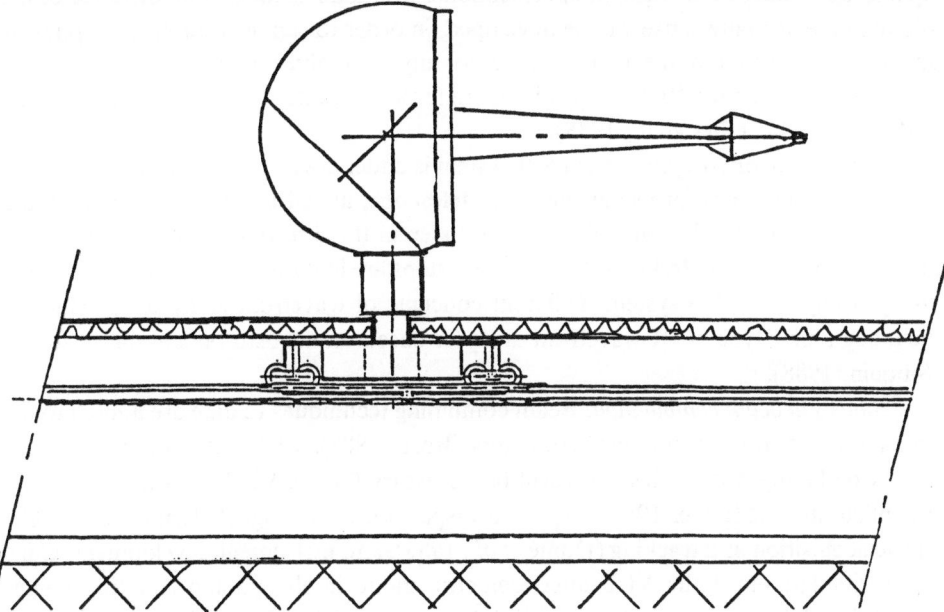

Fig. 2.   A preliminary study of 2 m-class auxiliary telescopes for the VLT (Plathner, 1988). The altazimuth telescope is continuously movable on an underground rail. Its very compact appearance limits its cross-section to wind.

mode (Figure 3). Although the final time-table is not fully determined yet, the major milestones are:

- first light on unit 8 m telescope staging from 1996 to 2001;
- final definition of interferometric mode in 1989;
- design and construction of auxiliary telescopes in 1990–1994;
- first interferometric operation in 1995–1996;
- progressive combination of large telescopes from 1997–1998 onward.

The main points of current developments are the following:

– *Studies on mechanical stability.* Interferometric requirements are indeed severe, since a constant optical path must be maintained between the entrance pupil of each telescope, i.e., the primary mirror, and the combined focus. The main sources of phase noise are indeed vibrations, excited by wind, drives or additional subsystems (fans, pumps, ...). It is not clear yet whether large inertia structures will in this respect be worse than small, current operated interferometric telescopes. The tolerable level of vibrations is set by the phase noise spectrum of the atmosphere itself, due to the beam propagation from the star to the telescope through the turbulent Earth's atmosphere. This phase noise, although site and seeing dependent, is known (ESO/VLT Working Group on Interferometry, 1980) and first experiments have already been carried (Bourlon and Léna, 1988) to examine the vibration behaviour of various telescopes at the level required for interferometric operation. The observed levels are not significantly above the required ones. Further studies are planned to feed back on the VLT design. It is already worth noting that interferometric operation at or above a wavelength of 2 µm significantly relaxes the required specifications. Methods of measuring distance of ca. 100 m in the absolute sense will be developed, in order to rapidly establish the phasing and the configuration of the interferometer for any given object in the sky. In this sense, the plan is to make the VLT an absolute interferometer, internally co-phased (Roddier and Léna, 1984; Léna *et al.*, 1988).

– *Studies in adaptive optics.* A current system is under development (Kern *et al.*, 1988) with $n = 19$ actuators for use at the 3.6 m telescope, aimed to give diffraction-limited images at 2.2 µm and beyond. This is a first step in the direction of equiping the VLT at these wavelengths (Merkle, 1986); the next steps are likely to include a $n = 64$ system and possibly a $n = 256$ system. Different concepts of wavefront analyser are studied, including a new type of IR-wavefront analyser based on acousto-optical phase sensor (Sinquin, 1988).

– *Studies in beam combination.* Beam combining techniques to date are limited to the operation of two-telescope interferometers (Bosc, 1988), where the task is relatively easy. Combining $N \geq 3$ telescopes will be necessary for the VLT. This includes pupil reconfiguration (Merkle, 1988a), spectral dispersion of the signal (Léna *et al.*, 1988), phase acquisition and tracking (Damé *et al.*, 1988), and instrument development. Computer simulations of the VLT interferometric mode in the photon-counting regime ($\lambda \leq 1.4$ µm) have been performed (Reinheimer and Weigelt, 1987) to evaluate the efficiency of speckle masking techniques in image reconstruction.

In conclusion, the VLT interferometric mode is a long-range program giving to this

Fig. 3. A view of the VLT. This mock-up shows at the forefront the interferometric tunnel where beam combination occurs. For colour reproduction of this figure see colour section.

array of telescopes an unprecedented capability of high-resolution imaging. Such performances, especially at infrared wavelengths, will not be surpassed by a space instrument for a long time, since efficient space operation in the infrared requires a significant telescope cooling.

References

Bosc, I.: 1988, in F. Merkle (ed.), 'The New Optical Table of GI2T', *High Resolution Imaging by Interferometry*, Garching, F.R.G.

Bourlon, P. and Léna, P.: 1988, in F. Merkle (ed.), 'Vibration Testing of Telescopes and Interferometers', *High Resolution Imaging by Interferometry*, Garching, F.R.G.

Damé, L. *et al.*: 1988, in F. Merkle (ed.), 'Active Stabilization in Stellar Interferometry', *High Resolution Imaging by Interferometry*, Garching, F.R.G.

ESO/VLT Working Group on Interferometry: 1986, 'Interferometric Imaging with the Very Large Telescope', *VLT Report 49*, European Southern Observatory, Garching, F.R.G.

European Very Large Telescope Proposal: 1987, *Interferometry*, Ch. 12, Garching, F.R.G.

Kern, P., Léna, P., Rousset, G., Fontanella, J. C., Merkle, F., and Gaffard, J. P.: 1988, 'Prototype of an Adaptive Optical System for Infrared Astronomy', in M.-H. Ulrich (ed.), *ESO Conference on Large Telescopes and their Instrumentation*, Garching, F.R.G.

Léna, P.: 1983, in J. P. Swings and K. Kjar (eds.), 'Aperture Synthesis in the Infrared: Prospects for a VLT', in *ESO's Very Large Telescope, Cargèse Workshop*, Garching, F.R.G.

Léna, P.: 1987, *Messenger* **50**, 53.

Léna, P., Ridgway, S., and Mariotti, J. M.: 1988, in F. Merkle (ed.), 'Interferometric Beam Combination at Infrared Wavelengths', *High Resolution Imaging by Interferometry*, Garching, F.R.G.

Merkle, F.: 1986, in S. d'Odorico and J. P. Swings (eds.), 'Adaptive Optics for the VLT', *ESO's Very Large Telescope*, Garching, F.R.G.

Merkle, F.: 1988a, *J. Opt. Soc. Am.* **A5**, 904.

Merkle, F.: 1988b, 'Adaptive Optics Activities at ESO', in *ESO Conference on Large Telescopes and their Instrumentation*, Garching, F.R.G.

Plathner, D.: 1988, in F. Merkle (ed.), 'A New Mount for Mobile Telescopes in an Optical Interferometer', *High Resolution Imaging by Interferometry*, Garching, F.R.G.

Reinheimer, T. and Weigelt, G.: 1987, *Astron. Astrophys.* **176**, L17.

Roddier, F. and Léna, P.: 1984, *J. Optics* **15**, 171, 363.

Sinquin, J. C.: 1988, Thèse de Doctorat, Université Paris VII.

Vivekanand, M., Morris, D., and Downes, D.: 1988, in F. Merkle (ed.), 'Continuously Movable Telescopes in Optical Interferometry', *High Resolution Imaging by Interferometry*, Garching, F.R.G.

INFRARED INTERFEROMETRY WITH SINGLE TELESCOPES*

ALAIN CHELLI

Instituto de Astronomia, Mexico

(Received 9 January, 1989)

Abstract. We examine the merits of three basic interferometric techniques using two-dimensional infrared arrays: speckle interferopmetry, pupil plane interferometry, and speckle holography.

1. Detector Requirements and Limiting Magnitudes

To realize two-dimensional infrared interferometry, we need arrays with a fast read-out, a good linearity and pixel uniformity, a small umber of dead pixels (critical in speckle interferometry) and a dynamics of the order or larger than $10^{-7} e^-$ (important when the sky noise dominates). A 128×128 two-dimensional infrared detector is sufficient to perform image or pupil plane interferometry with a 7.5 m telescope. Adopting a quantum efficiency of 50%, a read-out noise of 100 e^- and a seeing of 1.3″ in the visible, it will allow full imaging at magnitudes $K = 9$, $L = 7$, and $M = 5$.

2. Speckle Interferometry

Speckle interferometry allows to estimate the object visibility by analysing speckled images. The experiment is easy to implement and the signal processing is simple. However, because of redundancy of the input and the output pupil, the method is sensitive to speckle noise and the signal-to-noise ratio per interferogram saturates to a constant vlaue on bright sources. It is also sensitive to aberrations (hence, to non-stationary local turbulence and seeing variations) and suffers serious calibration problems. As a consequence, it is difficult to estimate the visibility with a precision better than 5%. Theoretically, image and pupil plane interferometry achieve the same limiting magnitudes.

3. Pupil Plane Interferometry

Pupil plane interferometers allow to estimate the object visibility by measuring the modulus of the complex degree of coherence in the pupil plane. Because the output pupil is not redundant, the method is insensitive to speckle noise and the signal-to-noise ratio per interferogram does not saturate to a constant value on bright sources. It is also insensitive to aberrations and apparently does not suffer serious calibration problems. In principle, the visibility can be determined with a very high precision on bright sources.

* Paper presented at the Symposium on the JNLT and Related Engineering Developments, Tokyo, November 29–December 2, 1988.

A pupil plane interferometer operated in flat tint has the great advantages that the signal processing is simple and the detector constraints are relaxed (dead ixels can be tolerated). Various kinds of experimental systems can be considered, but two are of particular interest. The 180° folding interferometer: it is simple implement, byt the low frequencies (codified in to the main mirror hole) are lost. The rotation shearing interferometer: it is complicated to implement, but all the frequencies are recovered and the spatial resolution can be controlled.

4. The Phase Problem

Several techniques have been proposed to recover the phase of the object spectrum, the most employed being the Know and Thompson method and the bi-spectrum analysis. The Know and Thompson method is based on the estimation of the phase-spatial gradient. It is sensitive to speckle noise and to aberrations, and achieve the same performances in the image and in the pupil planes. The bi-spectrum analysis, applicable in the image plane, has been recognized as a generalization in optics of the well-known phase closure technique currently used in radioastronomy. It is also sensitive to speckle noise, but it is insensitive to aberrations in the high-frequency domain. It achieves slightly better performances than the Know and Thompson method on bright sources, but it seems to be the inverse on faint sources. However, the bi-spectrum analysis needs much more computing time to recover the phase than the techniques based on the spatial gradient. Other alternatives could consist to estimate directly the closure phase. This can be done in the image plane by using a non-redundant pupil mak or in the pupil plane by using at least two rotation shearing interferometers. The two approaches give high signal-to-noise ratios on bright sources. In the first case, all the frequencies are not explored but the experimental system is simple, in the second case, all the frequencies are explored, but the experimental system is very complicated.

5. Holographic Method

The holographic method consists to record speckled images simultaneously with the use of a wavefront sensor in order to perform speckle holography. The reference wavefront, which can be used for adaptive control, may be analysed in a large optical bandwidth on an artificial star for the purpose (under study), a visible star in the infrared isoplanetic field (30% of the sky is covered at 2.2 μm, 100% at 5 μm), or the object itself (if sufficiently powerful). The holographic method allows to estimate the object complex spectrum (visibility and phase). It is insensitive to speckle noise and provides high signal-to-noise rations on bright sources. It is also insensitive to aberrations and does not suffer calibration problems. The holographic method presents all the advantages of the previous techniques and must be employed whenever possible. It is ideal to detect diffuse extended structures around bright sources.

6. Conclusions

It is not an easy matter to determine the optical method to do high spatial resolution. In fact, the best strategy may depend on the object we want to study and on what we expect to find. In the opinion of the author, the fundamental goal to achieve is not only to obtain a high signal-to-noise ratio, but also and perhaps chiefly, to avoid systematic errors in order for the latter to have a meaning. A self-calibrated method has also the interest to allow an automatic signal processing which can be used by any non-specialist in the field of high spatial resolution. The holographic method has all these characteristics and must be a current system installed in all large telescopes. If it is not possible to use it, it is preferable to determine the visibility in the pupil plane and the phase in the image plane by the mean of the bi-spectrum analysis and eventually using a non-redundant pupil mask. Finally, for objects at the limit of sensitivity and if we want to determine very simple geometries like multiplicity, classical speckle systems being more luminous than pupil-plane interferometers, may be employed.

ISOBE – I believe a limiting magnitude of speckle interferometer is strongly dependent on image size. Since the JNLT site on Mauna Kea has high quality of image, the limiting magnitude should be much fainter than the value you showed.

CHELLI – The limiting magnitudes I have given are for a 1 arc sec seeing in the visible. With a two-dimensional infrared array, they improve as r_0^2 for a fixed integration time per interferogram.

SPECTRAL SPECKLE INTERFEROMETRY*

NAOSHI BABA

Hokkaido University, Sapporo, Japan

(Received 2 December, 1988)

Abstract. A method of obtaining an objective prism spectrum of a stellar object with diffraction-limited spatial resolution is described.

1. Introduction

It is possible to incorporate a spectroscopic technique into speckle interferometry. The speckle spectroscopy is a method to get an objective prism spectrum of a stellar object with diffraction-limited spatial resolution. Weigelt *et al.* (1981) have proposed a speckle spectroscopic method based on the speckle masking algorithm. We (Baba *et al.*, 1988) have proposed a wideband speckle spectroscopic method relying on the shift-and-add (SAA) (Bates and Cady, 1980).

Since we described the principle of our method in Baba *et al.* (1989) and Baba and Tabata (1988), here, we only outline our method and the experimental setup. In this paper the emphasis is placed on the data reduction procedure.

2. Method and Experimental Setup

We divide the incident light from a stellar object into two parts. One beam forms a wideband speckle image and the other beam is used to form a dispersed speckle image. Because these two images have the fixed relation with the position, the SAA operation on the wideband speckle image leads to the SAA objective prism spectra.

Figure 1 shows our experimental setup, where the grating G is used for the division of the incident beam and for the spectroscopic device. We make an artificial double star of fiber-fed light sources. One example of the image observed at P is shown in Figure 2. (See colour plate.)

3. Data Reduction

The vital aspect of the SAA applied to our method is to find the speckle constructively formed by the wideband light. To this end, we operate the spatial filter in the wideband speckle image and find the location of its maximum value. The 5×5 spatial filter employed in our reduction procedure is shown in Figure 3.

The form of this filter is determined after several iterations. The initial form of the filter

---

* Paper presented at the Symposium on the JNLT and Related Engineering Developments, Tokyo, November 29–December 2, 1988.

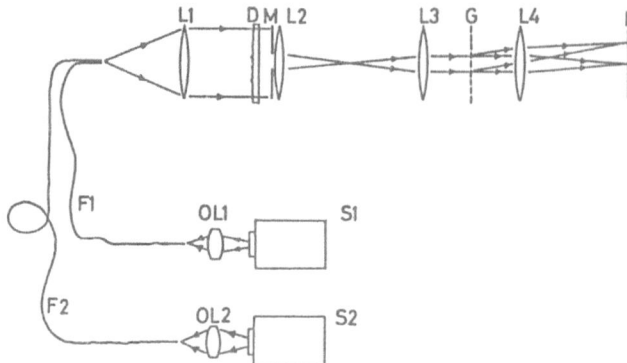

Fig. 1. Experimental setup, where D is a diffuser to simumate atmospheric turbulence and M is an aperture mask.

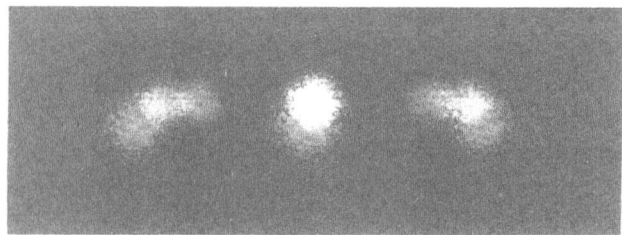

Fig. 2. Wideband speckle image and dispersed speckle image of two point sources (He–Ne laser and Hg lamp). For colour reproduction of this figure see colour section.

$$
\begin{array}{ccccc}
7 & 12 & 14 & 12 & 7 \\
12 & 17 & 18 & 17 & 12 \\
14 & 18 & 20 & 18 & 14 \\
12 & 17 & 18 & 17 & 12 \\
7 & 12 & 14 & 12 & 7
\end{array}
$$

Fig. 3. 5 × 5 spatial filter fo determine the location of the speckle with maximum output value.

is estimated from the image reconstructed by the simple SAA (centering the frame to the position of the maximum intensity without filtering operation).

When the object under observation has a spiky portion, such a filter is useful. However, if a binary star with similar intensities or a diffusely extended object is concerned, the suitable matched filter will be employed (Ribak et al., 1985).

Figure 4(a) shows the contour plot of the objective prism spectra based on the SAA operation. Forty-six speckle images are used in this reconstruction. As can be seen, the emission spectra of Hg source are not well resolved because of a strong background component. The contour plot in Figure 4(b) shows the ideal spectra, which are observed at P in Figure 1 without the diffuser D.

There may be various methods to estimate background components. Since we treat

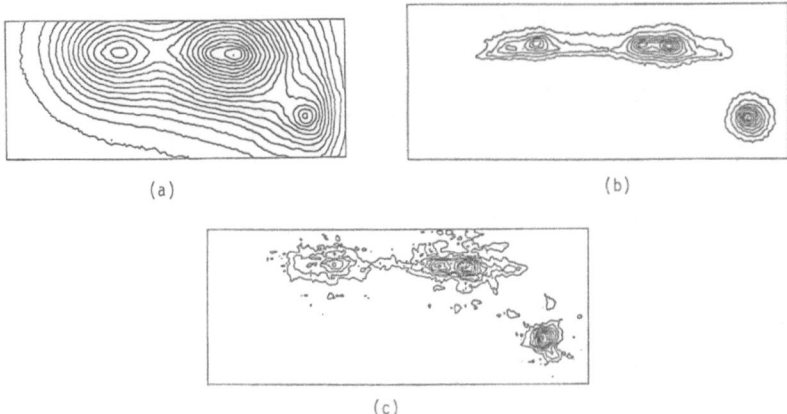

Fig. 4.   Objective prism spectra (a) by SAA, (b) without diffuser, (c) after background subtraction.

speckles holding spiky property, a median filter can be useful to reduce the spiky components and to estimate the background components. We operate 1-D median filter with window-size of 9 to each dispersed speckle image along vertical and horizontal directions. After median filtering, we further operate a 9×9 local averaging filter. A perspective view of a speckle image is plotted in Figure 5(a). After the smoothing operation described above, we can estimate its background component as shown in Figure 5(b). The subtraction of these images results the background-extracted image.

Fig. 5.   Speckle image and its estimated background.

When we use the background-extracted images at the stage of the addition operation in the SAA, we can reconstruct a bias-free image, as shown in Figure 4(c). For the case that there remains negative regions in the reconstructed image, we set zero to these regions.

4. Discussion

We are constructing a speckle spectrometer as shown in Figure 6(a). (See colour plate for Figures 6(a) and 6(b).) Though we used a Ronchi grating as a disperser in our previous experiments, a prism is employed in a new system. If we suitably choose transmittance to reflectance ratio of half mirrors, we can balance a wideband speckle image with a dispersed speckle image. Figure 6(b) shows one example observed in the

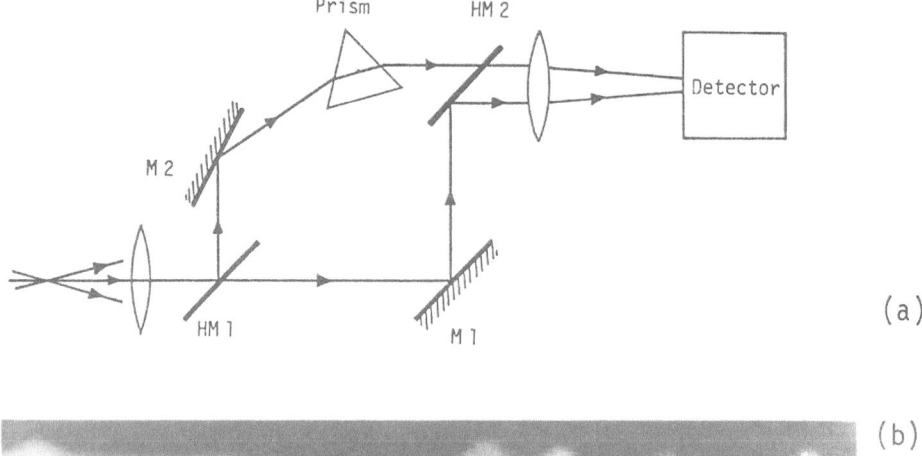

Fig. 6.   New version of speckle spectrometer and a sample image. For colour reproduction of this figure
see colour section.

new system. Since a wideband image is used to detect the brightest speckle, most of the
incident light can be directed to a spectroscopic device.

Our method will be useful to multiple-star systems. For general objects, the projection
speckle spectroscopy method by Grieger *et al.* (1988) will be applicable, because the
SAA works for projected speckle images.

References

Baba, N. and Tabata, M.: 1988, in F. Merkle (ed.), *High Resolution Imaging by Interferometry*, Garching.
Baba, N., Tabata, M., and Murata, K.: 1988, *Opt. Letters* **13**, 616.
Bates, R. H. T. and Cady, F. M.: 1980, *Opt. Commun.* **32**, 365.
Grieger, F., Fleischmann, F., and Weigelt, G.: 1988, in F. Merkle (ed.), *High Resolution Imaging by Inter-
ferometry*, Garching.
Ribak, E., Hege, E. K., and Christou, J. C.: 1985, *Proc. Soc. Photo-Opt. Instrum. Eng.* **556**, 196.
Weigelt, G.: 1981, *Proc. Scientific Importance of High Angular Resolution at IR and Optical Wavelength*, ESO
Conf.

SUBMILLIMETER INTERFEROMETRY*

M. ISHIGURO

Nobeyama Radio Observatory†, Nobeyama, Japan

(Received 28 February, 1989)

Abstract. The technical feasibility of submillimeter interferometry at Mauna Kea, Hawaii, by connecting existing and planned optical/IR telescopes as well as submillimeter telescopes is discussed.

1. Introduction

The subjects of fundamental scientific importance in submillimeter interferometry are star-formation, structure of galaxies, nuclei of active galaxies and quasars, etc. The emission from cool dust and molecules is stronger at submillimeter wavelengths than at millimeter and IR wavelengths. Due to higher receiver noise temperature and lower atmospheric transparency, a submillimeter interferometer should be located at high and dry site. In spite of these difficulties interferometry at submillimeter wavelengths is attractive because of the advantage in getting high resolution with relatively short telescope spacings.

The millimeter arrays at OVRO, Hat Creek and Nobeyama are in operation and the IRAM array will soon be in operation. Larger millimeter arrays have been proposed by NRAO and Nobeyama, independently. However, none of these arrays will not be able to work at short submillimeter wavelengths. The Smithsonian Astrophysical Observatory has already begun planning a submillimeter array consisting of six 6 m diameter telescopes operating in the wavelength range from 1.3 mm to 0.35 mm (Moran *et al.*, 1984; Ho, 1988, priv. comm.).

In this paper we investigated the technical feasibility of submillimeter interferometry by connecting optical/IR telescopes as well as submillimeter telescopes which are existing and planned at Mauna Kea, Hawaii, in relation to the JNLT project.

2. Heterodyne Interferometry

The instrument under consideration here uses a heterodyne interferometry technique because direct interferometry might be very difficult except for very short baselines. The heterodyne technique is also a necessary approach to obtain a frequency (velocity) resolution good enough for submillimeter wave spectroscopy.

Figure 1 shows a simplified block diagram of the heterodyne interferometer. At the mixer submillimeter wave signal received at each telescope is mixed with a local oscilla-

---

* Paper presented at the Symposium on the JNLT and Related Engineering Developments, Tokyo, November 19–December 2, 1988.
† Nobeyama Radio Observatory is a branch of the National Astronomical Observatory, the Ministry of Education, Science and Culture of Japan.

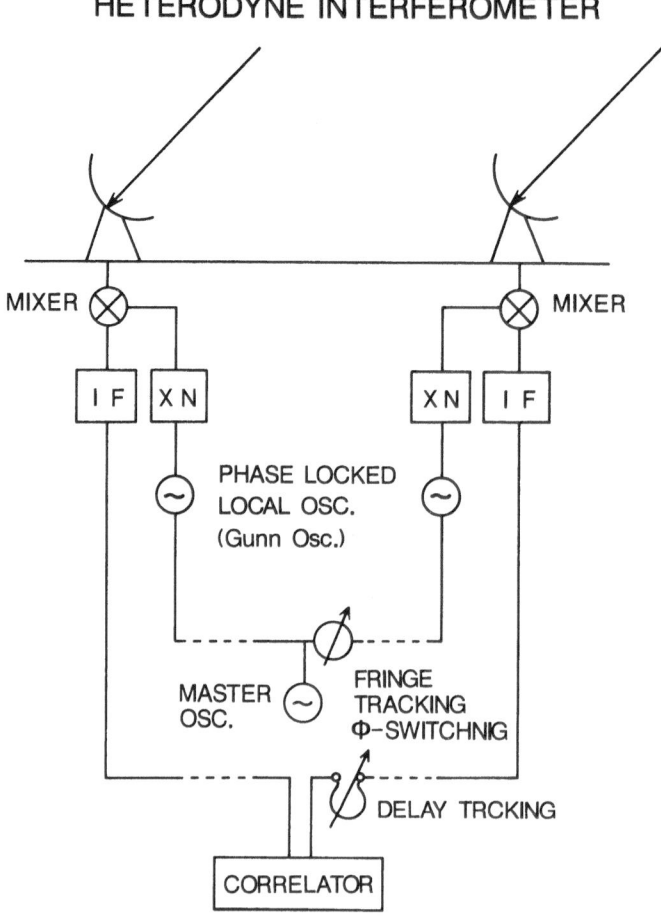

Fig. 1.   A simplified block diagram of the heterodyne interferometer.

tor (LO) signal to produce intermediate frequency (IF) signal. The mixers are usually cooled down to cryogenic temperature to obtain a low noise temperature. To preserve phase coherence between the received signals, the local oscillators are phase-locked to the common reference signal (∼ 1 GHz) distributed by under-ground cables. Fringe tracking and phase switching are usually performed through the LO signals. Delay tracking in IF is much easier than doing at submillimeter wavelengths. After delay compensation the IF signals are correlated.

3. Array Configuration

We will consider the array of four telescopes, JNLT 7.5 m, WMKT 10 m, JCMT 15 m, and CSO 10 m. Figure 2 shows a superficial view of the telescopes. The maximum and minimum baselines are 440 m (JCMT–WMKT) and 140 m (JNLT–WMKT, JCMT–CSO), respectively. The baseline is longer in the north to south direction than

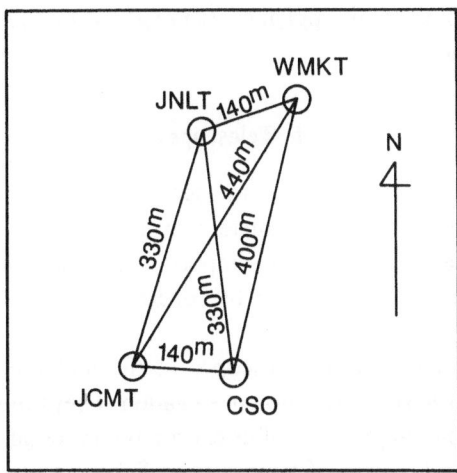

Fig. 2. A plan view of the telescopes at Mauna Kea which could be used as a submillimeter interferometer array.

in east to west direction, and can provide long N–S projected baselines for low declination sources. The array, however, is nearly a parallelogram and the $(u - v)$ coverage is necessarily redundant and lacks short baselines. The $(u-v)$ coverage is not favourable for mapping extended sources. If it is possible to locate a movable telescope close to one of these telescopes, the $(u - v)$ coverage will be greatly improved.

The field-of-view (primary beam) and the resolution of the (JNLT–WMKT) interferometer are summarized in Table I for various wavelengths. With 140 m baseline, 0.36″ resolution is obtained at $\lambda = 0.35$ mm. When telescopes of unequal size are used, the effective primary beam of the interferometer is given by the geometrical mean of the two different primary beams.

TABLE I

Field-of-view and resolution of the [JNLT–WMKT] interferometer

f (GHz)	λ (mm)	Field-of-view (″)	Resolution (″)
230	1.30	37	1.30
345	0.87	25	0.90
460	0.65	19	0.67
690	0.43	12	0.44
860	0.35	10	0.36

Atmospheric phase scintillation degrades the image quality of synthesized maps especially at short wavelengths. It can be expected to achieve 0.3″ radio seeing at Mauna Kea in very good weather condition (Bieging et al., 1984). Due to the atmospheric limitations, it will be better to do experiments at baselines shorter than 100 m at $\lambda = 0.35$ mm. Therefore, the baselines in Figure 2 probably be too long at short submillimeter wavelengths. If more than three telescopes are available, it will be possible to restore the degraded images by applying *closure-phase* or *self-calibration* techniques (Pearson and Readhead, 1984).

4. Telescopes

Optical and infrared telescopes can provide sufficient surface accuracy and pointing accuracy when they are used at submillimeter wavelengths. The mechanical stability of pathlength of each telescope is also an important problem in phase-coherent sub-millimeter interferometry. To guarantee the operation at $\lambda = 0.36$ mm, stability of 30 μm/30 min is required.

Change of pathlength for all six telescopes of the multiple mirror telescope was measured (Hege *et al.*, 1985) as a function of elevation. They found that the change was repeatable and was about 20 μm hr^{-1} after elevation correction. Pathlength difference between telescopes as a function of temperature difference was also measured to be 65 μm/°C. Pathlength difference of 50 μm corresponds to the phase error of about 1 radian at $\lambda = 0.3$ mm. If the mechanical and thermal stabilities are known and predictable, they will be corrected in fringe tracking. Even if they are unknown, they can be corrected through frequent calibrations if their time variation is slow.

By use of optical or infrared telescopes as radio telescope, the problem of optimum telescope illumination must be considered. Radio telescopes use horn antenna to illuminate the primary mirror. As shown in Figure 3, the illumination by the horn at the central portion of the primary mirror is higher than at the edge. A large central lockage due to secondary mirror in optical telescopes causes loss in aperture efficiency and high side-lobe level. The problem has been studied by Harris (1988) and he found that the optimum illumination edge taper for the large infrared telescope is near 10 dB. The maximum aperture efficiency achievable for telescopes of small central blockage like IRTF ($\sim 1\%$) is about 0.8.

Figure 4 shows the diameter of equivalent perfect mirror along wavelengths for JNLT, WMKT, and JCMT. The 'perfect mirror' means the mirror without blockage and surface error. We assumed the surface accuracy of 30 μm r.m.s. and $\eta_0 = 0.6$ for JCMT, and $\eta_0 = 0.8$ for JNLT and WMKT. At wavelengths shorter than 0.4 mm, WMKT and JNLT have effective apertures larger than JCMT. Therefore, WMKT and JNLT might be a best interferometer pair at short submillimeter wavelengths.

5. Receiver System

The IF system, delay lines, and correlator are the same as those for millimeter and centimeter interferometers. Mixers and LO's are the key components for the sub-millimeter interferometry.

Schottky diode mixer, InSb hot-electron bolometric mixer, and SIS (Superconductor-Insulator-Superconductor) mixer have been used as submillimeter mixers. The Schottky mixer can be operated at as low as $\lambda = 0.1$ mm. A heterodyne receiver with a Schottky diode mixer and an optically pumped submillimeter laser for the sub-

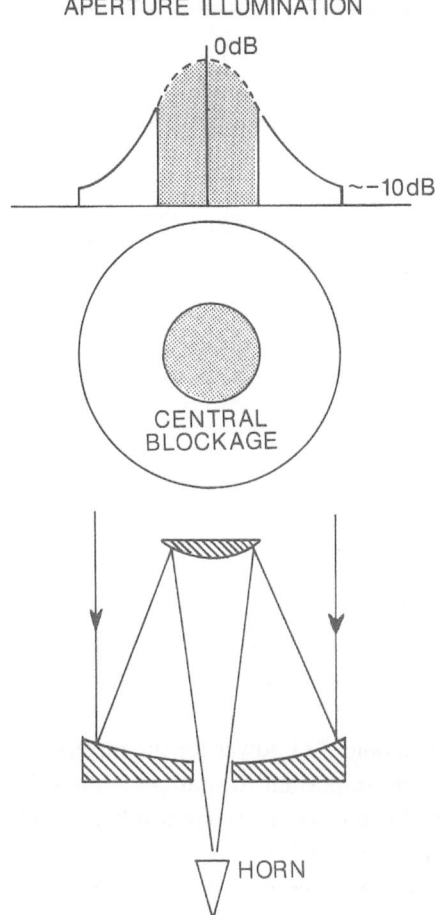

Fig. 3.   Aperture illumination of an optical/IR telescope when it is used as a radio telescope.

millimeter wavelength range from 100 μm to 1000 μm is described by Roeser *et al.* (1986). A similar heterodyne receiver system was successfully used in the Kuiper Airborne Observatory to detect the $J = 7–6$ rotational transition (372 μm) of carbon monoxide (Roeser *et al.*, 1987). The Schottky diode mixer, however, requires relatively high power (~ 0.1 mW) LO sources such as carcinotrons or lasers, and will not be practical especially for interferometer. The InSb bolometric mixer has low noise performances at submillimeter wavelengths, but the bandwidth is very limited (~ 1 MHz).

The SIS mixer is a promising device, because its noise temperature is lower than the Schottky diode mixer and requires very small LO power (~ 1 μW). As shown in Figure 1, millimeter oscillator at each telescope is phase-locked to the common reference signal. Then submillimeter LO signals are obtained by frequency-multiplying signals from millimeter oscillators. Klystrons or Gunn oscillators are used for the millimeter oscillators. The Gunn oscillators are preferable because it is compact and can be operated at low voltages. When the number of multiplication N is getting large, it will

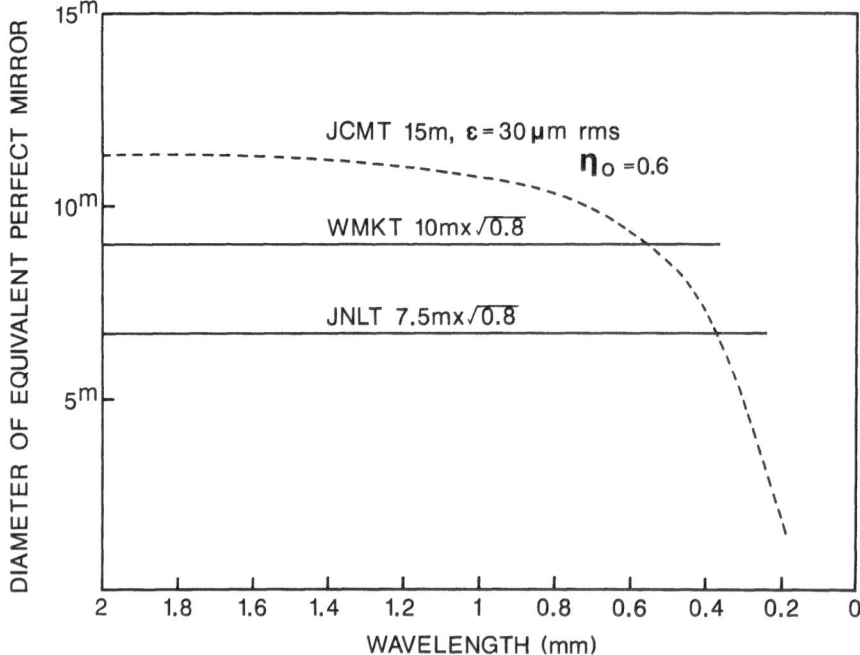

Fig. 4.   Diameter of equivalent perfect mirror for JCMT, WMKT, and JNLT telescopes.

be very difficult to obtain enough LO power for the mixers. So, the mixers working at small LO power level is very important for designing the LO system in submillimeter interferometry. A round-trip phase-lock system has been used to lock the phase of the reference signals (Ishiguro *et al.*, 1984).

A quasi-optical heterodyne receiver using a Pb alloy SIS mixer and a planer logarithmic spiral antenna has been developed and shown broadband performance from millimeter to submillimeter wavelengths (Büttgenbach *et al.*, 1988). The measured noise temperature at frequencies between 115 GHz and 761 GHz was 33 K to 1100 K (DSB). This receiver has been tested at 115, 230, and 345 GHz at the CSO on Mauna Kea. An array of these planer antennas (Rutledge *et al.*, 1984) will be very useful for extending the field-of-view of the submillimeter interferometer (see Table I).

The expected sensitivity of the [JNLT–WMKT] interferometer to point source using current technology is calculated. Following equations are used and the result is shown in Table II.

TABLE II

Sensitivity of the [JNLT–WMKT] interferometer

f (GHz)	λ (mm)	T_{rx} (K)	t	T^*_{sys} (K)	$\Delta S_{r.m.s.}$ (mJy)
345	0.87	400	0.1	550	4
860	0.35	2000	1.0	17 000	130

$$\Delta S_{\text{r.m.s.}} = \sqrt{2}\, kT^*_{\text{sys}}/[A_e\,\sqrt{(BT)}]\,,$$

$$T^*_{\text{sys}} = T_{rx}\, e^{t\sec z} + T_{\text{atm}}[e^{t\sec z} - 1]\,,$$

where $\Delta S_{\text{r.m.s.}}$ is the r.m.s. noise in Jy for two-element interferometer and T^*_{sys} is effective system noise temperature outside of the Earth's atmosphere. We assumed effective aperture $A_e = 47\,\text{m}^2$, integration time $T = 8$ hours, bandwidth $B = 1\,\text{GHz}$, zenith distance $Z = 60°$, atmospheric temperature $T_{\text{atm}} = 280\,\text{K}$. T_{rx} and t is receiver noise temperature and optical depth of the atmosphere, respectively.

6. Summary

The technical feasibility of submillimeter interferometry at Mauna Kea by connecting existing and planned optical/IR telescopes as well as submillimeter telescopes is investigated. A heterodyne interferometry using a broadband SIS mixer and phase-stabilized solid-state LO system will have great promise. The [JNLT–WMKT] might be the best interferometer pair at short submillimeter wavelengths, because they will have largest collecting areas and the baseline is not too long. Adding more telescopes is desirable to fill short spacings and to utilize *closure-phase* or *self-calibration* techniques. For successful achievement of submillimeter interferometry, investigations of mechanical and thermal stability of pathlength, effective coupling of submillimeter wave to SIS mixer, development of phase-stabilized LO system is very important.

References

Bieging, J. H., Morgan, J., Welch, W. J., Vogel, S. N., and Wright, M. C. H.: 1984, *Radio Sci.* **19**, 1505.

Büttgenbach, T. H., Miller, R. E., Wengler, M. J., Watson, D. M., and Phillips, T. G.: 1988, 'Caltech Submillimeter Observatory Astrophysics', preprint.

Harris, A. I.: 1988, *Int. J. Infrared Millimeter Waves* **9**, 231.

Hege, K. E., Beckers, J. M., Strittmatter, P. A., and McCarthy, D. W.: 1985, *Appl. Opt.* **24**, 2565.

Ishiguro, M. *et al.*: 1984, *Proc. U.R.S.I. Int. Symp. Millimeter and Submillimeter Wave Radio Astronomy*, Granada, p. 75.

Moran, J. M., Elvis, M. S., Fazio, G. G., Ho, P. T. P., Myers, P. C., Reid, M. J., Willner, S. P.: 1984, 'A Submillimeter-Wavelength Telescope Array: Scientific Technical, and Strategic Issues', *A Special Report of the Smithsonian Astrophysical Observatory*, Cambridge, MA.

Pearson, T. J. and Readhead, A. C. S.: 1984, *Ann. Rev. Astron. Astrophys.* **22**, 97.

Roeser, H. P., Wattanbach, R., Durwen, E. J., and Schultz, G. V.: 1986, *Astron. Astrophys.* **165**, 287.

Roeser, H. P., Schaefer, F., Schmid-Burgk, J., Schultz, G. V., van der Wal, P., and Wattenbach, R.: 1987, *Int. J. Infrared Millimeter Waves* **8**, 1541.

Rutledge, D. B., Neikirk, D. P., and Kasilingam, D. P.: 1984, in K. J. Button (ed.), 'Integrated-Circuit Antennas', *Infrared and Millimeter Waves*, Vol. 1, Academic Press, p. 1.

BECKERS – Normally in radio interferometers one uses dishes of equal diameters and, hence, equal sensitivity and individual beamwidths. Do you encounter any special problems in using telescopes of unequal size especially when, for example, studying extended objects?

ISHIGURO – Effective primary beam is a geometrical mean of the two different primary beams. Using telescopes of unequal size have been done in VLBI, and is not so serious problem because the VLBI field-of-view is much smaller than that of individual telescope. The process of image formation is not straightforward, as you pointed out, if you want to study extended objects. We can correct for this effect basically, because the visibility data is a convolution of the true visibility and the Fourier transform of the effective primary beam pattern.

LENA – Do you have expect to have a non-flat wavefront over the 10–15 m aperture due to index fluctuation (H_2O)?

ISHIGURO – We have ever experienced a rapid (time-scale $\sim$ 1 min) beam broadening of the Nobeyama 45-m millimeter telescope. This phenomena could be attributed to the non-flat wavefront over the aperture. Same sort of things would happen at submillimeter wavelength, but I have no estimate at this moment.

HALL – The Smithsonian Astrophysical Observatory proposed to locate a 7-element submillimeter array on Mauna Kea.

SUMMARY OF THE SYMPOSIUM

SUMMARY OF THE SYMPOSIUM*

D. N. B. HALL

University of Hawaii, Honolulu, U.S.A.

(Received 2 March, 1989)

I will start by summarizing the various international projects that were described during the conference, and the lessons that may be drawn for the JNLT as an international facility operating at a site outside the homeland of the astronomers. Terry Lee emphasized the experience with United Kingdom Infrared Telescope, and Bob McLaren described a somewhat different mode of operation for the Canada–France–Hawaii Telescope. Two presentations then highlighted the experience with the Anglo-Australian Telescope. The conclusion I draw from all three of those presentations is that, apart from the technical aspects of the site and the telescope quality, it is absolutely vital that one takes into account the human resource needed to make a facility like this work. Close attention needs to be paid to putting together a dedicated team and inspiring them to make sure that the conditions are right for the astronomers to get the very best observations. We heard a little about the ESO operation and their upcoming decision about whether to stay with a developed site or whether to undertaken the task of developing another new site in Northern Chile and going to operation of two separate facilities there. Finally there was the delightful presentation on supra-national coopera-tion and the way that is working out for the James Clerk Maxwell Telescope.

To me, one of the really significant developments at this conference is that, quite apart from 8- to 10-m class telescopes which are the primary focus of things, 4-m class altazimuth telescopes have really gained acceptance. The MMT has, of course, been in operation for some time and so has a 2-m class altazimuth telescope in Australia. However, in earlier days of applying modern technology to telescope construction there were widespread reservations about altazimuth telescopes, particularly whether they could ever be made to point and track to the requisite accuracy, and whether one could rotate fields or de-rotate images to the requisite level of performance. This week we have heard about four 3- to 4-m class altazimuth telescope projects, all of which are about to go into operation and where initial testing, or, in the case of the Herschel telescope, actual observing, have shown them to be resounding successes. Even in the context of the potential gain in light-collecting power that can probably be achieved with 8- to 0-m class telescopes, one should not overlook that these four telescopes are a very substantial augmentation in observing capability. I was very impressed by the New Technology Telescope (NTT) results on active control of the primary mirror and also the inclusion

* Paper presented at the Symposium on the JNLT and Related Engineering Developments, Tokyo, November 29–December 2, 1988.

of apparatus to allow active collimation and updating of the secondary alignment and focus. This appears to be one of the limiting factors in achieving high-resolution imaging with existing telescopes. This bodes well for 8-m telescopes but it also clearly shows that very high performance 4-m class telescopes, built on the best sites and with only modest extensions of technology, are in fact realizing major cost savings and seem destined to provide outstanding performance.

We heard a number of presentations on image quality and I will make a couple of points here. One is that, for many limiting observations on point sources, 4-m telescopes with $\frac{1}{2}$ arc sec images will perform just as well as 8-m telescope with 1 arc sec images. We have seen enormous gains in recent years in many areas of observational astronomy by really pushing the image quality toward the atmospheric limits. It is crucially important, if 8-m class telescopes are to achieve their potential, that they are able to get this level of imaging. Also, the problems that have long been anticipated in efficiently matching instruments to 8- to 10-m class telescopes are very much diminished because of the improvements in image quality. Matching an instrument to an 8-m class telescope with $\frac{1}{2}$ arc sec images is the same as matching instruments to existing 4-m class telescopes with 1 arc sec images. It also become very clear at this meeting that sensitive, large format array detectors are available in the near infrared. These will allow use of adaptive optics to realize truly diffraction-limited imaging throughout the near infrared. Simple diffraction-limited imaging under good seeing conditions at 10 microns and beyond are a major thrust of 10-m telescopes. If one can achieve diffraction-limited performance, one does realize the full (aperture)^2 improvement in signal-to-noise ratio in background-limited observing conditions. This is an area where there is enormous potential and where the technology is now in hand to make that possible.

The key challenge for large telescopes are achieving mirrors of adequate quality while minizing the weight and supporting them. The rest of the telescope weight scales with the mirror weight. The three technical areas are blanks, polishing, and support. The mount and drive must be matched to the image quality.

In the area of mirror blanks, we have heard presentations on a number of alternate approaches, to get to 8-m class telescopes, all of which seem feasible, for the Keck approach where one takes 2-m class segments and builds them up into a mirror surface; the blanks are well within conventional limits. We heard from Roger Angel that the Arizona group has successfully cast a 3.5-m borosilicate honeycomb and that things are progressing to 6.5- and eventually 8-m class borosilicate honeycomb blanks. Similarly, we heard from Schott that the spin-cast zerodur process also appears fully capable of 8-m class blanks. Schott plans to produce two 8-m blanks for their catalog inventory along with the four for the VLT. Finally, we heard a very convincing presentation from Corning on their approach using proven technology to fuse together ULE boules and then to slump the blank to the required meniscus.

As far as polishing, there are clearly still challenges to be overcome. In the Keck approach, we heard that the stressed mirror polishing has worked extremely well to produce off-axis parabolas, but that they have run into a problem associated with trimming the edges from the blanks in order to make them hexagonal (necessary to fit them together a close-packed mirror surface). The proposed 'fix' involves bending harnesses which appear feasible, although it sounds very much a matter of putting off difficulties from the development phase into increased complexity during the operational phase. We heard of the Arizona group's plans for stressed lag polishing to produce very fast local rations, notably the $f/1.2$ required for the 'Columbus' and 'Magellan' projects. We also heard from Zeiss regarding their active polishing lap and techniques available there for meniscus mirror polishing. Contraves-Goerz feels that their experience with more modest class mirrors is such that they do not anticipate major problems in simply taking existing polishing techniques and applying them to 8-m class mirrors.

We heard about a wide variety of support systems. There were a number of presentations on the various approaches to support of meniscus mirrors and techniques for compensating wind loading. I think a lot of us still have concerns about how effectively one will be able to support a relatively flexible meniscus mirror in an 8-m class telescope without a high degree of active support. In this area, I was particularly encouraged by the results presented from the NTT project at ESO where, at least on 4-m class blank of approximately the same thickness as being proposed for the 8-m blank, they are able to demonstrate an extraordinary degree of control and the ability to fine-tune the primary mirror into the shape that is required. Given that the stars required to provide the servo-signal for that process run down into approximately the same magnitude range as the guide stars for the Hubble Space Telescope and the availability of the guide star catalog, I think that it is clear that, so long as the technical side of things works well there, it is a relatively straightforward matter operationally to select stars for sensing the mirror surface. Overall, particularly based on the NTT results, I was encouraged about the possibility associated with active supports on meniscus blanks.

I will now turn to actual 8- to 10-m class telescope projects. As was emphasized earlier on in the conference, astronomy does not seem finally to have broken out of the long period where it was simply impossible to get any of these large telescope projects funded. In the United States, the Field Committee Report for Astronomy in the 1980's specifically recommended developing technology during the decade rather than funding telescope construction. This week's presentations demonstrate that the technology is now in hand. The goal for the 1990's should be to build telescopes, not to further study technology. We have heard here from the two projects that are actually funded and are under way, the ESO VLT project and the Keck project. Of a range of other projects, I emphasize those where funding is likely or where active efforts are being made to move ahead with the funding of the project. Notable are the JNLT, 'Columbus', 'Magellan', and the NOAO 8-m projects. If, in fact, one looks at the collecting area increase associated with these telescopes, over what is available today, they represent a revolution in terms of collecting area and capability for optical/infrared astronomy. I thought the last talk was delightful; Harlan Smith showed us that there is life after failure to fund your 10-m telescope project – you simply go off and build one anyway.

The main thrust of this symposium was, of course, the JNLT. It provided a valuable opportunity for people throughout the international community to interact wth the JNLT project staff and the broader Japanese community of optical and infrared astronomers. We have heard in a number of sessions the details of the telescope, the level of characterization of the site and the plans for instrument and detector development. I was particularly impressed by the fact that, both in the area of optical CCD's and in the area of infrared arrays, it seems very likely that the best detectors available to the JNLT will be fabricated here in Japan, an amazing change from a few years ago.

We also heard a great deal about what I consider the two key areas for future evolution of the 7.5-m JNLT. Jacques Beckers went through and showed us that, through the near infrared, adaptive optics technology as now being demonstrated by a number of projects has the potential to get down to the diffraction limit of the telescope at 7.5 m aperture. We also heard a great deal about array telescope possibilities and the way the JNLT might be combined with other nearby large telescopes, both at submillimeter and infrared, and eventually even optical, wavelengths.

Clearly one of the key drivers for building 8- to 10-m class telescopes is to achieve high angular resolution imaging at very high sensitivity. Beyond 10 microns, the JNLT at a good site should be diffraction limited. This is an area which, as we heard this morning, is receiving relatively little attention at the moment, yet it is one of the areas where these larger telescopes really do achieve huge advances over capabilities that e have in existing telescopes. And I again stress that adaptive optics has the potential to extend this diffraction-limited performance down to much shorter wavelengths. Jacques Beckers emphasized that even an imperfect adaptive optics correction can potentially give you very high angular resolution, but t the expense of some limitation in contrast. I would also emphasize that achieving this full angular resolution is again a key factor in sensitivity, that much of the gain in going to the 8- to 10-m class aperture is in getting this full angular resolution so that one can achieve the maximum intrinsic sensitivity of the telescope.

It is useful to highlight the real sensitivity issues. A lot has been made about the improvements in auxiliary instrumentation, detectors, control efficiencies and things, and the steady evolution of improvement of auxiliary detectors and instruments. Although we have not built an effective telescope much larger than 5 m in the last 40 years, there have been huge advances in the capabilities in optical and infrared astronomy. I think we need to pay attention to the fact that, in many areas, one is now veyr close to the theoretical limitations of performance, at least in terms of quantum efficiency and associated detector characteristics which allow one to really work down to the natural background. There are exceptions; it is still difficult to get ideal CCD's, and infrared arrays are only just now becoming available. I think it is very clear on the time-scale of the JNLT that for most applications one is not going to be able to foresee further major improvements associated with improvements in detector efficiency or detector characteristics. This will largely eliminate the area where one is limited by detector characteristics and where one gains very fast in performance for collecting aperture. There are, however, key areas, notably backround limited operation and sensitivity to point sources where in the diffraction-limited case one is still gaining in signal-to-noise ratio as the square of the diameter of the telescope. This is already achievable in the thermal infrared and can be extended down into the near infrared with adaptive optics techniques. These are areas where the JNLT will gain by a factor of four or so in sensitivity relative to the best 4-m class telescopes.

There were a wide range of views on how well-matched 8- to 10-m class telescopes are to working on sources where one is background-limited and not at the diffractin limit of the telescope. Here one gains in sensitivity only linearly with the diameter of the telescope. I think the answer here is that, for observations which involve relatively short integration times, one can well ask why use an 8-m class telescope when by integrating 4 times the exposure on a 4-m class telescope one can achieve the same result. It will come back to the point I made in my earlier talk that in the Northern Hemisphere for the spring extragalactic observing season out of the plane of our Galaxy, there are in fact only three dark runs or a total of about 30 dark nights each year. As those of you who schedule large telescopes are aware, there is incredible pressure for this observing time and within the context of that really severe limitation, the gain in sensitivity by a factor of two (or observing time by a factor of four) on an 8-m class telescope really does open up a range of problems which simply cannot be addressed today. We are at a point for many fundamental problems where people are doing 10- or 20-hour integrations on fields and are realistically considering trying to go to a factor of five or ten longer, in order to achieve another magnitude, particularly in the extragalactic cosmological area. It really does take a big telescope like the JNLT and the commitment to schedule it for those key projects where it alone can make further progress on those.

Let me finally turn to the science presentations of the opening session. We heard first about the activity here in Japan, in X-ray space astronomy and also radio astronomy. I think the key thing to be drawn those presentations and the experience of large telescopes in the past several decades is that every new space mission or every new wavelength region that is opened up increases the pressure on large optical/infrared telescope time. Discoveries of new objects open up new areas of astronomy where researchers have to get substantial amounts of time on large optical/infrared telescopes to do the follow-up studies and put the new results in the context of the wider body of optical/infrared experience. It is also very clear that 8-m telescopes like JNLT are entirely complementary to the facilities like the Hubble Space Telescope. The Hubble Space Telescope will produce superb images over a large field and will be very good at finding faint objects and looking at the morphology. Large telescopes such as the JNLT have the collecting area and the sensitivity essential for follow-up spectroscopic observations and will also have substantial resolution and sensitivity advantages through the infrared. In an era of Hubble Space Telescope I see the JNLT and other 8-m class telescopes doing a lot of the follow-up observations and spectroscopy which will then pin down the physics of the sources. It is also evident that other upcoming space missions such as ISO and X-ray missions, are again going to create a great deal of demand for observing time on these classes of telescopes.

In the key area of cosmology, as highlighted by Professor Hayakawa, it is very clear that one is looking at magnitude limits around 23 or 24 to high enough red-shifts that one is close to or going through the minimum size of galaxies before they start to expand in angular size again. We now know that the compact core of these galaxies at redshifts around 3 are in the $\frac{1}{2}$ to 1 arc sec range and that the more extended material around them is typically a few arc sec across. In this regime, JNLT at a good site will be ideally matched to achieving optimum sensitivity. It will be very much more sensitive than smaller space instruments and, of course, one can exploit the higher angular resolution. One point that I thought came through during the discussions is that as Japanese astronomers move from their urrent optical/infrared facilities to a 7.5-m telescope, there is likely to be a need for observing on intermediate size telescopes. It seemed to me that sort of access, which might make sense as part of international agreements, may be very useful in preparing for the best use of the JNLT.

In conclusion, I think the basic message of this conference is that the key technologies for 8- to 10-m class telescopes in general, and specifically the JNLT, are available and that as funding becomes available, this new class of optical/infrared telescopes will be built to carry out forefront research on into the next century. The JNLT is a well conceived and developed project which will allow Japanese astronomers to be at the forefront of optical/infrared observational astronomy.

CANNON – I would like to follow up the point which Don Hall made, about Japanese astronomers probably needing access to international-size facilities in other countries if they are to make full use of the JNLT. I think it will be very important for Japanese astronomers to make use of other large telescopes *before* the JNLT begins working in several years time. It will be especially crucial for young astronomers to spend time working abroad, both to get experience with the best modern instruments on large telescopes, and to keep abreast of the latest and most exciting fields of research. I am sure that the U.K. and Australia and other countries represented here will welcome visiting Japanese astronomers, and I hope that on the Japanese side it will be possible to enable young astronomers, especially those who are just completing their PhDs, to make such visits.

SMITH – I would like to draw everyone's attention to the size of 8-m telescopes. They are facilities of the size of conference room – very large indeed.

BOKSENBERG (to Smith) – Although 8-m telescopes are large, actually they are not much taller than current 4-m telescopes because they all have such faster primary mirrors and so they are shorter in proportion.

OBITUARY*

DAVID S. BROWN

(1927–1987)

It is highly appropriate that this volume, which catalogues the exciting progress made towards the realisation of the Japanese National Large Telescope, be dedicated to David Brown, an influential figure in the design of postwar astronomical telescopes, a friend and colleague to many contributors to these proceedings, and a pioneer in the never-ending campaign for larger, more cost-effective telescopes.

David Scatcherd Brown was born in 1927 at Coventry in England, graduated from Cambridge and worked with the firm of Sir Howard Grubb Parsons, at Newcastle, from 1950 to 1985. He was appointed optical manager in 1961 and technical director in 1975. This period marked the renaissance of British optical astronomy and Brown's team at Grubb-Parsons was responsible for a succession of telescopes including the 2.4 m Isaac Newton, the 3.9 m Anglo-Australian, the 1.2 m UK Schmidt, the 3.5 m UK Infrared Telescope and, most recently, the highly-acclaimed 4.2 m William Herschel Telescope.

In addition, the team produced many telescopes and optical components for foreign observatories. Brown's name was associated with the production of high quality mirror

Astrophysics and Space Science **160**: 393–394, 1989.
© 1989 *Kluwer Academic Publishers.*

surfaces, and his remarkable knowledge of all aspects of telescope manufacture was in great demand worldwide.

Brown's contributions to astronomy earned him an honorary doctorate from Durham University in 1981 and when Grubb-Parsons closed in 1985, he was made the Science and Engineering Council Grubb-Parsons Fellow at Durham. He had already begun directing his attention to numerous aspects concerned with the next generation of large telescopes via studies at the University of Arizona and here in Japan. He spent several weeks in Japan as an invited scholar of the Japanese Society for the Promotion of Science and was a key figure in the 1987 JNLT workshop. He became the primary technical consultant to the UK Large Telescope Panel when it was formed in 1986 and contributed substantially to that study reported in these proceedings.

David Brown died suddenly after a short illness on July 17, 1987, aged only 60 years. He had only just returned from a second extended visit to Tokyo during which time the UKLT Panel's made its first visit.

Those who knew David will remember him as a quiet but friendly man. He complained very rarely about anything, else than funding shortages and the frustrations and delays which these bring to all ambitious projects. Like all of us, he enjoyed the immense hospitality he received in Japan. He spoke frequently about his rides on the Shinkansen (and complained only about Japanese tables with six-inch legs!). His enthusiasm and wisdom is sorely missed in a subject where everyone shares knowledge to push the frontiers ahead.

I am happy that the organisers of this meeting decided to dedicate the proceedings to David. He would be proud of the progress shown in the many significant articles in this volume, and particularly those relating to the JNLT. With all of us, he would wish the JNLT great success in the coming years.

RICHARD S. ELLIS

ANNOUNCEMENT

JAPANESE NATIONAL LARGE TELESCOPE AND RELATED ENGINEERING DEVELOPMENTS

*Proceedings of the International Symposium on Large Telescopes,
held in Tokyo, Japan, 29 November – 2 December, 1988*

Edited by

T. KOGURE
Department of Astronomy, Faculty of Science, Kyoto University, Japan

and

A. T. TOKUNAGA
Institute for Astronomy, University of Hawaii, U.S.A.

Please note that a hardbound edition of this special of *Astrophysics and Space Science*, Vol. 160, Nos. 1–2 (October, 1989), is available from the publisher.

ISBN 0–7923–0561–2   Prices: Dfl. 285,–/US $149.–/£94.–

VOLUME CONTENTS

Volume 160 Nos. 1–2

ARTICLES

Fig. 6. Fabrication in progress-tube on temporary supports, front view (photo courtesy J. Steffey). (Medwadowski, p. 39).

Fig. 7. Active support experiment at a tilted orientation. (Iye, p. 154).

Fig. 5. Fabrication in progress-tube on temporary supports, side view (photo courtesy J. Steffey). (Medwadòwski, p. 38).

Fig. 6. Demonstration prototype of an inflatable dome. This prototype is being tested at La Silla Observatory. (Enard, p. 52).

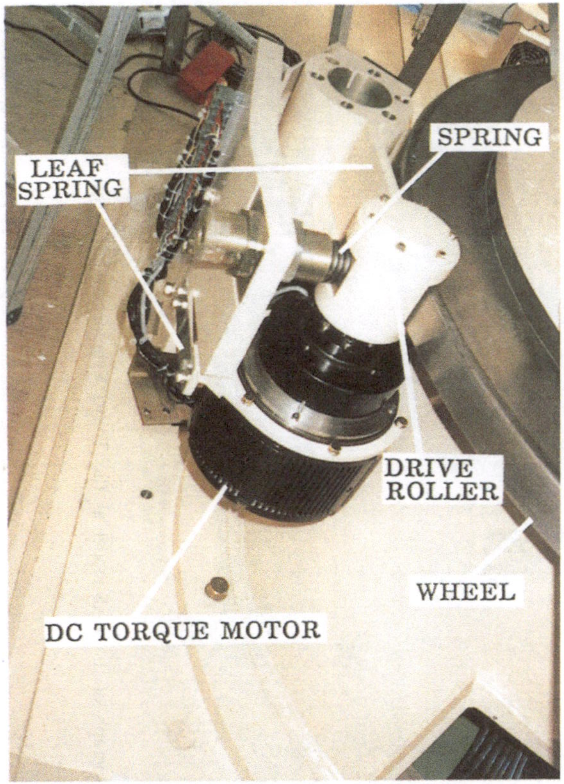

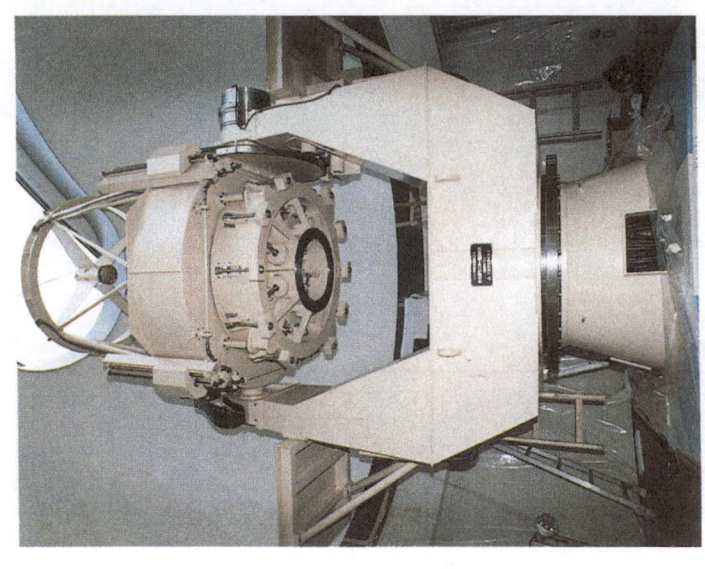

Fig. 8.   1.3 m telescope. (Itoh   *et al.*, p. 170).

SPRING

DRIVE
ROLLER

WHEEL

LEAF
SPRING

DC TORQUE MOTOR

Fig. 9.   AZ drive mechanism. (Itoh *et al.*, p. 171).

Fig. 3b. (Hügenell, p. 232).

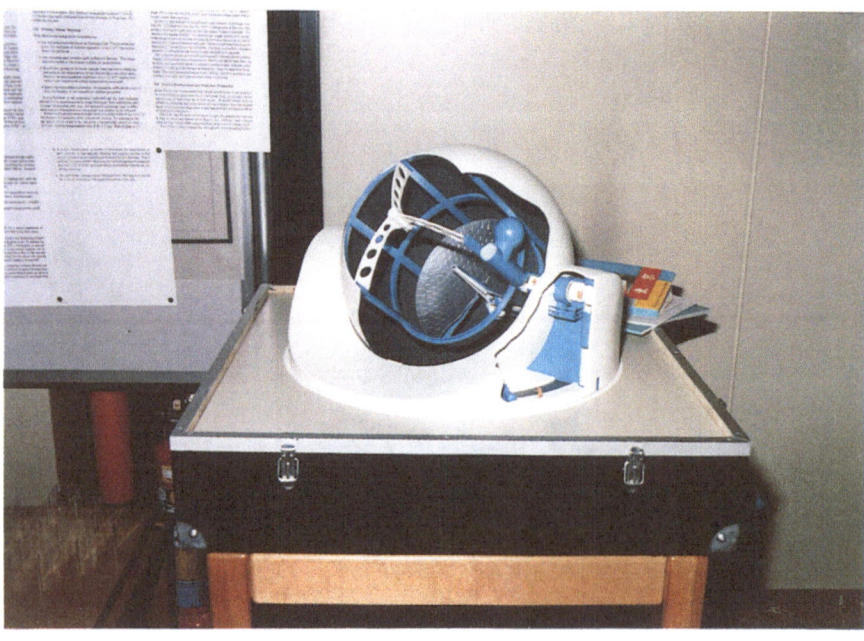

Fig. 14. Photograph of the ZAS-model at the Tokyo Symposium on Large Telescopes, 1988. (Hügenell, p. 239).

Fig. 5. Light weight Zerodur mirror made by 'Schott Glaswerke, Mainz'. (Hügenell, p. 233).

Fig. 3a.   (Hügenell, p. 232).